Disaster Risk Reduction

Methods, Approaches and Practices

Series Editor

Rajib Shaw, Keio University, Shonan Fujisawa Campus, Fujisawa, Japan

Disaster risk reduction is a process that leads to the safety of communities and nations. After the 2005 World Conference on Disaster Reduction, held in Kobe, Japan, the Hyogo Framework for Action (HFA) was adopted as a framework for risk reduction. The academic research and higher education in disaster risk reduction has made, and continues to make, a gradual shift from pure basic research to applied, implementation-oriented research. More emphasis is being given to multi-stakeholder collaboration and multi-disciplinary research. Emerging university networks in Asia, Europe, Africa, and the Americas have urged process-oriented research in the disaster risk reduction field. With this in mind, this new series will promote the output of action research on disaster risk reduction, which will be useful for a wide range of stakeholders including academicians, professionals, practitioners, and students and researchers in related fields. The series will focus on emerging needs in the risk reduction field, starting from climate change adaptation, urban ecosystem, coastal risk reduction, education for sustainable development, community-based practices, risk communication, and human security, among other areas. Through academic review, this series will encourage young researchers and practitioners to analyze field practices and link them to theory and policies with logic, data, and evidence. In this way, the series will emphasize evidence-based risk reduction methods, approaches, and practices.

Ayse Yildiz · Rajib Shaw
Editors

Disaster and Climate Risk Education

Insights from Knowledge to Action

Springer

Editors
Ayse Yildiz
School of Business
Institute for Environmental Futures
University of Leicester
Leicester, UK

Rajib Shaw
Graduate School of Media and Governance
Keio University, Shonan Fujisawa Campus
Fujisawa, Kanagawa, Japan

ISSN 2196-4106 ISSN 2196-4114 (electronic)
Disaster Risk Reduction
ISBN 978-981-97-5986-6 ISBN 978-981-97-5987-3 (eBook)
https://doi.org/10.1007/978-981-97-5987-3

This Springer imprint is published by the registered company Springer Nature Singapore Pte Ltd.
The registered company address is: 152 Beach Road, #21-01/04 Gateway East, Singapore 189721, Singapore

Preface

Education is the key to knowledge and action. While we talk about disaster education or climate risk education, many times, knowledge remains with the people, which gives them a certain level of awareness. However, there is always a gap between awareness and education. To address this issue, we have developed the KIDA (Knowledge Interest Desire Action) framework, which spans from knowledge generation to enhancing action. To implement the KIDA framework effectively, collaboration among schools, communities, and families is essential.

Local governments and the board of education need to play a role in providing support systems for this approach. The necessity for such an approach has been reflected through disaster experiences all over the world, including the recent tragic Turkey–Syria earthquake of 2023. Emerging risks of climate change have added dimensions to the so-called "disaster education," and understanding and disseminating disaster climate nexus as an education and learning process is very important. We can create a comprehensive and efficient disaster education system that allows us to deal with the changing risks of our world while nurturing a generation dedicated to creating a sustainable future by encouraging collaboration among all stakeholders and prioritising vulnerable groups' education and involvement.

Keeping this in mind, this book will cover the broader aspects of disaster and climate change risk, including school education, community education, teacher training, technology in education and educational governance. The book will include in-depth reviews and new data based on case studies across the globe involving multi-disciplinary research. Finally, the book suggests specific policy and action measures to enhance disaster and climate risk education and school and community resilience. We will be glad if the readers find the book useful.

Leicester, UK — Ayse Yildiz
Fujisawa, Japan — Rajib Shaw

About This Book

Education serves as a cornerstone for gaining knowledge and taking initiative. However, despite efforts in disaster and climate risk education (DCRE), a gap often exists between awareness and meaningful action. To bridge this gap and promote a progression from awareness to empowerment, we offer the KIDA (Knowledge Interest Desire Action) framework. The foundation of this framework is collaboration between schools, communities, and families, which calls for support from educational boards and local governments.

Following international tragedies like the Turkey–Syria earthquake of 2023, as well as the ever-changing difficulties posed by climate change, DCRE has to be given priority immediately. Our book proposes a comprehensive strategy that includes governance, capacity building, education in schools and the community, and technology integration. Our insights, analyses, and practical policy suggestions are based on multidisciplinary research and worldwide case studies, with the goal of strengthening resilience and cultivating a generation committed to sustainability.

Overall, this book is a call to action, rather than just a guide. It provides a comprehensive exploration of DCRE, including educational, governance, and technological aspects. It aims to prepare individuals and communities to face the challenges of a changing world head-on by improving knowledge and promoting preparedness.

Contents

Editors and Contributors

About the Editors

Ayse Yildiz is an Assistant Professor in Risk at the School of Business at the University of Leicester, UK. An expert in climate and disaster risk, resilience and disaster preparedness, she is an affiliate of both the University of Leicester's Institute for Environmental Futures and the Society of Risk Analysis. Ayse's research focuses primarily on community-based disaster resilience, risk perception and child-centred disaster risk reduction. She engages with diverse audiences to promote awareness and understanding of disaster-related issues, including government agencies, local teams, schools and students of varied cultural and religious backgrounds. Ayse has convened seminars and workshops alongside representatives of the UN Office for Disaster Risk Reduction and regularly contributes expert opinion to post-disaster media coverage.

Rajib Shaw is a professor in Graduate School of Media and Governance of Keio University, Japan. He is the Co-founder of a Delhi (India) based social entrepreneur startup Resilience Innovation Knowledge Academy (RIKA), and RIKA Institute. He is the Co-chair of the United Nations Asia Pacific Science Technology Advisory Group (AP-STAG) and was CLA for IPCC's 6th Assessment Report. Professor Shaw is the recipient of prestigious "Pravasi Bharatiya Samman Award (PBSA)" in 2021 for his contribution in education sector. He also received United Nations Sasakawa Award in 2022 for his lifetime contributions in the field of disaster risk reduction. He published 67 books and over 450 research papers. More about his work can be found in: www.rajibshaw.org.

Contributors

Rachel M. Adams Natural Hazards Center, University of Colorado Boulder, Boulder, Colorado, United States

Faduma Ali Scientific and Cultural Organization (UNESCO), United Nations Educational, Paris, France

Saki Arima-Shirai SEEDS Asia, Kobe City, Hyogo Prefecture, Japan

Ramazan Aslan Faculty of Health Sciences, Emergency Aid and Disaster Management Department, Ardahan University, Ardahan, Türkiye

B. Burcak Basbug Department of Statistics, Middle East Technical University, Ankara, Türkiye

Lee Bosher School of Business, University of Leicester, Leicester, England

Jolie Breeden Natural Hazards Center, University of Colorado Boulder, Boulder, Colorado, United States

Mehar Chauhan Department of Psychology and Neuroscience, University of Leicester, Leicester, UK

Vladimir M. Cvetković Faculty of Security Studies, University of Belgrade, Belgrade, Serbia;
Scientific-Professional Society for Disaster Risk Management, Belgrade, Serbia;
International Institute for Disaster Research, Belgrade, Serbia

Robert M. de Groot U.S. Geological Survey, Reston, Virginia, United States

Sanjit Dhami Division of Economics, School of Business, University of Leicester, Leicester, UK

Thoa Kim Do Earth Systems, Hanoi, Vietnam

Canay Doğulu Department of Psychology, TED University, Kolej, Çankaya, Ankara, Türkiye

Siva Ram Edupuganti Department of Architecture and Planning, Indian Institute of Technology Roorkee, Roorkee, India

Ozgur Erdur-Baker Faculty of Education, Middle East Technical University, Ankara, Turkey

Gul Ergun Burdur Mehmet Akif Ersoy University, Faculty of Health Sciences, Department of Nursing, Burdur, Türkiye

Lutfa Tilat Ferdous Business School, University of Leicester, University of Leicester, Leicester, UK

Yinan Gao Graduate School of Media and Governance, Keio University, Fujisawa, Japan

Rong Huang Graduate School of Media and Governance, Keio University, Fujisawa, Japan

Tomonori Ichinose Department of Education, National University Corporation, Miyagi University of Education, Sendai, Japan

Gözde Ikizer Department of Psychology, TOBB University of Economics and Technology, Sogutozu, Çankaya, Ankara, Türkiye

Masayoshi Inoue Ehime University, Matsuyama, Japan

Sadaf Ismail Scientific and Cultural Organization (UNESCO), United Nations Educational, Paris, France;
Graduate School of Advanced Integrated Studies in Human Survivability, Kyoto University, Kyoto, Japan

Tomoki Itamiya Division of Dental Education, School of Dentistry, Kanagawa Dental University, Yokosuka, Japan

Hiralal Kapar Kathmandu University, Lalitpur, Nepal

Ayse Nuray Karanci Department of Psychology, TOBB University of Economics and Technology, Sogutozu, Çankaya, Ankara, Türkiye

Tomo Kawane Graduate School of Media and Governance, Keio University, Fujisawa, Japan

Thi Kinh Kieu The University of Danang University of Science and Education, Da Nang, Vietnam

Kaori Kitagawa Faculty of Education and Society, IOE, UCL's, London, UK

Atul Kumar Department of Architecture and Planning, Indian Institute of Technology Roorkee, Roorkee, India

Tim Marjoribanks Water Engineering and Development Centre, School of Architecture, Building and Civil Engineering, Loughborough University, Loughborough, England

Edwin F. Marquez SEEDS Asia, Kobe City, Hyogo Prefecture, Japan

Patricia Mie Matsuo Institute of Advanced Studies, Universidade de São Paulo, Sao Paulo, Brazil

Sara K. McBride U.S. Geological Survey, Reston, Virginia, United States

Sefa Mızrak Faculty of Health Sciences, Emergency Aid and Disaster Management Department, Gümüshane University, Gümüşhane, Türkiye

Mahua Mukherjee Department of Architecture and Planning, Indian Institute of Technology Roorkee, Roorkee, India

Ha Thu Nguyen The Vietnam National Institute of Educational Sciences, Hanoi, Vietnam

Lyn Alexandria T. Novilla SEEDS Asia, Kobe City, Hyogo Prefecture, Japan

Yukihiko Oikawa Nara University of Education, Nara, Japan

Lori Peek Department of Sociology and Natural Hazards Center, University of Colorado Boulder, Boulder, Colorado, United States

Tikaram Poudel Department of Language Education, Kathmandu University, Lalitpur, Nepal

Fathima Roshan Rakeeb Business School, University of Leicester, University of Leicester, Leicester, UK

Amrita Sabhapandit Royal School of Architecture, The Assam Royal Global University, Guwahati, Assam, India

Bhola Saha Royal School of Architecture, The Assam Royal Global University, Guwahati, Assam, India

Dikshya Saikia Jogananda Deva Satradhikar Goswami College, Bokakhat, Assam, India

Aiko Sakurai Kobe University, Kobe, Japan;
Tohoku University, Sendai, Japan

Takeshi Sato Tohoku University, Sendai, Japan

Rajib Shaw SEEDS Asia, Kobe City, Hyogo Prefecture, Japan;
Graduate School of Media and Governance, Keio University, Fujisawa-shi, Kanagawa, Minato, Japan

Mitsuko Shikada-Otsuyama SEEDS Asia, Kobe City, Hyogo Prefecture, Japan

Rosana Louro Ferreira Silva Biosciences Institute, Universidade de São Paulo, Sao Paulo, Brazil

Vanja Šišović Scientific-Professional Society for Disaster Risk Management, Belgrade, Serbia

Laura Stamp Science Teacher in Oak Park Elementary School District 97, Oak Park, Illinois, US

Nana Sugaya SEEDS Asia, Kobe City, Hyogo Prefecture, Japan

Hans Svennevig Faculty of Education and Society, IOE, UCL's, London, UK

Salih Taskin Burdur Mehmet Akif Ersoy University, Institute of Social Sciences, Family Studies Master's Program with Thesis, Burdur, Türkiye

Jennifer Tobin Natural Hazards Center, University of Colorado Boulder, Boulder, Colorado, United States

Thi My Thi Tong Vietnam Institute of Economic, Vietnam Academy of Social Science, Hanoi, Vietnam

Gaye Tuncer-Teksoz Faculty of Education, Middle East Technical University, Ankara, Turkey

Lisa Wakefield Business School, University of Leicester, University of Leicester, Leicester, UK

Yafang Wen Graduate School of Media and Governance, Keio University, Fujisawa, Japan

Soichiro Yasukawa Scientific and Cultural Organization (UNESCO), United Nations Educational, Paris, France

Ayse Yildiz School of Business, Institute for Environmental Futures, University of Leicester, Leicester, UK

Elanur Yilmaz-Na Faculty of Education, Recep Tayyip Erdogan University, Rize, Türkiye;
Faculty of Education, Lakehead University, Orillia, Canada

Ztionhiying Zhao Graduate School of Media and Governance, Keio University, Fujisawa, Japan

Part I
School Education

Chapter 1
Overview of the Concept: Disaster and Climate Risk Education

Ayse Yildiz and Rajib Shaw

Abstract Disasters caused by natural hazards have become more frequent and intense in recent years, and these events are being exacerbated by climate change. As a result, the importance of disaster and climate risk education (DCRE) in building community resilience and preparedness all around the world has been increasingly recognised. Despite its significance, challenges to the effectiveness of DCRE implementation include socioeconomic factors, financial constraints, and governance structures. Innovative approaches are required to address these issues, with a focus on school education, community education, governance frameworks, and technology. Prioritising these areas will help to improve DCRE initiatives and build more resilient and adaptable communities in the face of increasing environmental challenges. The purpose of this chapter is to give a general overview of the idea of DCRE, including its importance, goals, and key components, as well as to summarise the book outline.

Keywords Disasters · Climate change · Risk · Education · Community resilience

1.1 Understanding Disaster and Climate Risk

Disasters and climate change cause a significant threat to communities, ecosystems, and economies all around the world. The pressing need for comprehensive risk reduction strategies is made evident combined by socioeconomic vulnerabilities and an increase in the frequency and intensity of extreme weather occurrences.

Disaster risk refers to the possible negative consequences that a hazardous event might have on people, assets, and the environment (UNDRR 2024). It includes the likelihood of a hazard occurring (Blaikie et al. 2014) and the vulnerability of exposed elements to the effects of that hazard (Birkmann et al. 2017). Climate risk, meanwhile,

A. Yildiz (✉)
School of Business, Institute for Environmental Futures, University of Leicester, Leicester, UK
e-mail: ay111@leicester.ac.uk

R. Shaw
Graduate School of Media Governance, Keio University, Minato, Japan
e-mail: shaw@sfc.keio.ac.jp

A. Yildiz and R. Shaw (eds.), *Disaster and Climate Risk Education*, Disaster Risk Reduction, https://doi.org/10.1007/978-981-97-5987-3_1

refers to the risks driven by climatic variability and change, such as an increase in the frequency and severity of extreme weather events, including storms, floods, heat waves, and droughts (IPCC 2022). Understanding the relationship between climate risk and disaster risk is essential to comprehending how societies and ecosystems are becoming more vulnerable to natural hazards. Although disasters caused by natural hazards have always occurred, the frequency, severity, and spatial distribution of these occurrences are changing due to climate change, which exacerbates pre-existing vulnerabilities and creates new risks. Understanding the relationship between disaster risk and climate risk is crucial for effective disaster risk reduction and climate adaptation efforts. Without using a comprehensive strategy, risk management in one area has the potential to increase risks in other areas. This is also highlighted in the UNDRR Strategic Framework 2022–2025 by addressing the integration of disaster risk reduction and climate change into relevant strategies and policies across and within sectors in result 1.2 (UNDRR 2021). As stated in Article 8 of the Paris Agreement, which emphasises the need for comprehensive risk management, comprehensive disaster, and climate risk management includes teamwork to improve knowledge, response, and assistance in preventing, reducing, and mitigating loss and damage.

In the face of rising disaster and climate risk, building resilience becomes essential to ensuring sustainable development and protecting lives and livelihoods. Resilience in the context of the disaster and climate risk includes taking proactive steps, to foresee, absorb, recover from, and adapt to challenges to improve people's, communities', and systems' capacity to prosper in a changing environment. While resilience refers to developing the ability and adaptability to face and overcome adversity, adaptation refers to proactive adjustment to changes in circumstances. Resilience and adaptation are two essential components of effective disaster and climate risk management strategies to create more sustainable and resilient futures. Education emerges as an important facilitator in promoting resilience and adaptation by building knowledge and awareness and empowering vulnerable groups at individual, community, and institutional levels.

1.2 Education for Disaster and Climate Risk

The foundation of climate action and disaster risk reduction initiatives is education. Studies have shown that education can enhance risk perception (Yildiz et al. 2023; Wachinger et al. 2013; Shaw et al. 2004), encourage proactive measures (Ronan et al. 2015; Yildiz 2022; Shaw et al. 2011), awareness (Yildiz et al. 2021; Shaw et al. 2009; Schiwaku et al. 2007), and facilitate effective responses (Torani et al. 2019; Shaw et al. 2013). Positive results have been demonstrated by educational efforts that incorporate disaster preparedness into both formal curricula and informal learning environments (Yildiz et al. 2023; Shaw et al. 2011). In the context of climate change, education can help to adopt cutting-edge innovative methods, tools, and regulations that support low-carbon development pathways and enhance climate resilience

(Rousell and Cutter-Mackenzie-Knowles 2020; Monroe et al. 2019; Stevenson et al. 2017).

Despite the widely recognised importance of disaster and climate risk education (DCRE), a number of challenges prevent its effective implementation. These include financial challenges, socioeconomic factors, a lack of access to education, and governmental frameworks. In addition, the complexity of the risk landscape and the rapid pace of environmental change pose challenges to effective DCRE efforts. In order to advance DCRE initiatives, it is essential to address current challenges and investigate innovative approaches. The future direction of DCRE should include strengthening interdisciplinary cooperation, such as DCRE, into more comprehensive agendas for sustainable development and utilising digital platforms and technologies for outreach and engagement.

In order to overcome the challenges and increase the effectiveness of DCRE, we focus on the areas: school education, community education, governance, and technology. Each of these priority areas is essential to enhancing DCRE initiatives and addressing current obstacles in disaster and climate risk education and awareness-raising in the face of increasing environmental challenges.

1.2.1 School Education for Disaster and Climate Risk

Schools are essential settings for disseminating information and developing an understanding of climate change (Kumar et al. 2023; Feinstein and Mach 2020; Anderson 2010) and disaster risks (Shaw et al. 2013; Shiwaku et al. 2007) among young people. Educators can promote awareness, preparedness, and resilience from an early age by including disaster and climate risk education (DCRE) in the school curriculum. Students can get a comprehensive understanding of the complex relationships between natural hazards, climate change, and societal vulnerabilities by integrating DCRE concepts into a variety of areas, including science, geography, and social studies.

Although there are many initiatives worldwide to promote disaster and climate risk education in school curricula, several challenges remain. Limited DCRE integration into current school curricula is one of the main issues (Nurdin et al. 2017; Bonifacio et al. 2010). Despite its significance, DCRE is frequently given insufficient consideration in educational frameworks, leading to a disjointed and superficial covering. Many teachers lack the resources and expertise needed to teach DCRE subjects in an effective manner. Insufficient assistance and opportunities for professional growth could make it difficult for teachers to provide comprehensive and engaging lessons on the risks associated with disasters and climate change (Shaw 2015). Given the interdisciplinary nature of DCRE themes, knowledge from a variety of disciplines, including science, geography, social studies, and environmental studies, is needed. However, attempts to implement a comprehensive approach to DCRE may be hampered by the topic of compartmentalisation found in school education systems. In addition, some communities do not have access to high-quality education about climate change and disaster risk. Inequalities in the allocation of resources, lack of infrastructure, and socioeconomic differences worsen disparities in access to schooling.

In order to address these challenges, there is a need for comprehensive and innovative approaches. This includes revision and integration of DCRE into a variety of subjects in the school curriculum. Curriculum developers and educational policymakers need to work together to find ways to incorporate DCRE subjects into already-existing courses. It is also essential to provide continual training and professional development opportunities to teachers to increase their ability to teach DCRE effectively. Promoting interdisciplinary learning is another aspect of overcoming the challenges of DCRE. This can be through cross-curricular activities, integrated lesson plans, and collaborative projects to establish a holistic understanding of the DCRE concept. It is also important to prioritise equal access to education, especially in vulnerable communities.

1.2.2 Community Education for Disaster and Climate Risk

Communities are essential in any disaster and climate risk reduction initiatives. Education in the community can help create awareness and understanding and effectively respond to disaster and climate risks. Community education places a stronger emphasis on local knowledge, community involvement, and collaborative decision-making processes, unlike the traditional top-down methods (Baudoin et al. 2016; Parsons et al. 2016; Gaillard and Mercer 2013). A variety of activities are usually included in community education initiatives, such as training sessions, workshops, awareness campaigns, planning exercises involving participation, and the distribution of educational materials. These initiatives aim to develop community resilience, increase hazard awareness, and encourage risk reduction practices and a sense of individual and group responsibility (Satterthwaite 2011; Shaw 2015).

Even though community participation is important, limited community engagement is one of the challenges of community education. Attempts to promote meaningful engagement can be hampered by low awareness, conflicting priorities, and power dynamics within communities. New approaches should include community members in decision-making processes from the beginning through inclusive and participatory techniques to improve legitimacy and ownership. To encourage involvement and collaboration, it is crucial to establish community-based organisations, facilitate dialogue, and develop trust.

Communities with diverse cultural and linguistic backgrounds can find it difficult to communicate effectively and share knowledge (Imperiale and Vanclay 2021; Cannon 2015). The accessibility and efficacy of educational activities can be limited by traditional communication methods and language challenges, especially in multicultural contexts. Hence, to make sure that educational messages are understood by a variety of audiences, it is crucial to adopt culturally and linguistically sensitive communication techniques. In addition to this, the scope and effectiveness of educational activities can be limited by a lack of funds, inadequate facilities, and logistical issues. Hence, it is recommended that grants, collaborations, and community-based funding opportunities be prioritised to enhance community education initiatives' resilience and sustainability.

1.2.3 Governance in Disaster and Climate Risk Education

We refer to governance in disaster and climate risk education as the broad set of institutions, policies, and decision-making processes that guide the integration and implementation of educational programmes meant to increase people's knowledge, readiness, and resilience to climate change and disasters. In DCRE initiatives at all levels, from national to local, and across diverse sectors, including education, the environment, and disaster management, effective governance should provide and support coordination, coherence, and accountability (Sakurai 2016; Forino et al. 2015). At the national level, governments play an essential role in developing policies, plans, and guidelines that support DCRE in educational institutions and broad frameworks for disaster risk reduction. This can be a plan of action for integrating climate change and disasters into formal curriculum, teacher training programmes, and community outreach activities. At the local level, decentralised governance structures, such as community organisations, school boards, and local governments, are essential for delivering DCRE programmes. Given that the local governments are often the ones closest to the impacted communities, they are well positioned to encourage community involvement, collect resources, and establish collaborations among multiple stakeholders.

One of the main challenges of DCRE's governance is policy fragmentation, which occurs when many government sectors and agencies create independent policies and projects without necessary coordination. This fragmentation can sometimes result in effort duplication and inconsistencies. To avoid this, establishing a system to ensure the consistency of DCRE initiatives, such as cross-sectoral cooperation and inter-ministerial coordination, is highly recommended. In addition, many nations lack the institutional capacity and funding required to implement efficient DCRE governance structures. Enhancing institutional capacity through programmes for knowledge exchange, technical support, and capacity building will be essential to the success of the DCRE governance. It is also important to note that the effectiveness of the DCRE governance depends on collaboration and interaction with non-state players, such as academic institutions, the commercial sector, and civil society organisations. Hence, establishing platforms for discussion, advice, and collaborative decision-making can enable various stakeholders to provide DCRE initiatives with their knowledge, assets, and viewpoints.

1.2.4 Capacity Building for Disaster and Climate Risk Education

In this book, the term "capacity building" refers to the activities used to improve the abilities; resources; and knowledge of people, groups, and organisations engaged in disaster and climate risk education initiatives. Building capacity effectively can help stakeholders better understand, anticipate, and respond to climate-related risks and disasters, which ultimately promotes resilience and sustainable development (McBean and Rodgers 2010). A wide range of stakeholders is the focus of capacity-building initiatives, including vulnerable populations, emergency responders, educators, policymakers, and community leaders. Building capacity can promote preparedness, adaptability, and resilience at different levels by providing these stakeholders with the information, skills, and tools they need.

One of the main challenges in capacity building for DCRE is the difficulty of maintaining long-term capacity development, as training initiatives can be discontinued when initial project funds or support end. This can lead to the benefits of capacity building eventually diminishing over time. Enhancing sustainability can potentially be achieved via institutionalising and mainstreaming capacity-building initiatives within current frameworks and processes, for example, including capacity building in professional development programmes, organisational policies, and formal education curriculum.

1.2.5 Technology for Disaster and Climate Risk Education

The use of technology can greatly enhance the effectiveness and accessibility of disaster and climate risk education programmes. A wide variety of technological tools and innovations can be used to disseminate information and raise awareness about the risks associated with climate change and disasters, ranging from digital platforms to remote sensing technology. Through the use of technology, stakeholders can work together across geographic boundaries, access instructional materials, and participate in interactive learning experiences.

Virtual reality (VR), augmented reality (AR), social media platforms, online learning, remote sensing, mobile apps, geographic information systems (GIS), and interactive multimedia tools are some of the technological applications that can be used in DCRE. With the use of these technologies, users may receive real-time information on risks and the effects of climate change, visualise geographical data, and simulate disaster scenarios. They can also engage in virtual training exercises.

One of the challenges of using the technology for DCRE is the digital divide and access issues. Low-income groups, remote locations, and disadvantaged people may not have access to the required Internet infrastructure, software, or technology. Another challenge in using technology for DCRE is misconception or confusion, which may result from inaccurate content, out-of-date data, and misinformation, which can undermine the efficacy of technology interventions. Fact-checking techniques, peer-review methods, and quality assurance mechanisms can all be used to confirm the accuracy and reliability of information published via online platforms.

1.3 Book Outline

We divided this book into five sections: 1. school education, 2. community education, 3. governance, 4. capacity building, and 5. technology. Each section focuses on a unique aspect of disaster and climate risk education.

Section 1: School Education

The chapters in this section explore how DCRE can be included in educational practices and school curriculum. They address educational methodology 'citizenship pedagogies (CP)', which embeds DRR education in the existing provision of citizenship education, earthquake early warning in the western United States: special considerations for schools, a reflection of school education, enhancing educational efforts within disaster-stricken areas, map the present status of disaster education in national curriculums, and enhancing the integration of DCRE principles into education practices.

Section 2: Community Education

The emphasis of this section is on community-based approaches to DCRE. The chapter covers the following topics: community education and engagement for disaster risk reduction; school community engagement for proactive risk reduction; climate change and the human response to climate change; and climate change education through local education programmes.

Section 3: Governance

The development of DCRE policies and initiatives is heavily influenced by governance. This section's chapters explore the governance of disaster and climate risk at the local and national levels, looking at the frameworks for governance in different countries, as well as the empowerment of vulnerable communities.

Section 4: Capacity Building

Building capacity is essential for the successful implementation of DCRE. This section's chapters include enhancing capacity building for disaster and climate risk education through teacher training and first responder education and analysing capacity-building initiatives in countries like Serbia and the Philippines.

Section 5: Technology

Technology provides innovative ways to improve DCRE initiatives. This section consists of chapters that explore the use of augmented and virtual reality for immersive learning experiences, the significance of community education and engagement in blue-green infrastructure projects in India, and the role of conventional and disruptive technologies in disaster and climate education in China.

Overall, this book offers an in-depth overview of DCRE, addressing a variety of subjects, including technological innovations, governance, and school education. Each section advances knowledge of the critical role that education can play in fostering preparedness and resilience against the effects of climate change and disasters at different societal levels.

References

Anderson A (2010) Combating climate change through quality education. Brookings Global Economy and Development, Washington, DC

Baudoin MA, Henly-Shepard S, Fernando N, Sitati A, Zommers Z (2016) From top-down to "community-centric" approaches to early warning systems: Exploring pathways to improve disaster risk reduction through community participation. Int J Disaster Risk Sci 7:163–174

Birkmann J, Sorg L, Welle, T. (2017). Disaster vulnerability. Palgrave Handb Unconv Risk Transf:329–356

Blaikie P, Cannon T, Davis I, Wisner B (2014) At risk: natural hazards, people's vulnerability and disasters. Routledge

Bonifacio AC, Takeuchi Y, Shaw R (2010) Mainstreaming climate change adaptation and disaster risk reduction through school education: Perspectives and challenges. Clim Chang Adapt Disaster Risk Reduct: Issues Chall 4:143–169

Cannon T (2015) Disasters, climate change and the significance of 'culture'. In: Cultures and disasters. Routledge, pp 88–106

Feinstein NW, Mach KJ (2020) Three roles for education in climate change adaptation. Climate Policy 20(3):317–322

Forino G, von Meding J, Brewer GJ (2015) A conceptual governance framework for climate change adaptation and disaster risk reduction integration. Int J Disaster Risk Sci 6:372–384

Gaillard JC, Mercer J (2013) From knowledge to action: Bridging gaps in disaster risk reduction. Prog Hum Geogr 37(1):93–114

Imperiale AJ, Vanclay F (2021) Conceptualizing community resilience and the social dimensions of risk to overcome barriers to disaster risk reduction and sustainable development. Sustain Dev 29(5):891–905

IPCC (2022) Climate Change 2022: Impacts, Adaptation and Vulnerability. In: Contribution of working group II to the sixth assessment report of the intergovernmental panel on climate change. Retrieved from https://www.ipcc.ch/report/sixth-assessment-report-working-group-ii/

Kumar P, Sahani J, Rawat N, Debele S, Tiwari A, Emygdio APM, Pfautsch S (2023) Using empirical science education in schools to improve climate change literacy. Renew Sustain Energy Rev 178:113232

McBean G, Rodgers C (2010) Climate hazards and disasters: the need for capacity building. Wiley Interdiscip Rev: Clim Chang 1(6):871–884

Monroe MC, Plate RR, Oxarart A, Bowers A, Chaves WA (2019) Identifying effective climate change education strategies: A systematic review of the research. Environ Educ Res 25(6):791–812

Nurdin N, Rafliana I, Hidayati S, Oktari RS, Djalante R (2017) Integrating disaster risk reduction and climate change adaptation into school curricula: from national policy to local implementation. Disaster Risk Reduct Indones: Prog, Chall, Issues:213–234
Parsons M, Glavac S, Hastings P, Marshall G, McGregor J, McNeill J, Stayner R (2016) Top-down assessment of disaster resilience: A conceptual framework using coping and adaptive capacities. Int J Disaster Risk Reduct 19:1–11
Ronan KR, Alisic E, Towers B, Johnson VA, Johnston DM (2015) Disaster preparedness for children and families: a critical review. Curr Psychiatry Rep 17:1–9
Rousell D, Cutter-Mackenzie-Knowles A (2020) A systematic review of climate change education: Giving children and young people a 'voice' and a 'hand' in redressing climate change. Children's Geogr 18(2):191–208
Sakurai A (2016) Governance of the education sector and disaster risk reduction. Disaster Resil Educ Syst: Exp Jpn:11–23
Satterthwaite D (2011) Why is community action needed for disaster risk reduction and climate change adaptation? Environ Urban 23(2):339–349
Shaw R (2015) Disaster and climate change education in Pakistan. Disaster Risk Reduct Approaches Pak:315–335
Shaw R, Kobayashi SH, K, Kobayashi M (2004) Linking experience, education, perception and earthquake preparedness. Disaster Prev Manag: Int J 13(1):39–49
Shaw R, Takeuchi Y, Rouhban B (2009) Education, capacity building and public awareness for disaster reduction. Landslides–disaster Risk Reduct, 499–515
Shaw R, Takeuchi Y, Shiwaku K (2013) Disaster education in schools. In: Disaster management. Routledge, pp 82–96
Shaw R, Takeuchi Y, Ru Gwee Q, Shiwaku K (2011) Chapter 1 Disaster education: an introduction. In: Disaster education. Emerald Group Publishing Limited, pp 1–22
Shiwaku K, Shaw R, Chandra Kandel R, Narayan Shrestha S, Mani Dixit A (2007) Future perspective of school disaster education in Nepal. Disaster Prev Manag: Int J 16(4):576–587
Stevenson RB, Nicholls J, Whitehouse H (2017) What is climate change education? Curric Perspect 37:67–71
Torani S, Majd PM, Maroufi SS, Dowlati M, Sheikhi RA (2019) The importance of education on disasters and emergencies: A review article. J Educ Health Promot 8(1):85
UNDRR (2024) Sendai framework terminology on disaster risk reduction. Retrieved from https://www.undrr.org/terminology/disaster-risk
UNDRR (2021) UNDRR strategic framework 2022–2025. Retrieved from https://www.undrr.org/publication/undrr-strategic-framework-2022-2025
Wachinger G, Renn O, Begg C, Kuhlicke C (2013) The risk perception paradox—implications for governance and communication of natural hazards. Risk Anal 33(6):1049–1065
Yildiz A (2022) Analysing children's natural hazard risk perception and preparedness. Doctoral dissertation, University of Portsmouth. https://pure.port.ac.uk/ws/portalfiles/portal/70394480/Thesis_YildizA_FINAL.pdf
Yildiz A, Dickinson J, Priego-Hernández J, Teeuw R, Shaw R (2023) Effects of disaster education on children's risk perception and preparedness: A quasi-experimental longitudinal study. Geogr Journal
Yildiz A, Teeuw R, Dickinson J, Roberts J (2021) Children's perceptions of flood risk and preparedness: A study after the May 2018 flooding in Golcuk, Turkey. Prog Disaster Sci 9:100143

Chapter 2
Preparing for Disasters Through Citizenship Pedagogies

Hans Svennevig, Kaori Kitagawa, and Masayoshi Inoue

Abstract The subject area which should include DRR education is an ongoing debate. At the same time, school hours are crammed without a margin in many national curricula. This chapter proposes an educational methodology, 'citizenship pedagogies (CP)', which embeds DRR education in the existing provision of citizenship education. The version of 'citizenship' we use works to ensure inclusion and social action by utilising theories of universal citizenship. The principle of CP is that children have agency and can make decisions through deliberation. Drawing on a case study from Japan where DRR education is implemented in CP as part of the Integrated Study in an elementary school, we explore implementation in Citizenship education through Key Stages 3 and 4 in England. This provides opportunity for policymakers and teachers to enact DRR into existing curricula, improving capacity and responses to disasters from collaborative and participatory student, school and community co-creation, and communication methods.

Keywords Citizenship pedagogies · Disaster education · Citizenship education

---

The original version of the chapter has been revised: Author provided correction has been updated. A correction to this chapter can be found at
https://doi.org/10.1007/978-981-97-5987-3_24

H. Svennevig (✉) · K. Kitagawa
Faculty of Education and Society, IOE, UCL's, London, UK
e-mail: h.svennevig@ucl.ac.uk

K. Kitagawa
e-mail: k.kitagawa@ucl.ac.uk

M. Inoue
Ehime University, Matsuyama, Japan
e-mail: inoue.masayoshi.xk@ehime-u.ac.jp

A. Yildiz and R. Shaw (eds.), *Disaster and Climate Risk Education*, Disaster Risk Reduction, https://doi.org/10.1007/978-981-97-5987-3_2

2.1 Introduction

This chapter proposes 'citizenship pedagogies (CP)' as a sustainable educational methodology for developing community preparedness and resilience against disasters triggered by natural hazards. CP is a work-in-progress methodology being developed by a group of researchers based on our work (Inoue 2021, 2020; Samaddar et al. 2022, 2021; Kitagawa 2016, 2020, 2021; Svennevig 2014).

The chapter broadly follows the definition of 'disaster education (DE)' offered by Shaw, Shiwaku, and Takeuchi: 'disaster education', 'disaster risk education', and 'disaster prevention education' are 'different expressions that essentially mean disaster risk reduction (DRR) education' (Shaw et al. 2011; Dorosz and Quinn 2019). The term 'education' tends to imply schooling, but we use disaster education (DE) to encompass broad learning activities of DRR in formal, informal, and non-formal settings. The importance of the role of 'Knowledge and Education' for DRR was highlighted in the Hyogo Framework for Action 2005–2015, which emphasised formal and non-formal education and awareness-raising as important components for DRR. The Sendai Framework for DRR 2015–2030 has taken a more embedded stance referring to education, training, and capacity building across all four priorities. As Shiwaku et al. (2016) put it, 'education' is regarded as 'a crosscutting issue to achieve the four priorities'.

The midterm review of the Sendai Framework issued in March 2023 reports that since 2015 many governments have made greater investments to engage the public in risk communication and recognised the significant role of DE in increasing resilience (UNDRR 2023a). The review introduces some good practices for the implementation of DE. To list a few, in Türkiye, the government declared 2021 a Disaster Training Year, providing training through online and in-person channels to more than 60% of the population to foster preparedness in them (UNDRR 2023a; Prambudi 2017). Education programmes for the general public were introduced in Cambodia to persuade them to invest in climate-resilient infrastructure, such as wind-resistant homes (UNDRR 2023a). To address gender gaps in DRR as well as political processes, Gender approach Advocacy Training Series for women was implemented in Columbia (UNDRR 2023b, Cornish et al. 2023). Mauritius focussed on the capacity building of communities by designing the Community Disaster Response Program, which supports volunteering for vulnerable populations and empowers communities to engage in DRR (UNDRR 2023b; Gong et al. 2021). Recognising such innovative achievements, the review also reinforces four priority areas stressing continuous investment in DRR training, education, and capacity building in fostering DRR culture through 'all-of-society approaches' (UNDRR 2023a).

This chapter first discusses the three major challenges of the provision of disaster education addressed in recent disaster risk reduction literature and how citizenship pedagogies respond to those challenges. The chapter also looks at citizenship education (CE) and particularly our conceptualisation of 'citizenship'. We then move on to elaborate on the CP framework that we propose. This is followed by an illustration of its practice in Japan and potential development in England. Our conclusion is that

CP can be used to address the shortfall in access to disaster education within school curricula. It can be used to engage both schools and the community in practice to improve disaster education communication.

2.2 Challenges of Disaster Education

We suggest citizenship pedagogies as one methodology that can respond to the current challenges of the provision of disaster education, which are threefold. Two of the challenges derive from the call for 'all-of-society engagement' in the Sendai Framework. Policy-makers and academics have mainstreamed 'participatory' and 'community-based' approaches to engaging as wide populations as possible for reducing disaster risks and adapting to climate change (Kitagawa 2019; McNamara and Buggy 2017; Samaddar et al. 2022, 2021). Despite the extensive number of participatory projects implemented, two problems remain: (1) people do not always wish to participate in DRR and climate change adaptation (CCA) initiatives and (2) many initiatives discontinue once the funding terminates (Kitagawa 2019; Paton et al. 2015; Shaw 2015; Yamori 2020). There is emerging evidence to show that even in the communities recognised for their 'good practice', a limited number of members participate regularly (Kitagawa and Samaddar 2022). This indicates the complexity of 'participation' and a need to rethink how best to improve the quantity and quality of participation in DE.

Our premise is that the above two challenges of participation can be mitigated by enabling agency and action in learners from a young age. Some DRR researches have already demonstrated that 'starting early' is an effective pathway for resilient communities (Luetz and Merson 2020; Prambudi 2017; Rofiah et al. 2021). Children growing up with high disaster awareness is significant in times of climate emergency. With its principle of diversity, CP positions children and young people to have a stake in society 'children are citizens' (Jerome and Starkey 2021; Kouno (2021); Osler and Starkey 2005). Existing social systems tend to treat children as subject to protection. Citizenship pedagogies agree with the importance of protection but focus more on children's agency and capacity to collaborate and make decisions. This involves action, which is facilitated through deliberation, co-creation, and co-decision-making processes in CP. Besides CP considers children as catalysts for the formation of solidarity in the community. Children bring their disaster learning to adults around them, building 'common values', e.g. preparedness as collective social action, in communities as participation becomes an everyday matter and, hence, sustainable.

The third challenge concerns school curricula—which subject area should include disaster education has been an ongoing debate. In 2012, UNICEF/UNESCO published a consultancy report, 'Mapping of Global DRR Integration into Education Curricula' (Selby et al. 2012). It illustrated various ways in which DRR was integrated into national curricula. The most identified approach was DRR-related topics being incorporated into certain subjects, particularly the physical and natural sciences. Few cases were found in which DRR was the major focus as a special subject area. The number of cross-curricular or interdisciplinary approaches to DRR was limited. The continuity of DRR learning from the primary-to-secondary phases was limited (Selby et al. 2012). In the past decade, since these findings were announced, DRR integration should have advanced in many countries. However, which subject is the most suitable area to teach DE remains a challenging question for many countries. Geography as the 'carrier' subject for DE has remained a popular option (Gong et al. 2021; Kamil et al. 2020). Some researchers have argued for a need for connected thinking between multiple subjects. Park (2020), for example, situates DE in science and technology studies and, at the same time, emphasises their connections with human aspects and social contexts. Disasters being 'failures of socio-technical systems', analysing both science and society is required to reduce future disaster risks (Park 2020). Ye and Xu (2023) discuss the importance of thematic learning and how this can be used to develop problem-solving skills and concepts such as critical thinking. In their example, they consider how students over a series of subjects can learn about weather-related disasters and how this enables creative thinking to be established to develop solutions to complex problems. The challenge is to ensure a curriculum can host this multi-modal exercise while supporting teachers to have the expertise and time to respond to it (Ye and Xu 2023). Such connectedness is valued in the Integrate Study, a cross-curricular hour at the primary level, which is widely used for DE in Japan (Kouno 2021; Todoroki and Yamashita 2022).

The question of which subject to host DE is conceptual, as well as practical, because school hours are crammed with many national curricula to cover without extra space to fit DRR topics. There is a substantial volume of anecdotal evidence on this point. For example, in a workshop which discussed pedagogical approaches in teaching climate justice at the Association for Citizenship Teaching annual conference in London, UK, June 2023, several attendees in conference sessions expressed the difficulty of including more topics in current timetables. In this regard, DE can be integrated into the present curriculum provision through CP without creating extra hours. If there is no extra burden on teachers and schools, it is likely CP will enable continuity of teaching disaster education.

2.3 Citizenship Education

Citizenship education has an extensive history as written about from Plato about Socrates to Marshall on Citizenship and Social Class (1950). The delivery of the subject in curricula is an often-contested ideal, depending on the nation in which it is delivered. Osler and Starkey (2005) define citizenship as a feeling, status, and practice, and it is in this form that we see citizenship pedagogies as a method in which to teach about disaster education. This form of citizenship education requires knowledge about a range of topics, including but not limited to human and child rights; peace education; wide overarching international agreements such as the UN sustainable development goals and political, legal, financial, and critical media literacy.

It requires students to be able to problem-solve, and critically engage with a wide range of complex themes and topics. Beyond this, it requires engagement with the community through active citizenship.

Citizenship education in England is about the development and teaching of social and moral responsibility, political literacy, and community involvement as expressed in 'The Crick Report', which led to the formation of the subject in the English national curriculum (Qualifications and Curriculum Authority 1998). The 'Crick Report' was influential in Japan, and the 2000s experienced many policy developments. In 2006, 'The Declaration of Citizenship Education' was announced by the Ministry of Economy, Trade and Industry (METI) as an outcome of one of the working groups that discussed the socio-economic benefits of citizenship education (Fujiwara 2008; Saito 2020). In parallel, MEXT piloted an initiative of the new subject 'Citizenship' at selected elementary (6–12 years old), lower (13–15 years old), and upper (16–18 years old) secondary schools across Japan. Although Citizenship has not been established as an independent subject area in the Courses of Study (national curriculum), citizenship education is promoted in the latest 2017/18 revised version of the Courses of Study for both elementary and lower secondary levels as follows:

> In providing moral education, each school should take particular care to ensure that it contributes to the development of Japanese citizens who would apply a spirit of respect for human dignity and reverence for life in specific activities at home, school and in other social situations, have rich humanity, respect traditions and culture, love one's country and hometown…, contribute to culture…, honor the public spirit, make an effort to develop society and the state, respect other countries, contribute to world peace and the development of the international community and the preservation of the environment, and have interest in exploring new possibilities for the future (MEXT n.d.).

Citizenship education tends to be covered under the theme 'civics' in the subject of Social Studies[1] in both elementary and lower secondary curricula. Some schools may spend time on citizenship education in the subject Integrated Study, which is a cross-curricular hour that enables to look at diverse current affairs, e.g. international understanding, environmental issues, applying a range of pedagogical approaches from problem-solving to fieldwork (MEXT n.d.).

In England, citizenship education was established as part of the national curriculum in 2002; since then, a few additional reports have been written to strengthen its scope. One of the recommendations of the 'Ajegbo Report' considering Diversity and Curriculum was for citizenship education to focus on local contexts and improve student voice (DfES 2007). Preston (2012) makes the relevant point that when using citizenship education in DE, we need to counter exclusionary and nationalistic practices that can exist in curricula. By utilising theories of universal citizenship from scholars such as Osler, Starkey, Jerome, and others while responding to national curricula citizenship pedagogies is designed to ensure inclusive community disaster education.

While the English curriculum was reviewed in 2013, Citizenship was retained with just under 22,000 students taking the full course GCSE examination in 2023

[1] Social studies comprise social studies, geography, history, and civics in the elementary level and geography, history, and civics in the lower secondary level.

(ACT 2023). In 2018, the House of Lords recommended in their Ties that Bind report on the state of Citizenship and civics that Citizenship should be strengthened as a subject in the English context, in response from 2019 OFSTED the school inspectorate in England required that:

> learners for life in modern Britain by: equipping them to be responsible, respectful, active citizens who contribute positively to society; developing their understanding of fundamental British values; developing their understanding and appreciation of diversity; celebrating what we have in common and promoting respect for the different protected characteristics as defined in law (OFSTED 2019)

While many in the subject did not feel this inclusion went far enough, it does demonstrate that citizenship education is required to be taught in schools and gives mention to participation in society. Considering both the inclusion of citizenship education in the English and Japanese contexts we propose that both methods have similar pedagogic engagement, and therefore draw them together when considering how they can be used to teach disaster education.

2.4 Citizenship Pedagogies

As CP is a form of participatory action research (PAR) for its emphasis on proactive *p*articipation, social *a*ction, and *r*esearch, it improves social capital in communities (Cornish et al. 2023; Kindon et al. 2007; Vaughn and Jacquez 2020). Participants will experience the cyclical PAR process demonstrated in Fig. 2.1.

As climate change affects the whole globe, we promote international engagement. The CP design enables an international exchange between different country contexts to share communities' experiences of and responses to climate-related disasters. Local children and young people may film their stories of disaster experiences and local DRR and CCA measures. Storytelling can be facilitated for the exchange, given that the effectiveness of storytelling for disaster recovery and preparedness has been well documented (Dorosz and Quinn 2019; Hou 2019; Jung et al. 2013; Mangione et al. 2013). Schools can create a space on their websites in which student ideas can be shared, and feedback can be exchanged between community groups. Through such exchange and reflection, children and young people can achieve increased climate and preparedness awareness as they become interested in not only self-help measures such as preparing disaster relief bags but also in mutual aid and public assistance measures like the development of disaster prevention warehouses. By enhancing children's awareness to the impact of reflecting on shared views and communicating with others, they improve their social participation and solidarity, which in turn can build and improve sustainable societies.

The implementation cycle of CP practice includes co-evaluation from all participant perspectives to ensure continuity. Co-evaluation can be undertaken in a workshop format as part of a school event to feedforward to the subsequent CP cycle. The storytelling exchange will be an opportunity for co-evaluation. Being part of the

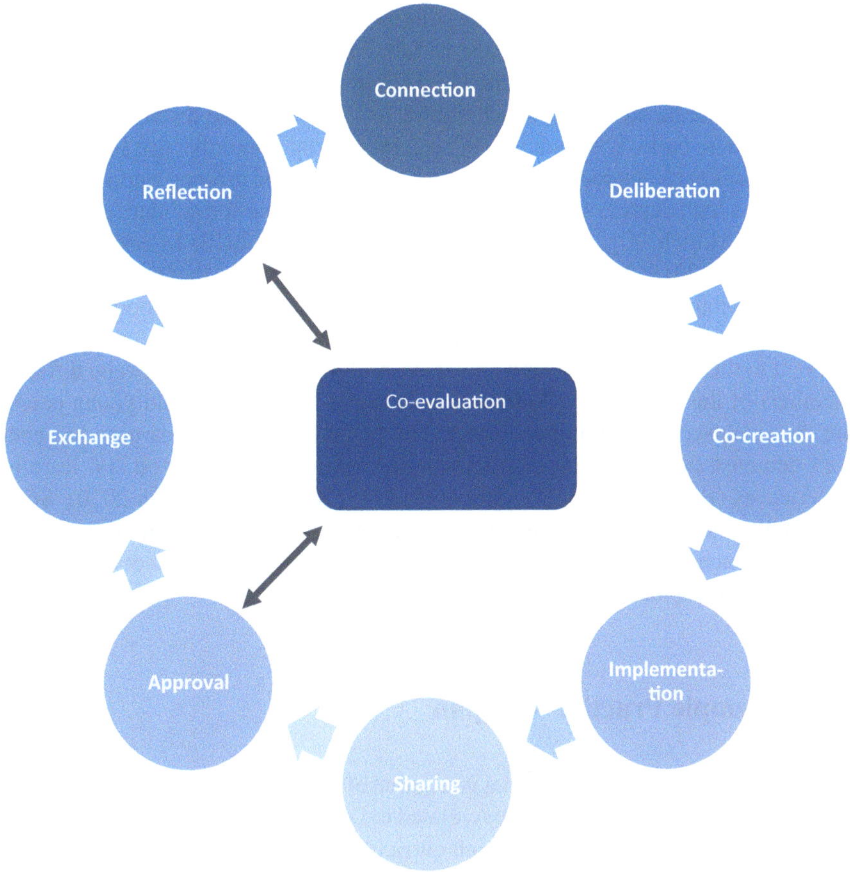

Fig. 2.1 CP procedure cycle (based on Inoue 2020). Connection: external human resources (e.g. academic expert) introduce the topic, and students connect the problem with their real-life situations; deliberation: students brainstorm in groups how to address the problem through deliberation; co-creation: external human resources and students co-create preparedness initiatives for the problem; implementation: students implement the initiatives and review the outcomes; sharing: students disseminate the initiatives; approval: community members' feedback to students; reflection: external human resources, students, and community members reflect on the counterparts' feedback on the initiatives using, e.g. school websites

evaluation allows the development of a sense of ownership in children and young people, which contributes to their sustainable proactive participation. Furthermore, schools could compare the evaluation outcomes of CP between different contexts to identify what can be learned from each other to advance CP and disaster preparedness. Meaningful reciprocal learning can be expected, particularly in schools' local areas that have high disaster risks.

We propose CP can address the two challenges for participatory projects identified in an earlier section (people do not always wish to participate in projects; many projects are temporary without funding) in the following ways. Firstly, CP enables agency in learners from a young age to familiarise themselves with participating in social actions; secondly, by embedding CP in the present educational provision, schools will not be burdened, which allows sustainability; and thirdly, by inspiring communities to build 'common values', e.g. preparedness as collective social action, which widens participation for continuity. As demonstrated in many studies, communities with a high degree of social capital exercise resilience and solidarity in disaster situations (Aldrich 2012; Kargillis et al. 2014; LaLone 2012; Li and Tan 2019; McEwen et al. 2018; Murphy 2007). Aldrich (2012) considers that the development of social capital is different between villages and cities and neighbourhoods, which is much to do with the organisation of the different community members—what appears relevant to them in their work and how they 'bridge' participation between different members of the community. Aldrich (2012) shows that children and young people are the catalysts for building solidarity. The power of vicarious learning has been well demonstrated—learning can occur by observing and sharing others' learning experiences (Bandura 1986; Bandura et al. 1963; Kuri and Kaufman 2020). Some community members may get involved in or even start leading community-based DE activities. Such inclusive and intergenerational interactions will further increase the level of social capital.

2.5 Example Practice in Japan

Inoue (2020, p. 84) summarises what is required of DE in Japan as follows: 'the development of lessons dealing with issues related to DRR measures' and 'the planning of DE imagining the association between curricular and extracurricular activities'. In response, he developed a form of DE incorporating the principles of CE that 'children are citizens' (Jerome and Starkey 2021). Conventionally, in Japanese school education, teachers who have knowledge instruct pupils who do not have the knowledge, and thus children are 'objects'. Supporting Osler and Starkey's perspectives that children have the right to be informed and participate in decision-making as citizens, Inoue argues for DE that fosters citizenship in pupils, focussing on the notion of 'social participation'. As opposed to 'passive social participation', which is limited to students' acquiring new knowledge and skills, he aims for pupils' 'simulated social participation' and 'proactive social participation' (Fujiwara 2011; Inoue 2020). The former involves external professionals coming to the classroom to interact with pupils—bringing society into school, simulating a public space. The latter indicates schools reaching out to society by pupils actively engaging in social action to make a positive change. Such a two-way relationship between school and society is critical in enabling children to formulate their responsible views and to collaboratively make decisions through deliberation.

Inoue's (2021) original form of CP comprises two broad phases, which include some sub-phases. 'The formation of own opinion' is the first phase in which pupils understand the problem, realise they can partake in tackling the problem, and develop their own views. They have an opportunity to interact with external DRR experts who offer their versions of solutions. The second phase is 'the reformulation of the opinion' whereby the pupils engage in deliberation with the experts to exchange perspectives. This process exposes pupils to new and different viewpoints, allowing them to reflect on their initial opinions. The experts will feedback to the pupils positively when encountering innovative thoughts. This assures the pupils that they are also the subject of social action and increases their motivation to engage more. Both parties then finalise their solution, bearing in mind the differences of the opinions. The most important point in this final part is the pupils become aware that the revised solution which they co-constructed with the experts will be implemented in the real world, i.e. their learning outcome is connected to society (Inoue 2021).

As one of his CP practices, Inoue collaborates with an elementary school in Seiyo City, Ehime Prefecture. Using the Integrated Study hours, he delivers the DE package lessons developed with city officials. For example, one of the lessons looked at the topic of the stockpile storage located in the school playground. The school and the city government needed to put an instruction displayed on the storage to inform evacuees what to do in an emergency. The problem was they could not agree on what to write on the note, and therefore decided to involve pupils. Deploying the above process of the CP phases, Inoue invited pupils to discuss in groups what they think should be written on the note. This was followed by the exchange of opinions between the groups and the external DRR expert—it was Inoue himself in this lesson. He introduced a need to consider non-Japanese residents in the town, although small in number. Incorporating such information, the groups revised their notes, which were then shared with a city official who was present at the lesson. He then offered feedback to the whole class appraising some of the unexpected ideas such as the creation of an audio version for people who have difficulty reading (Kitagawa and Samaddar 2022). Thus, CP requires collaboration and commitment between schools, the local government, and universities (or other expert organisations) with a belief that children do have a stake in society, and they should understand that.

2.6 Potential Application in England

Up to 2023 disaster education was not commonly taught or required to be taught in the UK. The content, concepts, knowledge, and teaching activities of citizenship education are often contested and lead to many scholars being engaged in improvements in this area; this means that while it has a rich history, CE is open to further development as a relatively new required subject in school curricula in England. The Association for Citizenship Teaching a membership body for educators engaged in citizenship education works to further spread awareness and opportunities of the delivery of citizenship education in schools and could additionally be used to develop curriculum materials in this field if it was used as a vehicle for DE. We propose that using CP to teach DE through the English citizenship education curriculum will enable the subject to be taught effectively within existing curricula and curriculum

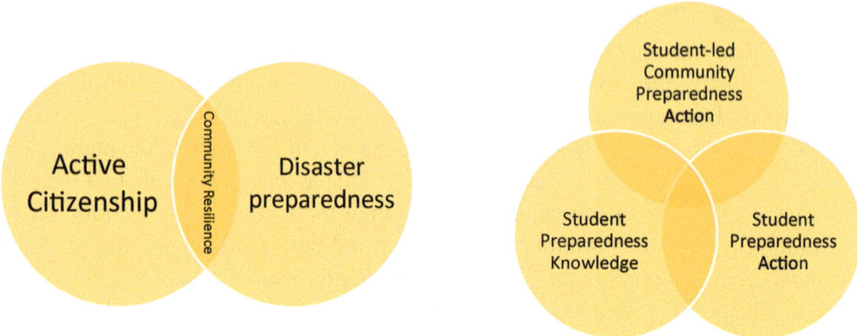

Fig. 2.2 Citizenship education teaching disaster education (Svennevig 2014)

time and improve community DE through community involvement and participation. For example, as demonstrated in Fig. 2.2, when considering one aspect of disaster education; disaster preparedness, active citizenship education can be used to build community resilience. To do this, students in schools need to be educated with disaster preparedness knowledge, the skills to act on their knowledge and the skills to disseminate this to communities, skills that should be taught through active citizenship as part of citizenship education.

We posit that the development of CP to teach DE should start with trainee teachers of citizenship teacher education programmes and with the introduction of continual professional development from organisations such as the Association for Citizenship Teaching to establish CP in teacher education. Thus, CP will trigger cascading impacts through participated trainee teachers' practice of CP at future schools, pupils influencing their peers, families, and neighbours and more schools playing a role of a hub in the region in building preparedness and resilience.

There are multiple areas of the current 2013 Key Stage 3 and 4 English curriculum including through examination-related content where DE via CP can be incorporated (Department for Education 2013). We identify some of these in Tables 2.1 and 2.2 mapping the curriculum to disaster education opportunities.

When considering managing 'risk' within the key stage 3 curriculum above, while this is often linked to financial and economic literacy, adaptations in this requirement to teach disaster education would be relatively straightforward and could further assist in the teaching of disaster education around climate change adaptation and sustainable futures. In 2023, some non-governmental organisations including Send my Friend to School, The British Red Cross, and Picture News include DE within their citizenship education programmes as supporting teachers to teach around this managing 'risk' element of the 2013 curriculum.

As a consequence of the range and scope of these curriculum requirements, disaster education can be inserted into curriculum content without more time being required in schooling. Research has shown children and young people are the drivers of developing resilience and solidarity in communities (Aldrich 2012). Deriving from local, universal, and global citizenship perspectives, it is important to centre

Table 2.1 Key stage 3 citizenship education and disaster education opportunities (adapted from Svennevig and Kitagawa 2023)

English citizenship national curriculum (Summary) key stage 3	
Extracted and adapted from curricula requirement (knowledge and skills)	*Disaster education opportunity*
UK political systems	Scrutinising policy, i.e. disaster mitigation
Precious liberties	Rights that citizens have before, after, and during disasters
Legal justice system	How to protect, scrutinise, and inform disaster policy
Role of community groups and participation in school	Learn about disaster practice and communication
Managing 'risk'	Risk management/perception

Table 2.2 Key stage 4 citizenship education and disaster education opportunities (adapted from Svennevig and Kitagawa 2023)

English citizenship national curriculum (Summary) key stage 4	
Extracted and adapted from curricula requirement (knowledge and skills)	*Disaster education opportunity*
Critical media literacy	Communicating/education around risk/disaster/mitigation
Holding those to power to account	Scrutiny of policy, select committees, prevention
Human rights, international law, and relationships	International agreements, COP, climate, treaties, etc. Campaigning/lobbying on a global scale. Climate science meets action
Inclusion and diversity	Making sure disaster communication is inclusive
Responsible participation (active Citizenship) and volunteering	Disaster preparedness/community action/communication/drills

DE through knowledge around CP and curricula and then develop practice related to active citizenship and community involvement. Teachers and students as the heart of this transformative practice will gain affirmation and confidence in 'being a citizen', while learning about disasters including those caused, for example, by climate change. Teachers, in particular, will be able to acquire new pedagogical techniques and curriculum tools for citizenship education and disaster education, some of which will be transferrable and provide thematic links in teaching other topics. Taking their experiences to future schools over their career and by being leaders in roles enable a cascade of the knowledge of disaster education to occur. Such positive experiences would be a trigger for further active participation in DE and beyond. This situates the school, teachers, and young people as hubs for passing on knowledge to other community members.

2.7 Conclusion

We conclude this chapter by summarising three areas of benefits that CP could bring to DE. Firstly, community participation models in DRR and CCA have largely remained top-down, determined by policymakers and experts. The recognition that local communities should define the concept of participation and its operational mechanism is on the rise. Drawing on the humanitarian conceptualisation of 'citizenship' introduced by Starkey and Osler, CP focusses on children, young people and their communities in enabling impactful and sustainable lifelong DRR. As a collaborative knowledge development tool, CP facilitates proactive participation and provides an alternative ontology for participation. This will directly contribute to the development of knowledge in the areas of household disaster preparedness, risk governance, participatory risk communication, and community-led DRR and CCA. We believe that one of the significant outcomes of CP is that it can contribute to the development of long-term sustainable societies.

Secondly, very few methodologies with theoretical underpinnings that enable effective participation are available in the learning and teaching of DRR and CCA. The majority of existing participation studies investigate whether the derived criteria or method of participation is empirically valid. Yet, it remains uncertain whether the criteria and collective knowledge of participation can be used to facilitate participatory DRR and CCA. Context and time dimensions are another methodological challenge in participatory approaches. CP aims to address these by investigating participation as a co-creation of knowledge and from the participants' perspectives. CP also has the potential to be implemented in the scientific methods of DRR, CCA, PAR, and connected wider fields. These opportunities could be used to further strengthen the incorporation of CP into curricula, and to clarify potential collaboration methods with interested organisations and institutions.

Thirdly, disaster education currently lacks effective participatory and co-creation tools where participants proactively define problems, solutions, and their applications. Citizenship pedagogies would fill this gap by developing a curriculum and pedagogy with a strong school-community collaboration ethos. CP can be embedded in the existing curricula, which means it will not burden schools that are already facing a high curriculum demand, this could be done within the English and Japanese contexts easily and extrapolated to other nation contexts in the UK or internationally where engagement with political literacy and active Citizenship occur, for example, in the USA as part of service learning, or in other nations worldwide as part of their own citizenship education. This innovation will benefit researchers and research-practitioners in the fields of disaster education, citizenship education, and teacher education and further promote interdisciplinary researchers and research-practitioners across these fields. As a participatory mechanism to enable co-creation and capacity building, CP could also be applied to other topics beyond DRR and CCA.

References

Aldrich DP (2012) Building resilience: social capital in post-disaster recovery. The University of Chicago Press, Chicago

Association for Citizenship Teaching (2023) GCSE Citizenship studies results 2023—The value of citizenship. Available at https://www.teachingcitizenship.org.uk/gcse-citizenship-studies-results-2023-the-value-of-citizenship/

Bandura A, Ross D, Ross SA (1963) Vicarious reinforcement and imitative learning. Psychol Sci Public Interest 67:601–607. https://doi.org/10.1037/h0045550

Bandura A (1986) Social foundations of thought and action: A social cognitive theory, social learning theory. Prentice-Hall, Inc, Englewood Cliffs, NJ, US

Cornish F, Breton N, Moreno-Tabarez U, Delgado J, Rua M, de-Graft Aikins A, Hodgetts D (2023) Participatory action research. Nat Rev Methods Primers 3:1–14. https://doi.org/10.1038/s43586-023-00214-1

Department for Education (2013) Citizenship programmes of study: Key stages 3 and 4 National Curriculum in England

Department for Education and Skills (DfES) (2007) Curriculum review: diversity and citizenship (Ajegbo report) PPSLS/D35/0107/14. London: DfES

Dorosz EJ, Quinn M (2019) Disaster simulation and storytelling: effective modalities for knowledge dissemination. Disaster Training in Nursing: An International Collaboration

Fujiwara T (2008) Possibilities for citizenship education in Japan: Evaluation of pilot practices. Annual Review of Academic Research, Doshisha Women's College of Liberal Arts 59:89–106

Fujiwara T (2011) Comprehensive cognition and social participation in social studies education. J Soc Stud:29–40. https://doi.org/10.18992/socialstudies.2011.113_29

Gong Q, Duan Y, Guo F (2021) Disaster risk reduction education in school geography curriculum: Review and outlook from a perspective of China. Sustain 13:3963. https://doi.org/10.3390/su13073963

Hou JZ (2019) Unlocking the power of digital storytelling for community resilience building in disaster risk communication: A case study from Australia. In: Presented at International Association of Media and Communication Research (IAMCR) conference 2019

Inoue M (2020) Principles of diaster prevention learning aiming to foster citizenship: focusing on the development of "let's think about a city that is resilient to disasters!" in a Civil Field Unit of Social Studies in junior high school. Disaster Educ Res 1:81–92

Inoue M (2021) Principles and methods of social studies lesson structure that encourage deliberation between external human resources and children: Through a development and practice of the geographical field unit "hazard maps for the region." Soc Stud Educ Res 95:1–12

Jerome L, Starkey H (2021) Children's rights education in diverse classrooms. Bloomsbury Publishing, London

Jung JY, Toriumi K, Mizukoshi S (2013) Neighborhood storytelling networks, Internet connectedness, and civic participation after the Great East Japan Earthquake. Asian J Commun 23:637–657. https://doi.org/10.1080/01292986.2013.819930

Kamil PA, Utaya S, SumarmiDH U (2020) Strengthen disaster preparedness for effective response on young people through geography education: A case study at school in the Tsunami affected area of Banda Aceh City, Indonesia. IOP Conf. Ser: Earth Environ. Sci. 412:012016. https://doi.org/10.1088/1755-1315/412/1/012016

Kargillis C, Gillham D, Kako M (2014) Disaster survivors: A narrative approach towards emotional recovery. Aust J Emerg Manag, 29:25

Kindon S, Pain R, Kesby M (ed) (2007) Participatory action research approaches and methods: Connecting people, participation and Place. Routledge, London. https://doi.org/10.4324/9780203933671

Kitagawa K (2016) Situating preparedness Eeucation within public pedagogy. Pedagogy. Cult Soc https://www.tandfonline.com/doi/full/10.1080/14681366.2016.1200660

Kitagawa K (2019) Exploring 'everyday-life preparedness': Three case studies from Japan. Int J Disaster Risk Reduct 34:265–274. https://doi.org/10.1016/j.ijdrr.2018.11.025

Kitagawa K (2020) Disaster risk reduction activities as learning. Nat Hazard https://doi.org/10.1007/s11069-020-04443-5

Kitagawa K (2021) Conceptualising 'disaster education'. Educ Sci 11(5):233. https://doi.org/10.3390/educsci11050233

Kitagawa K, Samaddar S (2022) Widening community participation in preparing for climate-related disasters in Japan. UCL Open Environ. https://doi.org/10.14324/111.444/ucloe.000053

Kouno T (2021) Contents and characteristics of elementary school disaster prevention education in the integrated learning period—Toward the development of disaster prevention education across subjects. Bull Osaka Christ Coll Semin 62:15–33

Kuri SK, Kaufman EK (2020) Leadership insights from hollywood-based war movies: An opportunity for vicarious learning. J Leadersh Stud 14:53–61. https://doi.org/10.1002/jls.21682

LaLone MB (2012) Neighbors helping neighbors: An examination of the social capital mobilization process for community resilience to environmental disasters. J Appl Soc Sci 6:209–237. https://doi.org/10.1177/1936724412458483

Li Z, Tan X (2019) Disaster-recovery social capital and community participation in earthquake-stricken Ya'an areas. Sustain 11:993. https://doi.org/10.3390/su11040993

Luetz JM, Merson J (2020) Climate change and human migration as adaptation: Conceptual and practical challenges and opportunities climate action. In: Leal Filho W, Azul AM, Brandli L, Özuyar PG, Wall T (ed). Encyclopedia of the UN sustainable development goals, Springer, Cham

Mangione GR, Capuano N, Orciuoli F, Ritrovato P (2013) Disaster education: a narrative-based approach to support learning, motivation, and students' engagement. J E-Learn Knowl Soc 9 https://doi.org/10.20368/1971-8829/837

Marshall TH (1950) Citizenship and social class and other essays. Cambridge University Press, Cambridge

McEwen L, Holmes A, Quinn N, Cobbing P (2018) 'Learning for resilience': Developing community capital through flood action groups in urban flood risk settings with lower social capital. Int J Disaster Risk Reduct 27:329–342. https://doi.org/10.1016/j.ijdrr.2017.10.018

McNamara KE, Buggy L (2017) Community-based climate change adaptation: a review of academic literature. Local Environ 22:443–460. https://doi.org/10.1080/13549839.2016.1216954

MEXT (n.d) Elementary school, lower secondary school: general provisions, national curriculum standards (2017–2018 Revision) improvement of academic abilities (Courses of study) Available at https://www.mext.go.jp/en/policy/education/elsec/title02/detail02/1373859.htm

MEXT (n.d) Hours for integrated study (Inquiry) Available at https://www.mext.go.jp/a_menu/shotou/sougou/main14_a2.htm

Murphy BL (2007) Locating social capital in resilient community-level emergency management. Nat Hazards 41:297–315. https://doi.org/10.1007/s11069-006-9037-6

OFSTED (2019) School inspection handbook. Crown Copyright

Osler A, Starkey H (2005) Changing citizenship: democracy and inclusion in education. Open University Press, Maidenhead

Park W (2020) Beyond the 'two cultures' in the teaching of disaster: or how disaster education and science education could benefit each other. Educ Philos Theory 52:1434–1448. https://doi.org/10.1080/00131857.2020.1751126

Paton D, Anderson E, Becker J, Petersen J (2015) Developing a comprehensive model of hazard preparedness: Lessons from the Christchurch earthquake. Int J Disaster Risk Reduct 14:37–45. https://doi.org/10.1016/j.ijdrr.2014.11.011

Prambudi, DI (2017) Integrating disaster mitigation education in the elementary school curriculum. In: Presented at the 1st international conference on social sciences education—"Multicultural transformation in education, social sciences and wetland environment" (ICSSE 2017), Atlantis Press, pp 79–82. https://doi.org/10.2991/icsse-17.2018.19

Preston J (2012) Disaster education: 'Race.' Sense Publishers, Equity and Pedagogy

Qualifications and Curriculum Authority (1998) Education for citizenship and the teaching of democracy in schools: Final report of the Advisory Group on Citizenship 22 September 1998 (No. QCA/98/245). London

Rofiah NH, Kawai N, Hayati EN (2021) Key elements of disaster mitigation education in inclusive school setting in the Indonesian context. Jàmbá: J Disaster Risk Stud 13:1–8 https://doi.org/10.4102/jamba.v13i1.1159

Saito M (2020) The characteristics of the concept of "citizen" in Japanese citizenship Education: drawing on METI's report and Shinagawa Ward's "Citizenship" subject. Seijo Communication Studies Research 11:21–39

Samaddar S, Oteng-Ababio M, Dayour F, Ayaribila A, Obeng FK, Ziem R, Yokomatsu M (2021) Successful community participation in climate change adaptation programs: on whose terms? Environ Manage 67:747–762. https://doi.org/10.1007/s00267-020-01421-2

Samaddar S, Si H, Jiang X, Choi J, Tatano H (2022) How participatory is participatory flood risk mapping? Voices from the flood Prone Dharavi Slum in Mumbai. Int J Disaster Risk Sci 13:230–248. https://doi.org/10.1007/s13753-022-00406-5

Selby D, KagawaUNESCO, UNICEF, F (2012) Disaster risk reduction in school curricula case studies from thirty countries. UNESCO and UNICEF, Paris

Shaw R (2015) Disaster risk management at local and community levels. In: Davis I (ed) Disaster risk management in Asia and the Pacific. Routledge; Asian Development Bank Institute, Oxford, pp 211–231

Shaw R, Takeuchi Y, Ru GQ, Shiwaku K (2011) Chapter 1 disaster education: an introduction. In: Shaw R, Shiwaku K, Takeuchi Y (eds) Disaster Education. Emerald Group Publishing Limited, Community, environment and disaster risk management, pp 1–22

Shiwaku K, Sakurai A, Shaw R (eds) (2016) Disaster Resilience of Education Systems: Experiences from Japan, Disaster risk reduction, methods, approaches and practices. Springer, Tokyo

Svennevig H, Kitagawa, K (2023) Integrating disaster education with citizenship education. In: Presented at disaster education for a just and resilient future: an international conference, University of Southampton, UK

Svennevig, H (2014) How could citizenship education be used to teach disaster preparedness? (Dissertation), University of Manchester

Todoroki Y, Yamashita N (2022) Disaster prevention education for elementary school students: From the perspective of multicultural society. Bull Educ Res Teach Dev Kagawa Univ 45:49–55

UNDRR (2015) Sendai framework for disaster risk reduction 2015–2030. In: Third world conference on disaster risk reduction Sendai, Japan

UNDRR (2023a) The report of the midterm review of the implementation of the sendai framework for disaster risk reduction 2015–2030. Midterm Review of the Sendai Framework. Geneva, Switzerland

UNDRR (2023b) Midterm review of the implementation of the sendai framework for disaster risk reduction 2015–2030: Good practices in disaster risk reduction, Extracted from formal submissions to the Midterm Review of the Sendai Framework. Geneva, Switzerland

UNISDR (2005) Hyogo framework for action 2005–2015: Building the resilience of nations and communities to disasters. In: World conference for disaster reduction. Kobe, Japan

Vaughn LM, Jacquez F (2020) Participatory research methods—Choice points in the research process. J Particip Res Methods 1. https://doi.org/10.35844/001c.13244

Yamori K (2020) Disaster risk communication: a challenge from a social psychological perspective, Integrated disaster risk management ser. Springer, Singapore

Ye P, Xu X (2023) A case study of interdisciplinary thematic learning curriculum to cultivate "4C skills." Front Psychol 14:1080811. https://doi.org/10.3389/fpsyg.2023.1080811

Chapter 3
Earthquake Early Warning in the Western United States: Special Considerations for Schools

Lori Peek, Rachel M. Adams, Jolie Breeden, Jennifer Tobin, Robert M. de Groot, and Sara K. McBride

Abstract Schools in the western United States are vulnerable to damaging earthquakes, leaving children and staff at risk to death, injury, and disruption. ShakeAlert—the U.S. Geological Survey-managed earthquake early warning system—has the potential to enhance school safety by providing rapid mass notification of an earthquake before shaking arrives. The extra seconds of warning can trigger automated system shutdowns and prompt people to drop, cover, and hold on. To explore how earthquake early warning can be more effectively integrated into schools, our research team conducted 118 interviews with school leaders, teachers, students, and other members of the school community in earthquake-affected school districts in Alaska and California. The interviews revealed that interviewees are enthusiastic about the potential of ShakeAlert to enhance school safety efforts. They also asked questions and provided feedback in the following six areas of concern: (1) awareness and knowledge; (2) funding and diversion of resources; (3) alert channels; (4) alerting threshold and frequency of alert delivery; (5) alert message content and drills; and (6) school infrastructure. This chapter offers recommendations to address these considerations and to help support schools in adopting this potentially life-saving technology.

Keywords Children · Schools · Earthquakes · Preparedness · Drills · Earthquake early warning · Alaska · California · Oregon · Washington · United States · Social science · ShakeAlert®

L. Peek (✉)
Department of Sociology and Natural Hazards Center, University of Colorado Boulder, Boulder, Colorado, United States
e-mail: lori.peek@colorado.edu

R. M. Adams · J. Breeden · J. Tobin
Natural Hazards Center, University of Colorado Boulder, Boulder, Colorado, United States

R. M. de Groot · S. K. McBride
U.S. Geological Survey, Reston, Virginia, United States

A. Yildiz and R. Shaw (eds.), *Disaster and Climate Risk Education*, Disaster Risk Reduction, https://doi.org/10.1007/978-981-97-5987-3_3

3.1 Introduction

Earthquakes rank among the costliest, deadliest, and most disruptive natural hazards in the United States and globally (FEMA 2023; Ritchie and Rosado 2024). This is due, in part, to their widespread occurrence, varying intensity levels, and unpredictability (McBride et al. 2022). Earthquakes can also generate devastating cascading hazards such as tsunamis and landslides. These events pose extreme risks to life safety in heavily populated places with aging building stock, unenforced building codes, or substandard construction practices (FEMA 2020; FEMA 2023; Kenny 2009). When earthquakes occur in locations where critical infrastructure—such as schools, hospitals, and fire stations—is not resistant to hazards, loss of life can be extensive (Motamedi et al. 2012; Peek 2008).

Earthquake early warning (EEW) systems represent a potential life-saving intervention (de Groot et al. 2022). First adopted in Mexico City in 1991, EEW can allow for automated shutdown of vital infrastructure, such as trains, and provide critical seconds of warning to people who need to take protective action (Reddy 2023). EEW systems began to proliferate in the early 2000s and are now operational in several countries. According to McBride and colleagues (2022), Mexico, Japan, South Korea, Taiwan, China, and the United States offer public alerting, while Canada and countries throughout Europe, Latin America, and Asia provide limited alerts or are testing and developing new systems.

The ShakeAlert® EEW System, which is also referred to as the ShakeAlert System and ShakeAlert, is managed by the United States Geological Survey (USGS) and entails a collaboration with academic and state partners. It is the first earthquake public alerting system in the United States. The system relies on an extensive network of ground sensors to rapidly detect earthquakes; estimate the amount of shaking; and, in an event of sufficient magnitude, rapidly issue official ShakeAlert Messages (Minson et al. 2019). ShakeAlert began delivering alerts to people who downloaded a cellular phone application and those who can receive Wireless Emergency Alerts (WEAs) on their devices in 2019 in California, followed by Oregon and Washington in 2021 (McBride et al. 2022).

3.1.1 Schools and ShakeAlert

School buildings are often deemed critical infrastructure within communities because they are fundamental to social life. They are where children receive a formal education, gain access to physical and mental health care, and are socialized by their teachers and peers (Meece and Eccles 2010). School facilities serve as hubs within communities for adults and young people, and during disasters they often are used as emergency shelters for the public (Lai et al. 2016). Despite the vital role they play in society, schools can be more vulnerable to damage in earthquakes and other natural hazards than other building types, leaving school children and staff especially at risk to injury and disruption (Applied Technology Council 2017; Fothergill and Peek 2015).

ShakeAlert-powered alerts can be delivered within schools and other institutional settings via licensed technical operators. In 2015, Eagle Rock High School, which is in the Los Angeles Unified School District in Southern California, became the first school in the nation to test ShakeAlert. During the early implementation phase, 10 of the high school's science classrooms were connected to the system, and alerts were programmed to go off on the teacher's computer (Xia 2015). In 2019, Stanwood Elementary became the first school in Washington to connect to ShakeAlert. According to a case study produced by the USGS (2022b), the Stanwood-Camano School District added alerting capability to all 13 of its buildings and connected them through an upgraded, centralized public address (PA) system that is automatically activated when shaking is expected. The PA system sounds an alert tone followed by a spoken message: "Earthquake! Earthquake! Expect shaking soon. Drop, cover, hold on. Protect yourself now" (USGS 2022b: 2). The total cost of upgrading the district PA network was under $100,000, with grants covering most of the initial expenses. Ongoing implementation costs are about $150 per month (ibid: 2). In 2022, the Medford 549C School District became the first in Oregon to adopt ShakeAlert. The district worked with a licensed technical operator to integrate ShakeAlert-powered alerts into their existing intercom system. In the event of shaking, the intercom is programmed to deliver spoken alerts telling students to expect shaking and to drop, cover, and hold on (Medford Alert 2022).

An estimated 5.5 million children attended K-12 public schools in California in 2022, with an additional 1.1 million in Washington public schools and 582,000 in Oregon public schools (United States Department of Education 2023). There are 977 school districts in the California public school system, 306 in Washington, and 197 in Oregon, with these districts encompassing thousands of elementary, middle, and high schools across the three-state region (United States Department of Education 2023). As of March 2024, however, only four school districts (one in California, two in Washington, and one in Oregon) and one private school (in California) had adopted the ShakeAlert system. While the diffusion of any innovation can take time (Rogers 2003), the adoption rate of ShakeAlert among school districts in the western United States has been slow. In response, the USGS has taken steps to provide detailed information, case studies, and bilingual resources to the education sector (Crayne et al. 2023; Sumy et al. 2022; USGS 2023).

Given the low uptake of ShakeAlert in U.S. schools and the lack of available research on EEW in schools (Adams et al. 2022), the four primary authors of this chapter conducted 118 interviews with school staff and others in four earthquake-affected school districts to explore the following questions: How can ShakeAlert be most effectively integrated into schools in the western United States? What concerns do school leaders, teachers, emergency managers, students, and other people with a stake in school safety have about the adoption and use of EEW? How might those concerns or considerations influence their willingness to use EEW in schools?

3.2 Research Sites

This chapter draws on interviews gathered in the city of Anchorage and the borough of Matanuska-Susitna in Alaska, and the city of Ridgecrest and unincorporated community of Trona in California. These sites were selected because they experienced damaging earthquakes that disrupted school districts in the affected regions. In the case of Alaska, the November 30, 2018, M7.1 Anchorage earthquake damaged all 96 Anchorage School District buildings and forced the closure of two schools. Anchorage school leadership estimated that repairs to damaged schools and necessary seismic upgrades could exceed $150 million (Tunseth 2019). In the nearby Matanuska-Susitna Borough, 47 school buildings were damaged. The district incurred more than $1.8 million in repair costs with millions more needed to replace a school that was closed due to significant structural damage (Early 2019; Rodgers et al. 2021). In California, the July 4 and 5, 2019, earthquake sequence (M.6.4, M5.4, and M7.1) caused structural and non-structural damage to schools in both the Sierra Sands Unified School District in Ridgecrest and the Trona Joint Unified School District in Trona. Sierra Sands delayed the start of the 2019–20 school year because two of the 10 schools in the district sustained substantial damage. The neighboring Trona Joint Unified School District was home to two schools before the 2019 earthquakes. After the earthquake sequence, the high school had to close due to severe damage to the building, damage to gas and water lines, and hazardous soil conditions (Ciani 2023).[1]

Because schools represent a key sector that the USGS is seeking to reach, our university-based research team added a component to our post-earthquake field research to also learn about the adoption of EEW at a pivotal time. When fieldwork was initiated in early 2020, California, which has the highest number of people and most infrastructure exposed to earthquakes nationally, had only recently gained widespread public access to the ShakeAlert system (FEMA 2023; McBride et al. 2022). Alaska, which is the most earthquake-prone state in the nation, does not yet have access to ShakeAlert, but may in the future (Salisbury et al. 2020).

The diversity of the four districts informed our site selection criteria. As shown in Table 3.1, the districts represented in the sample are in different geographical locales and of widely varying sizes, enrolling between 273 and more than 43,000 students and employing between 18 and more than 2,500 teachers. The total number of schools within each district ranged from two in Trona to 96 in Anchorage. The districts also educate and serve racially and socioeconomically diverse students and employ staff members who have a range of earthquake experiences. These and other factors made these sites sociologically important for further investigation (Phillips 1997; Peek et al. 2021).

[1] In December 2023, Trona broke ground for a new high school to replace the building that was permanently closed after the 2019 earthquakes. The new school is estimated to open in 2026. Until then, high school students will continue to attend classes at the nearby elementary school (Ciani 2023).

Table 3.1 Research sites: geographic and demographic information by school district, 2021–2022 (*Source* U.S. Department of Education 2023)

School district	Location	2021–2022 school enrollment	Full-Time Equivalent (FTE) classroom teachers	Total schools	U.S. Department of Education Location Population Classification
Anchorage	Anchorage, Alaska	43,054	2,510.26	96	City: Large
Matanuska-Susitna Borough	Houston, Palmer, and Wasilla, Alaska	19,319	993.62	49	Town: Distant
Sierra Sands Unified	Ridgecrest, California	5,100	233.72	11	Town: Remote
Trona Joint Unified	Trona, California	273	18.6	2	Rural: Remote

3.3 Methods

In-person one-on-one as well as small group interviews were conducted with 88 participants in Alaska from January 20 to 25, 2020. Thirty additional interviews with participants in California were gathered from February 17 to 20, 2020. A purposive sampling approach was used to selectively recruit participants with knowledge of earthquakes and schools for this study (Phillips 2014). The Institutional Review Board (IRB) protocols and recruitment materials for each research site are published and publicly available (Tobin et al. 2021a, b).

Each of the four school districts approved the research in their schools. Within the schools, superintendents and other district officials, principals, teachers, school building officials, school emergency managers, and students participated. Earthquake engineers, government officials, journalists, and parents were also interviewed. Participants were identified through several means. USGS collaborators in both Alaska and California were contacted. Interviewees were also identified through reviewing publicly accessible email and phone lists of superintendents and principals and by reading media coverage of the impacts of the Anchorage and Ridgecrest earthquakes on local schools (Tobin 2020). After consulting this coverage, school emergency managers, principals, and others who were named or quoted were contacted. While in the field, participants were asked for recommendations of others who could inform the research, and contacts were followed up with by phone or email. IRB-approved semi-structured interview guides were used to inform data collection efforts (for the published research instruments, see Tobin et al. 2021a, b). Participants were asked questions about the damaging earthquakes and their effects on the educational system and broader community; their experience with drills and disaster preparedness; and their perceptions of, questions about, and concerns regarding EEW systems.

At the end of each interview, participants were invited to complete a close-ended demographic information form. All interviews were audio recorded.

When in Alaska and California, the four primary authors of this chapter conducted observations at local community events, educational training sessions, school board meetings, and within structurally damaged, as well as undamaged, school facilities. Hand-written notes were taken to capture initial observations, which were later transcribed and analyzed to provide additional context. Numerous photographs were taken while in the field, as well. All personally identifiable interview and observational data were uploaded and stored on a password-protected computer nightly.

Upon returning from the field, the qualitative audio files were professionally transcribed. After checking the transcripts for accuracy and completeness, documents were uploaded into ATLAS.ti, which is a qualitative software analysis program that allows for team-based coding. The four primary authors developed a draft codebook organized by the topics addressed in the interview guides, knowledge of the literature, and data collection efforts. Three interviews were coded by the primary authors to test the group coding process for intercoder reliability (MacPhail et al. 2016). Once comfortable with the process, 3 months, and a total of 207 work hours, were spent on open coding (searching for patterns and updating the codebook with emergent codes), axial coding (searching for generalizable themes), and representative coding (selecting interview quotes that represent relevant findings) (Marshall and Rossman 2011). The final codebook, which was added to and refined throughout the coding process, included 11 groups of generalizable themes, and 144 distinct open codes within those 11 groups. Detailed descriptions were written for each code. For this chapter, we focus on the group of codes associated with earthquake early warning, specifically. We use pseudonyms throughout the chapter and have changed some details to protect the identities of study participants.

3.4 Earthquake Early Warning: Special Considerations for Schools

We organize our findings around six key thematic considerations that the interviewees raised in response to the potential adoption of EEW in schools: (1) awareness and knowledge, (2) funding and diversion of resources, (3) alert channels, (4) alerting threshold and frequency of alert delivery, (5) alert message content and drills, and (6) school infrastructure. Because none of the interviewees in the study worked at, attended, or were associated with a school that had adopted EEW, much of the interview time centered around questions that the participants had about the system and concerns that they felt would have to be overcome before ShakeAlert could be adopted and implemented within their local school district. As such, the findings presented below represent their questions and concerns about the system.

3.4.1 Awareness and Knowledge

Our first key finding was that there was a lack of awareness of EEW among participants. Once those in the study learned more about the value of EEW, they agreed that stakeholders in the education sector would greatly benefit from more information about the system.

Of the 118 people in the study, less than 5% indicated that they understood EEW and how it might help to protect children and staff in schools during an earthquake. Therefore, after the topic was introduced in the interviews, the lead interviewer offered to provide a brief description of ShakeAlert before asking subsequent questions. Almost all participants requested further explanation so they could grasp how the system works. After listening to the explanation, some participants expressed confusion and asked for further clarification because there are a variety of cellular phone applications that provide earthquake information *after* events. Only ShakeAlert, however, provides early warning using the sensor network *before shaking* from an earthquake already in progress. Once participants grasped this fact, most responded highly enthusiastically with comments such as "wow" or "that sounds amazing!" A building manager for the Anchorage School District was also excited to learn about the technology but more circumspect than some in her assessment of EEW: "I think it would be very useful, but I think people need to understand what it means." Other interviewees echoed this sentiment when they underscored that awareness of the system and knowledge about how it functions would be key to its adoption in schools.

As the interviewees processed the potential impacts of EEW, several offered hopeful assessments of the difference that such a system could make for their students. In particular, they emphasized that EEW could enhance students' sense of safety while also making the threat of earthquakes feel less overwhelming. For example, an assistant elementary school principal in Alaska said, "I think it would be great. It would be great. It would provide, for earthquakes, a sense of control over something that feels so uncontrollable." A high school special education teacher in California thought that EEW could make a "big impact." She continued: "Because right now, it's so scary because it can happen whenever. I think that if they had even a couple seconds of warning, they'd be like—that makes more sense to them." A school emergency manager in Alaska thought that EEW would be especially helpful for younger children with less earthquake experience:

> I can see in schools, especially with little kids, if you have that 10 seconds, you can get everybody under their desks a lot quicker than waiting to see.... Thinking about kids who are kindergarten, first, second, third [grades]—they might be a little bit more scared, nervous, not know what to do. That 10 seconds helps that teacher get those kids safe and secure. I think that's very beneficial.

Several interviewees raised questions about EEW that they would want to have answered by the USGS or its partners: "what is the science behind the system?" "how much warning time would you get?" and "what would you do with the time available?" Because the interviewees in the study had all recently experienced damaging earthquakes, several processed their new understanding of EEW through that lens. For example, a school building official in Alaska said:

> It's just hard for me to wrap my mind around what would I do with 10 seconds? What would I have done differently [in 2018] if I'd known 10 seconds before the earthquake hit…. Would I have done something different? I would have gone under my desk sooner, maybe, but I'm not going to leave the building, so I'm not sure. Not that it's not beneficial, but people should know what they are supposed to do, rather than just know an earthquake is coming.

Others in the study—especially those who were in administrative leadership positions or who had responsibility for school building safety—were resolute that people who work in schools should be consulted about the design and adoption of EEW. They felt this would help make the system more effective for the unique school environment and would encourage wider understanding and adoption. One high school principal in Alaska was especially wary of his district purchasing any new technology without proper consultation. As he noted, "Oftentimes, things are implemented, and the question is why we didn't talk to this obvious group of people over here that need to use this darn thing."

3.4.2 Funding and Diversion of Resources

Interviewees were concerned with how much it would cost to implement and maintain ShakeAlert. Indeed, financial concerns were top of mind for school employees because of already overstretched budgets and the reality of aging technical systems. An emergency manager in Alaska pondered how EEW could be installed in their district: "I wonder how that would work because our intercom systems are so old, and it costs so much money to redo all of them or even get that feature." A building manager in the same district agreed and said: "That's what I was thinking for tying it into our systems. I don't even see how that would be possible, honestly. We have five different intercom systems. Some are so old we are getting parts off eBay for them. We're in the process of updating them, but I don't know how that would tie in. Do we have to pay for that?"

One of the technical experts in our study—a licensed structural engineer—had deep familiarity with EEW. His concerns were more broadly focused on the cost of installing the sensors across the state of Alaska to make the system operational. In his opinion, the tens of millions of dollars it would take to install and maintain the sensors (Salisbury et al. 2020) would be better spent on structural mitigation:

> Those numbers are huge for Alaska. To come up with that kind of funding would be very difficult in this state with our current budget situation. And being an engineer, I take the approach that it is better to design up front to survive the earthquake than to warn people afterwards that it's coming. With that type of funding, how many schools could you go and retrofit? How many buildings could you make safe so that when the earthquake comes you don't have to worry about the earthquake warning because the building is going to be fine.

The father of a student at a school that was badly damaged in the 2019 Anchorage earthquake agreed. He said: "I think the early warning system sounds like a great idea. But I think the more important thing is just to realize this is earthquake country. We need to plan for it to be earthquake country. The buildings need to be built for earthquake country." He, like other respondents in the study, worried that funding would become a "zero-sum game," with any available funds for building safety being diverted to fund a future early warning system.

3.4.3 Alert Channels

Channels refer to the ways that alerts or other messages can be sent to the public, school staff, parents, or students. Channels are important because they determine the content and type of information, such as the sounds or number of words an alert can contain. In schools, the primary channel for the centralized receipt of EEW is through a PA or intercom system, although in California, Oregon, and Washington, students and staff can also receive direct ShakeAlert-powered alerts through WEAs or downloadable applications on their phones (USGS 2024).

Many respondents saw the public availability of ShakeAlert messages through cellular phones as potentially problematic. On the one hand, the widespread free delivery of these alerts via cellular phone applications could address the cost concerns of the system. On the other hand, many study participants discussed strict "no cell phone" policies for both students and staff in their schools and wondered how this would affect their ability to receive alerts. A middle school principal in California said: "Our policy is to not have cell phones during instructional days because they tend to be instructionally distracting." A high school teacher in a neighboring school agreed: "You're not supposed to have your phones out in class. A lot of students do comply and keep their phones put away or on silent." A group of students in Alaska emphasized that if EEW was only available via mobile alerts, it would not be "useful":

> For students like us, it's not as useful because many classes, many professors, many teachers, are like, 'No phones out, no matter what.' The teachers and professors are probably going to get the notifications, but they are less likely to check their phones because of the interruption to the class.

Some of our participants were concerned that the public release of EEW in the western United States could contribute to "panic" or "chaos" in schools, which don't yet have a plan for centralized adoption. These participants emphasized that regardless of "no cell phone" policies, the actuality is that students use their phones throughout the day. A journalist whom we interviewed worried that in some cases, students with cellular phones could receive alerts unbeknownst to their teacher:

> They've all got iPhones. Imagine thirty alerts that say, "We're going to have an earthquake" go off and the teacher's still teaching because she's not looking at her phone. Do those kids politely say, "Miss, we have an earthquake…." Or do they just devolve into a panic?

Even though ShakeAlert is not yet available in Alaska, an elementary school principal there said that, eventually, "if you tell school staff they have to have [EEW] on their phone, then people are going to say, 'You have to pay for my phone.'" She argued, therefore, that a "Siren makes sense because sirens are consistent." Although most of the participants in our study agreed, they also drew on their recent earthquake experience to share important cautionary tales. For example, as we toured damaged buildings in both Alaska and California, we learned that fire alarms went off in many of the schools during the active shaking. In fact, one school emergency manager showed us video footage of high school students and staff running out of a high school where a wall collapsed, sprinklers were activated, and fire alarms sounded during the earthquake. An Alaska-born, long-time elementary school teacher at another facility described the reaction that the fire alarm bells triggered: "A lot of schools evacuated because of the fire alarm, but then they're evacuating through halls that have water and fallen light fixtures and things. They said they really should have stayed in place, but how do you know?" This teacher and her colleagues worried that if schools introduced earthquake sirens, in addition to fire alarms, it might leave people more confused and at risk: "They've been discussing that with the fire department, about what to do in a case like that. Is it safer to stay in the building and ignore the fire alarm, assuming that it was just tripped by the earthquake? But how do you know? Maybe there really is a fire?"

Teachers, who have the most direct contact with students throughout the school day, were especially animated when they discussed how they thought any PA or intercom message should be designed. For example, several teachers worried that if the EEW was simply a sound, students could become confused given the litany of bells that ring throughout the day and different alarms that signal in emergencies. Those who worked with special education students were especially opposed to using an alarm sound for the EEW. A California teacher summed up this feeling:

> I wouldn't want it to be a loud alarm because loud alarms just throw them into chaos. It's awful to have an unexpected alarm in here. I think if there was an unexpected alarm and then all the earth was shaking, I think somebody would get hurt in here. It [would be better to have] a computer voice or something. Not anything really terrible and scary. I don't think it should be an alarm at all.

3.4.4 Alerting Threshold and Frequency of Alert Delivery

School administrators and teachers are responsible for many things, including keeping students engaged and learning throughout the school day. Given this, school staff were especially eager to learn more about the alert threshold for ShakeAlert (i.e., the point at which the earthquake reaches an intensity that people are notified through an EEW provider). In particular, interviewees were concerned with the alert threshold being set too *low*, therefore triggering alerts too frequently. Just as respondents worried that the lack of a centralized system could cause disorder, teachers also thought that too frequent alerts could be highly disruptive:

> It's tough. It would be really easy to cause chaos in the classroom. If one student gets an alert for an earthquake, that student for sure is like, 'Earthquake is coming, everybody be ready!'

> Then, all of a sudden, you've lost. Everything for the day is done. There is no coming back from that. Everybody is already under their desks or outside… I think it's great, but I think there would have to be a pretty well-defined threshold because I know I would be annoyed if every time there was a tremor, everyone was like, 'Oh, there is an earthquake!'

Another high school teacher in California agreed:

> I think if it were coming too often…. I think the other teachers would be bothered by it because they are pushing to the bell to teach every single minute…. I do think teachers will be like 'Is it going to interrupt my class time? Are [the students] going to be scared after the announcement goes off?'

The need for trust and confidence in the system was emphasized by many of the respondents. As an engineer and seismic safety advocate in Alaska noted, "I think the biggest challenge would be finding… the right level of alert. Because what if you have false alarms? That then degrades the public trust of the system." An assistant high school principal in Anchorage who had dealt with the disruption of false alarms in her school similarly emphasized: "There's the cry wolf scenario I could imagine, too. The app told us we need to go, then it fizzles into nothing. Now we're kind of in the mindset, 'Do we trust it? Do we not?'"

Interview respondents in California and Alaska varied in terms of what they saw as the ideal alerting threshold. In California, most respondents thought the M4.0 to M5.0 range would work well, as described by this middle school principal:

> Anything 4.0 and above would probably be valuable because that's when people start noticing…. Anything less than 2.0's, I think most people do not even pick up on. On a personal level, I would really only want 4.0's and above, because that's when you start worrying about things happening [and when you would best] be able to support kids.

Those in Alaska tended to hold the opinion that ShakeAlert should be set at an even higher alerting threshold. For example, one elementary school teacher described how she had signed up through a commercial application for post-earthquake alerts after the 2019 Anchorage earthquake. She felt that the post-event alerts at M3.0- and M4.0-level were too frequent, and, therefore, she was concerned that anything at that level would not work in the context of Alaska, which regularly experiences lower magnitude seismic activity:

> I would want to know is it a 7.0 or above. Minimum 6.0. I would want to hear it. Even after all those aftershocks, it just became so annoying to me when it was M3.0s and M4.0s and stuff and my phone…. it was letting us know. I actually started getting very twitchy. I had to turn the alerts off.

3.4.5 Alert Message Content and Drills

Participants in the study were eager to understand what the EEW alert would say—either via a written message on the cellular phone application or a spoken message on a PA system or intercom—and how it might help prompt protective actions among school staff and students. Generally speaking, interviewees advocated for more information. For example, the structural engineer in Alaska quoted earlier said:

> The early warning system, as I understand it, in Mexico says, 'earthquake coming.' It doesn't tell you locality, magnitude, or what to do. If we get an early warning system, I think it has to indicate locality, magnitude, and what to do. 'M7.0 earthquake coming, originated in Mat-Su, drop, cover, hold on.' It has to be very specific, because people can only process things so fast. If we only get 10 seconds of warning, and we have to get that notification, look at that notification, recognize that notification, do something about that notification, I mean, it has to be a lot clearer than 'earthquake coming.' 'Earthquake coming' doesn't feel like a useful warning at all.

A high school teacher in Alaska agreed that more information would be especially helpful in schools. He argued that reminding people to take the recommended protective action—drop, cover, and hold on—would be the most important thing in ensuring that people don't "panic" and "run for the door" instead.

> It would be useful as long as those in charge—basically teachers, classroom teachers, responded calmly. I could see some people getting an alert, 'Big one coming in 30 seconds,' and panicking.... As long as our response to that was, 'Get under the tables now....' As long as we responded well to that information, then it could be very helpful.... As long as it was really clear this doesn't mean run for the door. This means get under a table. Because I would hate to see people running for the door in that sort of a situation and then it really starts shaking.

The participants in this study were in near unanimous agreement that, regardless of the alert channel or message content, the most important action schools could take would be to integrate EEW into existing drills. They asserted that this would both reduce potential inappropriate behavioral reactions while also helping students and staff know they should use the precious seconds of warning to drop, cover, and hold on. A public information officer in Alaska felt that the schools were well-positioned to integrate ShakeAlert into their existing preparedness activities.

> Our schools are really accustomed to drills. I mean, they do so many drills every semester around.... earthquake, or fire, or ALICE [Alert Lockdown Inform Counter Evacuate] drills, or whatever it may be. I think about that, and I just see we just need another practice in place, another set of protocols. If we [had earthquake early warning], we would want our staff, our teachers, and other people to know what to do so they would respond appropriately in the situation. To me, it's about designing protocols and practicing those at the schools to see what works and what doesn't and communicating that to families.

A former military member and current high school teacher in California agreed that holding drills and tabletop exercises and engaging in other preparedness activities would help ensure the effective adoption of EEW. However, she worried that school officials do not have the time—or organizational culture—to carry through with the changes that would be required to appropriately train and prepare:

> The military's pretty good about training and following up training and checking up on training and continuing training because there are lives at stake. Well, there are [lives at stake] in earthquakes, too, but nobody seems to see it that way. What I've seen in the civilian world.... it's like, implement and forget. It's like, talk about and forget. There's a lot of that. There's a lot of jaw flapping. Sounds good now, we'll see in a year from now. I'll check back in with you. [laughs]

A school building manager in Alaska worried that students and staff would become too reliant on the new technology and disregard existing drills or environmental queues such as the first signs of shaking. Furthermore, based on his experience in the 2018 Anchorage earthquake, many adults did not take the right protective action, leaving him wary of how EEW—without further education—would be of help:

> I think these things are great, but when people depend on something and that's the only way that they're going to know an earthquake's coming…. People sometimes rely on technology so much that when it's not there, they don't know what to do. We saw from the [2018 Anchorage] earthquake that adults definitely did not know what to do. They're running out of buildings. They're standing in doorways…. With 10 seconds, maybe if people knew that you're not getting in the doorway, that you're actually getting under your desk…. I think that would be very beneficial, but I think we need to teach them, keep teaching them what they need to do, first.

3.4.6 School Infrastructure

School buildings are of special concern in natural hazard events (Applied Technology Council 2017). These facilities tend to remain in use for longer than other building types and may not receive consistent capital renewal funds. They also often have large assembly rooms, such as auditoriums and gymnasiums, which are more vulnerable to structural and non-structural damage in earthquakes and other natural hazards due to their design. For these and other reasons, the interviewees in the study were concerned that EEW messaging and future drills consider the special nuances of school infrastructure.

One of the high school teachers in California taught theater arts and coached student athletes. He therefore spent much of his day outside of a formal classroom and in spaces such as the school theater and gymnasium without ready access to desks to drop, cover, and hold on. When asked if he saw the value in EEW, he enthusiastically replied:

> Oh, heck yeah, oh, heck yeah! Especially because, I mean, think about it. If I was in the gym, is there a place that the kids need to go? If you had the few extra seconds where you had the kids in showers, getting them safe. If we've got kids out on the field, get that designated place [set up] so that we can increase that potential so that we don't just have kids wandering and not near an adult voice to make sure they're okay.

Another teacher in the same school noted that the rhythms of the school day and the movement of children and staff through the hallways might make EEW an especially important technology for schools. This teacher noted that if an earthquake happened during "passing period," when hundreds of students are walking outside of classrooms, an early warning coupled with appropriate training would encourage kids to move quickly to the safest nearby space.

An elementary school teacher in Alaska had experienced several earthquakes, and she described how much time she had spent contemplating the potential dangers in her classroom. She thought that EEW was especially important in schools because of the potential for injury from non-structural components: "Definitely, if you had one of those heavy overhead [projector] things, you could move kids away from that area. Here, I'd probably get kids away from these cupboards."

A school emergency manager and team of building officials were especially enthusiastic about the potential of EEW in schools to trigger automatic gas and water shut off valves. They also recognized how important it would be to ensure that key personnel were trained to manage these systems. As a seasoned building manager emphasized: "Like shut off valves and stuff—depending on how much time it alerts you—then all of our operations custodians are familiar with where the shut off valves are in all the schools." Her co-worker agreed and underscored that the introduction of ShakeAlert "means more people need to be trained to know how to do this."

3.5 Discussion

This study drew on 118 interviews and observations in 4 earthquake-affected school districts in Alaska and California to glean critical insights about the challenges and opportunities for integrating the ShakeAlert EEW system into schools. We learned that school leaders, teachers, emergency managers, and others are enthusiastic about the potential of EEW to enhance school safety efforts. They also asked questions and provided feedback in the following six areas of concern: (1) awareness and knowledge, (2) funding and diversion of resources, (3) alert channels, (4) alerting threshold and frequency of alert delivery, (5) alert message content and drills, and (6) school infrastructure.

We found that there was very low awareness and knowledge of what EEW is and how it works. Other research has also reported low levels of public awareness of the system (Bostrom et al. 2022), although ours is the first study to specifically focus on knowledge among school officials with recent earthquake experience. After the system was described, nearly all respondents were universally excited about this potentially life-saving technology. They were also eager to ask follow-up questions so they could understand its capabilities, including what information alert messages could provide and how much lead time they would have in an earthquake. This positive reception suggests that more outreach could ensure that diverse members of the educational sector are aware of EEW and its potential as well as limitations (Crayne et al. 2023). Outreach for ShakeAlert could be focused on key decision-makers, such as school district superintendents, school principals, school board members, and others who will be involved in supporting its successful implementation, such as teachers and school emergency managers (Adams et al. 2022). Educating and actively including different school district personnels in decision-making can promote acceptance and buy-in among key stakeholders (Lai et al. 2016). Further, given the importance of buy-in at every level—from teachers and students in the classroom to leaders within the community—these findings suggest that the perspectives of those who will use the system should be actively sought as EEW is implemented across the western United States.

Another finding was that participants were particularly concerned about how much it would cost to adopt and maintain ShakeAlert, and whether their current PA systems could support EEW. The number of questions surrounding cost suggests that informational campaigns about EEW could provide more clarity on how funding for the system works, including what resources are available to offset costs. For instance, an option for adopting ShakeAlert involves having a school or school district become a USGS-licensed technical operator. This option allows for free access to the ShakeAlert Message data feed and all testing and development tools, but it requires considerable time, technical expertise, and an ongoing need to meet USGS-mandated performance standards. Alternatively, schools or school districts can become the customer of an existing licensed operator that sells and/or distributes ShakeAlert-powered products and services (de Groot et al. 2022). In this scenario, schools are responsible for paying for the installation and integration of the system, as well as the recurring alert delivery subscription fees to ShakeAlert technical partners (USGS 2022a). If a school or school district follows this second option, decision-makers need to be aware of potential mitigation or preparedness grants, as well as receive assurances that this will not lead to a diversion of funds away from critical building improvements and mitigation efforts (Applied Technology Council 2017).

The study highlighted participant concerns with alert channels, including the potential for mobile EEW alerts to disrupt the classroom or missed alerts due to "no cell phone" policies. This finding indicates that members of the school community need clarification on how alerts are issued, including awareness that cellular phones in California, Oregon, and Washington are already equipped to receive public alerts via WEA or other free downloadable applications. As participants described the numerous issues that could arise from non-centralized mobile alerting, many recommended that ShakeAlert be adopted through a school-wide PA or intercom system so that everyone gets the same clear message at once. Participants also suggested that any centralized system would need to be coupled with regular drills and semi-annual tabletop exercises so that students and staff know what to do as soon as they receive an alert. Such regular drills can allow students and staff to build behavioral self-efficacy (Adams et al. 2017) by developing muscle memory so that they can act quickly and instinctually as soon as they receive an alert (Adams et al. 2022; McBride et al. 2019).

Alert threshold was also a concern for many respondents, with several noting that if the threshold is set too low, it could lead to disorder in the classroom, missed instructional time, and a lack of trust in the system. The USGS has already established ShakeAlert delivery thresholds for different technical partners, such as a M5.0 set for WEA-capable devices and M4.5 set for smart phone applications downloaded by the public (USGS 2023). While individual schools or school districts would not have the ability to alter current thresholds, this finding emphasizes the importance of establishing confidence in the system (McBride et al. 2020). This is especially relevant for dealing with false alerts, which, as one respondent mentioned, could lead to a "cry wolf scenario" and degrade trust in ShakeAlert (Sutton et al. 2023). These findings suggest that communication strategies should both explain the possibility of false alerts and quickly send messages following a false alert that clearly underscore that recipients are safe and can return to regular activities (Bostrom et al. 2022; McBride et al. 2022).

In terms of content, participants generally agreed that alert messages should be specific and provide actionable information, such as reminding people to take the recommended protective action—drop, cover, and hold on (McBride et al. 2022). This aligns with empirically grounded best practices that have emphasized a shift from communicating *about risk*, to clearly communicating *what to do* about the risk itself (Wood et al. 2012). Further, participants remained resolute that the uptake of EEW in schools should be coupled with drills and tabletop exercises to help ensure that people follow instructions and know the role they are to play during an earthquake.

Finally, participants expressed concerns about the unique infrastructure of schools, including the areas where there are no desks or places to take cover. Therefore, schools could implement drills and exercises that adapt to the unique needs of the school environment. For example, if an EEW alert is issued between classes when students are in the hallways, they should be trained to use the advance warning to act in alternate ways to stay safe. Practicing drills that account for different scenarios—such as an earthquake happening before the school day begins when some students are still on school buses—and locations—such as when students are in classrooms, in the gymnasium, in the lunchroom, and on the playground—can help reduce confusion and promote effective protective actions. While ShakeAlert is a potential life-saving technology, it can only be effective if students and staff know what to do when they receive an alert under a variety of different conditions.

3.5.1 Limitations

Despite the novel contribution of this work, there are certain limitations that must be acknowledged. Purposive sampling was used to recruit participants from four U.S. school districts—two in California and two in Alaska. The sample is therefore not representative of all school districts in earthquake-prone regions in the United States, and the findings are not generalizable within the focal states or in other states with access to ShakeAlert. Nevertheless, the rich insights from this qualitative study paved the way for quantitative research involving an online survey of all school district superintendents across Alaska, California, Oregon, and Washington (Adams et al. 2024; Peek et al. 2022). Analyses of this survey data confirm that many of the key concerns highlighted in our qualitative work are relevant across a broader spectrum of school district leaders (Adams et al. 2024). As ShakeAlert becomes more widely available across earthquake country, additional research in this area can help to further refine strategies to increase ShakeAlert adoption in schools and keep school children and staff safe.

3.6 Conclusion

This is the first available study specifically designed to understand how the ShakeAlert earthquake early warning system could be most effectively integrated into schools in the western United States. The interviews revealed key questions and concerns that school staff and others have about the adoption and use of EEW in school buildings. Further, the work resulted in focused recommendations for how the USGS and its partners might best address those concerns and considerations. This research has implications for the future adoption and implementation of EEW in schools. Moving forward, it will be critical to continue to learn from the adults who work in and support schools as well as from students, their parents, and others with a stake in school safety.

Acknowledgements We are grateful for the participants in Alaska and California who contributed to this study. Lauren Dent, Ryan Gold, Brian Shiro, and Emily Wilkins offered careful reviews of earlier versions of this chapter. We also thank the volume editors, Ayse Yildiz and Rajib Shaw, for their editorial leadership and the invitation to share our research in this volume.

Funding This research was funded by U.S. Geological Survey (USGS) through intergovernmental personnel act assignments and supplemental support to the National Science Foundation (NSF Award #1635593). The interview data collection described in this article was conceptualized and implemented by the four primary authors at the University of Colorado Boulder and was not conducted by or on behalf of the USGS. Any opinions, findings, conclusions, or recommendations expressed in this material are those of the authors and do not necessarily reflect the views of the NSF.

Data Availability Statement The research protocols, recruitment materials, and interview guides that informed this work are published and available for download (Tobin et al., 2021a, b). The interview data are not publicly available due to the need to protect participant confidentiality. For more information on the data or other research materials, please contact the lead author and study principal investigator at Lori.Peek@colorado.edu.

Non-Endorsement Disclaimer: Any use of trade, firm, or product names is for descriptive purposes only and does not imply endorsement by the U.S. Government. ShakeAlert® is a registered trademark of the USGS and is used with permission.

References

Adams RM, Karlin B, Eisenman DP et al (2017) Who participates in the Great ShakeOut? Why audience segmentation is the future of disaster preparedness campaigns. Int J Environ Res Public Health 14(11):1407. https://doi.org/10.3390/ijerph14111407

Adams RM, Tobin J, Peek L et al (2022) The generational gap: Children, adults, and protective actions in response to earthquakes. Australas J Disaster Trauma Stud 26(2):67–82

Adams RM, Davies H, Peek L et al (2024) ShakeAlert® and schools: incorporating earthquake early warning in school districts in Alaska, California, Oregon, and Washington. Int J of Disaster Risk Reduct 112:104735. https://doi.org/10.1016/j.ijdrr.2024.104735

Applied Technology Council (2017) FEMA P-1000. Safer, stronger, smarter: A guide to improving school natural hazard safety. https://www.fema.gov/sites/default/files/2020-07/fema_earthqu

akes_p-1000-safer-stronger-smarter-a-guide-to-improving-school-natural-hazard-safety_aug 2017.pdf. Accessed 15 Jan 2024
Bostrom A, McBride SK, Becker JS et al (2022) Great expectations for earthquake early warnings on the United States West Coast. Int J Disaster Risk Reduct 82:103296. https://doi.org/10.1016/j.ijdrr.2022.103296
Ciani JV (2023) Trona breaks ground for new high school. The Daily Independent, December 4. https://www.ridgecrestca.com/news/trona-breaks-ground-for-new-high-school/article_1f0783dc-92f1-11ee-8060-2b99de156c82.html. Accessed 15 Dec 2023
Crayne J, Herrán C, Sumy DF et al (2023) Public education about ShakeAlert® earthquake early warning: Evaluation of an animated video in English and Spanish. Int J Sci Educ, Part B: Commun Public Engagement. https://doi.org/10.1080/21548455.2023.2238872
Early W (2019) Anchorage and Mat-Su school districts report millions in post-quake repair costs, with millions more to come. Alaska Public Media, February 7. https://alaskapublic.org/2019/02/07/anchorage-and-mat-su-school-districts-report-millions-in-post-quake-repair-costs-with-millions-more-to-come/. Accessed 12 Dec 2023
Federal Emergency Management Agency (FEMA) (2020) Building codes save: A nationwide study—Losses avoided as a result of adopting hazard-resistant building codes. https://www.fema.gov/sites/default/files/2020-11/fema_building-codes-save_study.pdf Accessed 1 April 2024
Federal Emergency Management Agency (FEMA) (2023) Hazus estimated annualized earthquake losses for the United States. https://www.fema.gov/sites/default/files/documents/fema_p-366-hazus-estimated-annualized-earthquake-losses-united-states.pdf Accessed 1 April 2024
Fothergill A, Peek L (2015) Children of Katrina. University of Texas Press, Austin, TX
de Groot RM, Sumy DF, McBride SK, et al (2022) ShakeAlert: A people-focused earthquake early warning system. Paper presented at the 12th National Conference on Earthquake Engineering. https://www.shakealert.org/wp-content/uploads/2024/04/1663488902_10608_de_Groot_Final_ShakeAlert_and_People.pdf Accessed 4 April 2024
Kenny C (2009) Why do people die in earthquakes? The costs, benefits, and institutions of disaster risk reduction in developing countries. World Bank Policy Research Working Paper No. 4823. https://ssrn.com/abstract=1334526. Accessed 12 Dec 2023
Lai BS, Esnard AM, Lowe SR et al (2016) Schools and disasters: Safety and mental health assessment and interventions for children. Curr Psychiatry Rep 18(12):109. https://doi.org/10.1007/s11920-016-0743-9
MacPhail C, Khoza N, Ranganathan M et al (2016) Process guidelines for establishing Intercoder Reliability in qualitative studies. Qual Res 16(2):198–212. https://doi.org/10.1177/1468794115577012
Marshall C, Rossman GB (2011) Designing qualitative research, 5th edn. SAGE Publications, Los Angeles, CA
McBride SK, Becker JS, Johnston DM (2019) Exploring the barriers for people taking protective actions during the 2012 and 2015 New Zealand ShakeOut drills. Int J Disaster Risk Reduct 37:101150. https://doi.org/10.1016/j.ijdrr.2019.101150
McBride SK, Bostrom A, Sutton J et al (2020) Developing post-alert messaging for ShakeAlert, the earthquake early warning system for the West Coast of the United States of America. Int J Disaster Risk Reduct 50:101713. https://doi.org/10.1016/j.ijdrr.2020.101713
McBride SK, Smith H, Morgoch M et al (2022) Evidence-based guidelines for protective actions and earthquake early warning systems. Geophysics 87(1):WA77–WA102. https://doi.org/10.1190/geo2021-0222.1
Meece JL, Eccles JS (eds) (2010) Handbook of research on schools, schooling, and human development. Routledge, New York, NY
Medford Alert (2022) Medford implements first school district earthquake early warning system. Medford Alert, December 5. https://medfordalert.com/2022/12/05/medford-implements-first-school-district-early-warning-earthquake-system/. Accessed 17 Jan 2024

Minson SE, Baltay AS, Cochran ES et al (2019) The limits of earthquake early warning accuracy and best alerting strategy. Sci Rep 9:2478. https://doi.org/10.1038/s41598-019-39384-y
Motamedi MHK, Sagafinia M, Ebrahimi A et al (2012) Major earthquakes of the past decade (2000–2010): A comparative review of various aspects of management. Trauma Monthly 17(1):219–229. https://doi.org/10.5812/traumamon.4519
Peek L (2008) Children and disasters: Understanding vulnerability, developing capacities, and promoting resilience. Child, Youth, Environ 18(1):1–29. https://doi.org/10.1353/cye.2008.0052
Peek L, Wachtendorf T, Meyer, MA (2021) Sociology of disasters. In: Caniglia BS, Jorgenson A, Malin SA et al (eds) Handbook of environmental sociology. Springer, Cham, Switzerland, pp 219–241
Peek L, Tobin J, Adams R et al (2022) 2022 ShakeAlert earthquake early warning and schools survey, in ShakeAlert Earthquake early warning and schools in the United States. (DesignSafe-CI PRJ-2958; Version 1). https://doi.org/10.17603/ds2-09yf-w512
Phillips B (1997) Qualitative methods and disaster research. Int J Mass Emergencies Disasters 15(1):179–195. https://doi.org/10.1177/028072709701500110
Phillips BD (2014) Qualitative disaster research. Oxford University Press, Cary, NC
Reddy E (2023) !Alerta! Engineering on shaky ground. MIT Press, Cambridge, MA
Ritchie, H, Rosado, P (2024) Natural disasters: Where and from which disasters do people die? What can we do to prevent deaths from natural disasters? https://ourworldindata.org/natural-disasters. Accessed 2 April 2024
Rodgers J, Hassan W, Thornley J et al (2021) Impacts of the 2018 M7.1 Anchorage earthquake on schools. Earthquake Spectra 37(3):1849-1874. https://doi.org/10.1177/8755293020988022
Rogers E (2003) Diffusion of Innovations, 5th edn. Simon and Schuster, New York, NY
Salisbury JB, Straight SH, Natalia R (2020) Earthquake early warning system for Alaska: Fact sheet. Alsk Div Geol & Geophys Surv Inf Circ 88. https://doi.org/10.14509/30454
Sumy DF, Jenkins MR, Crayne J et al (2022) Education initiatives to support earthquake early warning: A retrospective and a roadmap. Seism Res Lett 93(6):3498–3513. https://doi.org/10.1785/0220220159
Sutton J, Wood MM, Crouch S et al (2023) Public perceptions of U.S. earthquake early warning post-alert messages: Findings from focus groups and interviews. Int J Disaster Risk Reduct 84:103488. https://doi.org/10.1016/j.ijdrr.2022.103488
Tobin J, Adams R, Breeden J et al (2021a) 2018 Anchorage earthquake: Qualitative fieldwork, in ShakeAlert Earthquake Early Warning and Schools in the United States. (DesignSafe-CI PRJ-2958; Version 1). https://doi.org/10.17603/ds2-zfxv-9n62
Tobin J, Adams R, Breeden J et al (2021b) 2019 Ridgecrest earthquakes: Qualitative fieldwork, in ShakeAlert earthquake early warning and schools in the United States. (DesignSafe-CI PRJ-2958; Version 1). https://doi.org/10.17603/ds2-qyt3-a527
Tobin J (2020) A template for multidisciplinary virtual reconnaissance research. CONVERGE extreme events research check Sheets Series. (DesignSafe-CI PRJ-3158; Version 1). https://doi.org/10.17603/ds2-6h08-0p18
Tunseth M (2019) Quake-related repairs and improvements at Anchorage schools could top $150 million. Anchorage Daily News, July 29. https://www.adn.com/alaska-news/anchorage/2019/07/30/quake-related-repairs-and-improvements-at-anchorage-schools-could-top-150-million/. Accessed 15 Jan 2024
United States Department of Education (2023) Data and statistics. https://www2.ed.gov/rschstat/landing.jhtml?src=ft. Accessed 20 Dec 2023
United States Geological Survey (USGS) (2022a) FAQ: How to become a ShakeAlert® technical partner. https://www.shakealert.org/wp-content/uploads/2024/04/FAQ_-How-to-Become-a-Technical-Partner.pdf. Accessed 3 April 2024
United States Geological Survey (USGS) (2022b) ShakeAlert® earthquake early warning in schools: Stanwood-Camano School District, WA. https://www.shakealert.org/wp-content/uploads/2024/04/ShakeAlert-Case-Study-Education-Stanwood-Camano_August-2022.pdf. Accessed 4 April 2024

United States Geological Survey (USGS) (2023) Alert delivery thresholds. https://www.shakealert.org/system-information/alert-delivery-thresholds/. Accessed 10 Jan 2024

United States Geological Survey (USGS) (2024) How do I sign up for ShakeAlert® earthquake early warning system? https://www.usgs.gov/faqs/how-do-i-sign-shakealertr-earthquake-early-warning-system. Accessed 16 Jan 2024

Wood MM, Mileti DS, Kano M et al (2012) Communicating actionable risk for terrorism and other hazards. Risk Anal 32(4):601–615. https://doi.org/10.1111/j.1539-6924.2011.01645.x

Xia R (2015) First public test of earthquake alert system rolls out at L.A. school. Los Angeles Times, October 23. https://www.latimes.com/local/education/la-me-quake-early-warning-20151024-story.html. Accessed 15 Jan 2024

Chapter 4
Reflections on Disaster and Climate Education in the Japanese Education Sector

Aiko Sakurai and Takeshi Sato

Abstract In aiming for a carbon–neutral society, there is a call for the integration of disaster education and climate education in Japan. While in Japan's educational administration, disaster education and climate education are handled by different departments, their integration is being sought through revisions to the national curriculum guidelines. To achieve practical and effective disaster education tailored to the local disaster risks, efforts are being made in school settings, such as in-service teacher training simulating multi-hazards such as landslides and floods due to heavy rainfall during school hours, and improvements to disaster response manuals. Attempts from the perspective of mitigating damage from climate disasters, not limited to earthquakes and tsunamis, are being observed.

Keywords School disaster safety · Teacher training · Local disaster risk · Carbon–neutral society · Sustainable development

4.1 Introduction

In Japan, disaster risk governance and disaster risk reduction (DRR) policy in the education sector have evolved due to lessons learnt from past disaster experiences, such as the 1995 Hanshin-Awaji Earthquake and the 2011 Great East Japan Earthquake (GEJE) (Sakurai 2016). In the 13 years since the 2011 GEJE, in addition to earthquakes, Japanese schools have been experiencing climate-related natural disasters more frequently. The Japan Meteorological Agency (JMA) stipulates names for natural phenomena that have caused notable disasters to pass on the experience and valuable lessons learned from the disaster to future generations. After the 2011 disaster, three cases of earthquake disasters were added to the list of the named

A. Sakurai (✉)
Kobe University, Kobe, Japan
e-mail: sakuraia@people.kobe-u.ac.jp

A. Sakurai · T. Sato
Tohoku University, Sendai, Japan

A. Yildiz and R. Shaw (eds.), *Disaster and Climate Risk Education*, Disaster Risk Reduction, https://doi.org/10.1007/978-981-97-5987-3_4

notable natural phenomena including the 2024 Noto Peninsula Earthquake. On the other hand, nine climate-related disasters were named. Eight out of the nine disasters were caused by heavy rainfalls, and two out of the nine were typhoon-related disasters. Flooding and landslides have been more been more frequent due to heavy rainfalls and typhoons. Based on the lessons learnt from these past disaster experiences, Japanese disaster and climate education has been shifting towards a more participatory and risk-informed approach.

Furthermore, disaster education and climate education in Japan have been developed separately. Disaster education has been seriously addressed since the 1995 Great Hanshin-Awaji Earthquake and has been comprehensively promoted as school disaster safety, along with ensuring the safety of the learning environment and disaster management. Furthermore, disaster safety is positioned and promoted as one of the three areas of school safety alongside traffic safety and daily life safety. On the other hand, climate education in Japan has been positioned as environmental education since the 1990s, along with pollution education and nature experience education, as Education for Sustainable Development (ESD). Climate education has been originally implanted in the UNESCO Associated Schools as a part of ESD in Japan.

The 2011 GEJE and the Fukushima Daiichi Nuclear Power Plant Accident had a major impact on Japanese environmental education. Recognising that natural disasters are a crucial aspect to consider when examining the relationship between humans and the natural environment and sustainability, the Science Council of Japan has proposed the idea that disaster education and environmental education should be carried out in a mutually complementary manner (Science Council of Japan 2016). In June 2021, following the revision of the Act on Global Warming Countermeasures aimed at achieving a carbon–neutral society by 2050, MEXT sent notifications to all prefectural boards of education and indicated points to enhance environmental education at each school and the community. It can be said that demands for the integration of disaster and environmental education are increasing in Japanese society.

Thus, this chapter reflects on the evolution of Japanese disaster and climate education in Japan mainly from the perspective of school disaster safety by discussing ways to create more disaster-resilient schools.

4.2 Evolution of Japanese School Safety Governance Since the 2011 GEJE

(1) Institutional and Legal Arrangements

The School Health and Safety Act of 2009 describes the division of responsibilities associated with school safety among the educational administrative units. Five-year plans for the promotion of school safety have been formulated based on the School Health and Safety Act and have been promoted at the school level since the school year (SY) 2012. The third five-year plan (2022–2026) is currently being implemented.

The MEXT, as a central educational authority, has a responsibility to support and monitor safety efforts at the school level by collecting data, conducting surveys and research, providing safety-related information and guidance and publishing reports in each sector of school safety. This information is disseminated by the MEXT via its website.

Since the 2011 GEJE, the location of the division that manages school safety in the Ministry of Education, Culture, Sports, Science and Technology (MEXT) has been changed three times (Sakurai 2021). Previously, school safety was placed under the School Health Education Division of Sports and Youth Affairs Bureau in MEXT. This was because safety education had been dealt with health and physical education before disaster education was added as one of the fields of safety education in 2009. With the establishment of the Japan Sports Agency in 2015, disaster safety was moved under Health Education and Dietary Education Division of the Elementary and Secondary Education Bureau. Since environmental education has been under the Curriculum Division of the Elementary and Secondary Education Bureau, integration of both safety education and environmental education was expected. However, with the Ministerial restructuring in 2018, disaster safety was moved under a newly established Gender Equity, Cohesive Society Learning and Safety Division of the Education Policy Bureau. In addition, the Department of Facilities Planning and Administration under the Minister's Secretariat is responsible for the safety of school facilities.

(2) **Five-Year Plan on Promotion of School Safety**

The Five-Year (SY2012–SY2018) Plan on Promotion of School Safety was initiated 1 year after the GEJE, and reviewing the three five-year plans that have been implemented to date could improve our understanding of the overall achievements and issues associated with the Japanese school DRR policy since the 2011 GEJE (MEXT 2022).

The first five-year plan emphasised the lessons learnt from the GEJE experience, including the importance of (i) developing an attitude of independence among students, (ii) promoting practical safety education throughout school educational activities and (iii) implementing measures (e.g. robust school facilities and disaster prevention manuals) based on local characteristics to prevent damage caused by natural disasters. In the second five-year plan (SY2017–SY2021), school safety was promoted via the improvement of school safety plans and crisis management manuals. The Courses of Study were revised in 2017, and systematic and practical safety education was to be managed within the curriculum. In terms of safety management, regular inspections of school facilities, equipment and routes were carried out. Furthermore, as safety issues became more complex and diverse, cooperating and collaborating with families, communities and relevant institutions was promoted.

Although much progress was made, issues remained at the end of the second planning period in SY2021. First, various plans and manuals that had been prepared were not necessarily linked to effective initiatives. Second, there were differences in the awareness of school safety initiatives among communities, educational authorities and schools. Third, it was necessary to keep the memory of the GEJE alive and to promote practical disaster education nationwide in preparation for potential large-scale disasters that could occur in the future. Fourth, the actual situation in schools did not reflect the positioning of teachers and staff as the core proponents of school safety and the enhancement of their training. Fifth, various data and research results were not being utilised in schools. Lastly, sufficient progress had not been made due to insufficient follow-up on the plans themselves.

Considering the above issues, the third five-year plan (SY2022–SY2026) includes the following six basic directions: (i) establish a cycle for reviewing school safety plans and crisis management manuals and improve the effectiveness of school safety; (ii) promote safety measures from the perspective of the children and in close collaboration with various local actors; (iii) promote practical and effective safety education in all schools; (iv) implement practical disaster education and training based on local disaster risks; (v) visualise school safety by using data, such as accident information and the status of school initiatives and (vi) raise awareness of school safety (i.e. cultivate a culture of safety in schools). To achieve the goals of the third five-year plan, it is necessary to overcome the major challenges of making school safety efforts more effective and practical, which could be partly accomplished by basing them on local disaster risks and collaborating with the local population. These challenges could be addressed by applying a participatory risk-informed approach, which the revised Comprehensive School Safety Framework emphasises (GADRRES 2022).

4.3 Evolution of Disaster and Environmental Education

(1) Redefining the Aims of Disaster Education

Many lives were lost during the 2011 GEJE tsunami because of delays in activating evacuation procedures. Therefore, in terms of disaster education, the first five-year plan emphasised enhancing the ability of children to predict and avoid dangers on one's own. In 2013, the MEXT redefined the aims of disaster education in a revised reference document for disaster education entitled 'Development of Disaster Prevention Education to Cultivate Zest for Living' (MEXT 2013). The new definition, listed below, places more emphasis on the link between understanding, decision-making and action, implying that disaster education should be more practical to close the gap between understanding and changing behaviour. The definition also covered all hazard types, not only earthquakes, as well as daily life, not just school life. The definition was as follows:

(i) To deepen understanding of the current situation, causes and mitigation of natural disasters and other disasters, and to be able to make appropriate decisions and choose actions based on accurate thinking and judgement in the face of current and future disasters.

(ii) To be able to understand and predict the dangers associated with earthquakes, typhoons and other natural disasters and take actions to ensure one's own safety. At the same time, being able to prepare for your daily life.

(iii) To recognise the importance of respecting other lives and creating a safe and secure society, be able to participate, cooperate and contribute to the organisation's safety activities.

Disaster education in Japan does not depend on drills and/or specific subjects but is integrated into the existing curriculum. Therefore, schools are required to implement cross-curricular disaster education through scattered references to natural disasters and disaster management within each subject. This has been one of the major challenges faced by school teachers. Therefore, in the 2013 reference material, the goals of disaster education were also presented at each school level according to the children's developmental stage. The goals at the elementary level were to produce children who understood the hazards of disasters in various aspects of everyday life, could act safely and were attentive to the safety of others. The goals at the lower secondary level were to produce students who could act independently based on daily preparedness and sound judgement, understood the importance of local disaster prevention activities and of helping each other in times of disaster and were willing and able to take action. The goals at the upper secondary level were to produce students who were aware of their participation in the creation of a safe and secure society, could independently decide on their own appropriate role in local disaster prevention activities and could support activities in the event of a disaster.

(2) **Revision of the Courses of Study and Disaster Education and Environmental Education**

The Courses of Study were revised for the first time since the 2011 GEJE in 2017. The new Courses of Study aim to enhance not only disaster education but also environmental education. The general rules in the 2017 Courses of Study indicate that children are expected to become creators of a sustainable society. The following statements in the general rules are regarded as foundations of both disaster education and environmental education.

> 'Each school takes into account the actual situation of the students, school, and community, as well as the child's developmental stage, and aims to help them realize fulfilling lives, overcome disasters, and form the next generation of society, and that each school is required to organize an educational curriculum that develops the qualities and abilities necessary to respond to the various contemporary issues facing the world from a cross-curricular perspective' (MEXT 2017).

At the elementary level, the revised Courses of Study were implemented across all grades in SY2020, and natural disasters and environment are dealt with in social science, science, home economics, physical education, ethics, student-led activities, periods for integrated studies and life studies. At the lower secondary level, the revised Courses of Study were fully implemented in SY2021, and natural disasters and environment are dealt with in social science, science, health and physical education, ethics, periods for integrated studies, special activities, and technology and home economics. Disaster education and environmental education are overwrapping.

In science, natural hazards are covered primarily to convey an understanding of natural phenomena, including the mechanisms underlying earthquakes, tsunamis, volcanoes, rainfall, flooding and other natural events. Whereas in social science, which has been positioned as a core subject, natural disasters are presented to understand the interface between human activities and nature, including the relationship between a nation's natural environment and its population and the activities performed to protect human lives against natural disasters. In addition, in the 2017 revision, the natural disaster and disaster preparedness content within this subject was expanded.

Recently, a study was conducted to clarify the state of disaster education materials in Japan (Ikeda et al. 2021). The authors analysed 2,217 teaching plans that were developed and edited by Japanese schools and boards of education and found that 50.6% of the plans were made for schools located in major disaster-affected prefectures, including Hyogo Prefecture, Niigata Prefecture and Miyagi Prefecture. Most of the plans were developed after the 2011 GEJE, and 51.7% of the plans dealt with earthquakes, followed by 29.9% with tsunamis, and 57.3% were made for elementary school children. The majority of the plans were not developed for specific teaching subjects. Even though school-based disaster education increased in popularity after the 2011 GEJE, the study revealed that disaster education is mainly delivered outside of teaching subjects at the elementary level, that there are still only limited efforts in areas that have experienced disasters and that meteorological disasters are rarely covered by teachers (Ikeda et al. 2021).

(3) Geography as a Compulsory Subject in Upper Secondary Education

At the upper secondary level, the Courses of Study were revised in 2018, and the revisions are to be fully implemented in SY2024. Disaster-related content is included in geography and history, civics, science, health and physical education, art, home economics, periods for integrated studies and special activities.

Geography will be compulsory and a core subject of disaster education at the upper secondary level. Its content is divided into the following five major themes, based on the 1992 International Charter for Geography Education: 'location and distribution', 'place', 'interdependence between humans and the natural environment', 'spatial interdependence' and 'region'. These themes serve as a guideline for constructing the geography curriculum, and the relationships between the five themes and the learning aims are clearly stated in the revised 2018 Course of Study. First, the learning of skills is supported by the inclusion of specific content that teaches skills such as map reading and understanding geographic information systems. Second, international understanding and international cooperation are taught with an emphasis on not only acquiring knowledge of daily life and culture but also using the acquired knowledge and skills to work towards solving global issues. Third, the subject has been designed to teach students how to go beyond understanding current situations and develop future thinking. Lastly, geography is strongly related to junior high-school geography and the high-school elective of advanced geography (Ida 2019).

The natural environment and DRR are covered within the topic of sustainable development in the new geography course, and students are expected to obtain knowledge of the following issues: (i) natural disasters that have occurred in their local area, in Japan and worldwide; (ii) the relationships between features of the local natural environment and disaster preparedness and response efforts; (iii) the scale and frequency of natural disasters and (iv) the importance of disaster preparedness and response based on regional characteristics. Regarding the development of skills, students are expected to utilise various geographic information, such as hazard maps and topographic maps, and to collect, read and summarise this information. In terms of developing the ability to think, judge and express, students are expected to select a theme, consider the disaster preparedness and response based on regional characteristics and sustainable development from multiple angles and express their thoughts on these factors (MEXT 2018).

It is anticipated that making geography compulsory at the upper secondary level will result in more people being better equipped to manage daily life and the risks faced by society. However, at the same time, making geography compulsory will likely create a challenge for the teaching community because there is a long history of low rates of completion for geography courses during pre-service training and fewer social science teachers who majored in geography than in history. The introduction of a support system, training opportunities and materials may not necessarily ensure that the demand for qualified geography teachers is met. In addition, there is a high demand for practical training on how to develop teaching materials for natural disasters and disaster preparedness based on local cases. It has also been found that to improve disaster education in geography subjects, there is an urgent need to enhance training opportunities for teachers to improve their teaching abilities (Oda et al. 2023).

4.4 School Disaster Preparedness Against Meteorological Disasters

(1) Increasing Meteorological Disaster and Climate Change Adaptation in Japan

In recent years, climate change has led to more frequent and severe weather disasters worldwide. Increases in average annual temperatures and abnormal weather phenomena, such as extremely hot days and tropical nights, strong typhoons, short periods of intense rainfall and rising sea levels, are also predicted to occur in Japan. With the assumption that natural disasters may occur due to the effects of climate change and cause damage, the efforts made to reduce such damage are known as 'adaptation' measures to climate disaster risks. In Japan, the Climate Change Adaptation Act was promulgated in 2018 as part of the societal attempt to prevent and mitigate disasters through the introduction of hard measures (e.g. levees, seawalls and erosion control dams) and soft measures (e.g. increased disaster- and weather-related information dissemination, training, education and evacuation).

(2) Revision of Legislation and Guidelines on Evacuation

With the revision of the Flood Prevention Law and the Sediment Disaster Prevention Law in 2017 and 2021 and the introduction of the Act on Development of Areas Resilient to Tsunami of 2011, authorities are obliged to prepare an evacuation assurance plan, conduct evacuation drills against any relevant hazard (including floods, landslides and tsunamis) and report the results to the municipal government for all facilities under their responsibility that are designated as used by persons requiring special consideration in the local disaster management plan of the municipality. These arrangements are under the jurisdiction of the Ministry of Land, Infrastructure, Transport and Tourism (MLIT) (MLIT homepage).

In addition to the revision of the Flood Prevention Law and the Sediment Disaster Prevention Law, the Basic Act on Disaster Countermeasures was partially revised in May 2021. The Guidelines on Evacuation Information were subsequently revised to reflect that residents should take evacuation action based on their judgement without waiting for the evacuation information issued by their municipal government. In the revised guidelines, there are five alert levels that are each allocated a different colour (Table 4.1).

The evacuation information in the guidelines has been designed to urge the residents in the hazard area to take appropriate evacuation action. As shown in Table 4.1, residents should evacuate at alert Level 4 (purple; high risk of disaster), because at alert Level 5 (black; disaster is occurring or imminent), safe evacuation becomes difficult. At alert Level 3 (red), people who need time to evacuate (e.g. elderly people, infants and pregnant women) are urged to evacuate (Cabinet Office homepage). The revised guidelines are linked to disaster prevention and weather information released by the JMA and other disaster prevention agencies for each type of disaster, which is equivalent to the five alert levels (JMA homepage).

Through the abovementioned changes to evacuation policies and guidelines, the Japanese government is attempting to nudge residents into becoming more aware of their local disaster risk and into evacuating when disaster risk arises. Mass and social media are also used to alert residents when there is a significant risk of disaster.

(3) Importance of Risk-Informed Decision-Making by Schools

As the 2011 GEJE occurred on a Friday afternoon while students were at school, how schools can protect the lives of children from natural disasters has become an important issue in school management.

Okawa Elementary School in Ishinomaki, Miyagi Prefecture, was located on a river, four and a half kilometres inland from the mouth of the Kitakami River. At the GEJE, the school was hit by the tsunami, and 74 of its 108 students and 10 of its teachers died under school supervision. In subsequent court proceedings, it was determined that the school could have foreseen the tsunami arrival risk and evacuated the students and teachers to the location specified in their disaster response manual, based on articles 26 and 29 of the School Health and Safety Act of 2009. Although the school was not located in a tsunami inundation area on the prefecture's tsunami hazard map, the court concluded that the map did not exclude the risk of a tsunami reaching

Table 4.1 Revised evacuation information released by the Japan Meteorological Agency (JMA) for weather-related hazards

Alert information issued by municipalities		Evacuation information	Disaster prevention and weather information announced by the JMA		Colour
Alert level	Situation		Equivalent to alert level		
			Flood related	Landslide related	
5	Disaster occurring/ impending	Emergency safety measures should be taken to protect own life	Information on flooding Heavy rain (inundation) emergency warning	Heavy rain (landslide) emergency warning	Black
4	High risk of disaster	Evacuation instruction	Information on potential flood hazards	Landslide alert information	Purple
3	Risk of disaster	Evacuation of elderly and other vulnerable people	Flood warning Information to provide a warning on flooding	Heavy rain warning (landslide)	Red
2	Weather worsening	Heavy rain, flood or storm surge advisories (issued by the JMA)	Flood advisory	No announcement	Yellow
1	Risk of weather worsening	Probability of warnings (issued by the JMA)	No announcement	No announcement	White

Note The table includes warning information released by the JMA only, but does not include storm surge-related warning information. Other information that is not listed in the table includes information on river water levels announced by the Ministry of Land, Infrastructure, Transport and Tourism and prefectural governments
(*Source* Created by the author based on information from the Cabinet Office of Japan and JMA)

the school by considering other available information, such as the school's location. The court also determined that the school principal was required to accumulate more disaster risk information than residents to secure student safety based on the School Health and Safety Act (Sakurai 2021).

Following the judgement of the second instance court in the tsunami accident lawsuit involving Okawa Elementary School, schools across Japan are required to have highly effective school disaster response and preparedness plans in place, which have been developed in cooperation with local authorities, DRR-related organisations and experts and residents, that take into account the local characteristics and disaster risk of the school district.

In December 2019, after the finalisation of the court's decision, the MEXT issued a notification letter to all boards of education to improve school safety against natural disasters and promote practical disaster education (MEXT 2019). Through the letter, the MEXT requested that the educational authorities urge all schools to review, check and improve their school safety management and disaster education to prepare for future natural disasters, including earthquakes, tsunamis, volcanic eruptions and meteorological disasters. The MEXT also asked the boards of education to support all schools as they conduct evacuation drills and provide disaster education in collaboration with households, the community and related agencies. In the letter, the MEXT referred to the School Health and Safety Act, the Act on Development of Areas Resilient to Tsunami of 2011, and the revised Flood Prevention Law and Sediment Disaster Prevention Law. The preparation and dissemination of this notification letter could be regarded as a turning point in Japanese school disaster safety to move further forward beyond their experiences of the 2011 GEJE and tsunami disaster.

(4) **Current Situation and Challenges Faced by Schools**

As the risk of meteorological disasters increases, schools have a more important role to play in protecting children from disasters. The leaders of a school located in a hazardous area (with a flooding, landslide or tsunami risk) are obliged to prepare an evacuation assurance plan if they are specified as facilities for persons requiring special consideration in their municipal disaster management plan, according to the Flood Prevention Law and the Sediment Disaster Prevention Law. As mentioned above, schools are also required to have crisis management manuals in place, as stipulated in the School Health and Safety Act of 2009. The MEXT states that it is to include the evacuation assurance plan in existing crisis management manuals, instead of preparing another evacuation plan (MEXT 2021a).

However, this is 'easier said than done' for schools. For example, not all municipal disaster management plans position schools located in hazard areas as facilities for use by persons requiring special consideration. Therefore, regardless of whether schools are required to create an evacuation assurance plan, it is recommended that the school's leaders check whether the school is located in a hazardous area of any type. In addition, it is recommended that the school's leaders define the emergency evacuation sites, routes and means of transport to be utilised in the case that a Level 3 alert is issued and/or significant weather/disaster prevention information equivalent to a Level 3 alert is announced. Furthermore, it is a considerable amount of work for school leaders to obtain and understand the large volumes of information on disaster prevention, weather and evacuation, to apply it to their own school districts and to use it to formulate appropriate evacuation plans.

4.5 In-Service Teacher Training Programme for Climate Change Adaptation

How can schools determine the timing of evacuation based on disaster prevention, weather and evacuation information, and how can schools evacuate their students to a safe emergency site when the risk of natural disasters increases due to heavy rainfall while the students are at school? In Ishinomaki, where the authors have been conducting action research since 2012, in-service teacher training was developed and started in 2019. The aim of the training is to enhance the capacity of the teachers in charge of DRR; one such teacher is assigned to each school in the Ishinomaki municipality, and they make risk-informed decisions about their school's disaster risk management. The training is available in person and online (Tohoku University 2020) and involves exercises based on school cases. The training programme has been running for 4 years and has been progressively updated over time. Bi-annual training is provided for the teachers in charge of DRR at the school, and training for school administrators was launched in 2023.

(1) **Module 1: Understanding the Relationships Between Hazards and Landforms**

The current training course is composed of two modules. The first module involves reading hazard maps, as well as a topographic map and a geomorphological map of a case school. Learning about the topography of an area is key to understanding the disaster characteristics of a school district, and the topography is highly indicative of land conditions for various natural disasters. The relationship between topography and natural disasters is common knowledge among geographers, engineers and other natural scientists and is used as the basis for various hazard maps produced by the national and local governments.

Even though many teachers may be able to understand the colour coding on hazard maps, it is difficult for teachers to understand the relationship between topography and disaster risk in terms of why disaster risk is higher in some areas. Furthermore, they do not have the time to actively utilise publicly available disaster and weather information for school disaster management. Therefore, the training programme uses Web GIS layered hazard maps (Geospatial Information Authority of Japan Homepage) to compare topographical maps, topographical classification maps and other maps with flood and landslide hazard maps and makes the results available to teachers so that they can prepare their schools' plans and protect their students in the event of a natural disaster. The aim is to enable participants to understand, step-by-step, the assumptions that are made when hazard maps are prepared so that they can understand the relationships between hazards and landforms. Based on this understanding, the participants also examine unexpected disaster risks (beyond the use of hazard maps) by considering landforms at the end of the first module (Sakurai et al. 2022).

(2) **Module 2: Examining Multiple Evacuation Sites and the Timing of Evacuation**

The second module involves examining multiple evacuation sites, including vertical evacuation sites within the school and horizontal evacuation sites outside the school, as well as the timing of emergency evacuation; this is achieved by utilising knowledge and information about local disaster risks and real-time information on weather, disaster prevention and evacuation (Sakurai et al. 2023). For example, one of the included school cases examines the use of horizontal evacuation sites and routes in the event of the simultaneous occurrence of both flooding and landslides during heavy rainfall. This is because the school is located in an area that has two hazards (flooding and landslides) and has its second floor and above designated as an emergency evacuation site in case of flooding; however, the upper floors are not designated as an evacuation site in case of landslides. In addition, there are no designated emergency evacuation sites around the school that can be used in the event of flooding and landslides occurring almost simultaneously. By analysing this case, the participants learn that the government designates emergency evacuation sites and shelters for each type of disaster and that such sites are not always designated with complex disasters in mind. Hence, they learn that schools should not rely on the emergency evacuation sites designated by the government but should determine where the most appropriate evacuation sites are for different and complex disasters.

In another school case study in the second module, the timing of evacuation is examined at a two-storey school located in a flood hazard area that does not have roof access. This case is examined in the training for school administrators. The school is not designated as an emergency evacuation site in the event of flooding because the surrounding area is expected to be inundated to a depth of 3–5 m due to the proximity of two nationally designated rivers. Most of the land within the school district is floodplain and back marshes, and the school is surrounded by rice paddies. The nearest designated evacuation site is located 500 m southeast of the school and on a hill about 120 m above sea level. The participants are presented with a scenario in which it is noon and a heavy rain warning (landslide) was issued at 8.30 am after the students had arrived at school, and a Level 4 alert with evacuation instructions (landslide) was issued at noon for the area where the school is located, while the rain caused by a frontal depression continues. They are asked to determine whether the school should be evacuated at noon. The exercise emphasises the need for school administrators to be able to use such information to make decisions about the appropriate timing for evacuation. In this case, school administrators must start considering evacuation while there is still time remaining before actual flooding occurs, regardless of whether evacuation information is issued. Regarding warning information on flooding, as shown in Table 4.1, school administrators must collect real-time information on the local situation from residents, on the school situation from teachers and staff and on the weather, river levels and evacuation instructions.

There is no single correct answer that can be derived from the presented cases and exercises and applied to real situations because each situation is unique in terms of the type, location, extent and complexity of the disaster. However, it is vital that school administrators can accurately understand real-time information relevant to their school's location. For example, as shown in Table 4.1, there are different types of heavy rain warnings and evacuation information, such as 'heavy rain warning (landslide)' and 'flood warning' as well as 'evacuation instruction (landslide)' and 'evacuation instruction (flood)' and school administrators must be aware of what these warnings and evacuation information mean and how they differ. The above-mentioned scenario includes a 'heavy rainfall warning (landslide)' and Level 4 alert (landslide), not a 'flood'.

A questionnaire survey answered by the participated teachers in the training revealed that (i) it is important to understand the normal situation in the area where their school is located, as this will be a major factor when making decisions during an emergency; (ii) it is important to add local information to the information collected to determine when to evacuate and (iii) the information provided by various sources should be scrutinised to determine what is relevant to their own school. Others have stated that it is difficult to collect various pieces of information and make decisions in a limited amount of time and that it is necessary to make efforts to obtain information on their own to supplement the information provided.

4.6 Discussion

This chapter provides an overview of how school disaster safety has evolved in Japan since the GEJE in 2011 with the increase in meteorological disasters, and the following points have become clear as a result of this process. First, the third five-year plan for promoting school safety has been implemented, and there is now a need to expand the introduction of practical and effective school safety measures. To achieve this, it is essential to implement disaster management and disaster education programmes that are based on local disaster risks. Furthermore, crisis management manuals, evacuation drills and disaster education should be planned, implemented, reviewed and improved periodically via a Plan-Do-Check-Action (PDCA) cycle. This will result in schools working in collaboration with parents, residents, DRR experts, agencies and others to enhance school disaster safety and ultimately protect children from the effects of disasters.

Second, with regard to disaster education, revision of the curriculum guidelines in 2017 and 2018 resulted in an expansion of the volume and depth of DRR-related content in social studies subjects at the elementary level and in geography at the upper secondary level, and social science and geography were both designated as core subjects of disaster education. However, it will take time for the revised curriculum guidelines to be reflected in the practice of disaster education at the school level. There is a particular concern about the shortage of educators who can teach geography, and there is an extensive need to expand pre-service and in-service teacher training.

Third, as meteorological disasters become more frequent and severe due to climate change, there is an increasing need for schools to plan for and respond to not only earthquakes and tsunamis, but also floods, inundation and landslides caused by heavy rain. As a result of legislative changes, schools—and indeed all residents—are required to take appropriate and prompt evacuation actions based on weather/disaster prevention information and evacuation information, taking into account their own specific situation. Although the government is making efforts to make the relevant information understandable, the content is complex, and it is also making efforts to disseminate the information as widely as possible. As the risk of natural disasters increases, schools are encouraged to collect specialised information from various organisations on their own. However, it is not easy for school leaders to collect and interpret such information and to make evacuation decisions based on the local situation by themselves.

Therefore, in this chapter, we have introduced a teacher training programme that has been developed and implemented in Ishinomaki to raise awareness of disaster preparation and responsiveness among teachers and school administrators. The participants use school case studies to examine how schools should collect information and make decisions about evacuation routes, sites and timing when students are at school and faced with a natural disaster. This type of situation-based training is recommended by the MEXT (MEXT 2021b).

There is also an increasing need for both schools and students to create their own disaster prevention 'timeline', which is a chronological arrangement of disaster prevention action plans and their implementation bodies based on assumptions and situations that may occur during a disaster. Preparing a timeline helps one learn basic knowledge about DRR and disasters, consider potential responses while having a concrete image of a disaster (e.g. a typhoon and heavy rain) and visualise and share who will do what and when which allows each person to proactively respond in the event of an emergency. Students could learn how to create such a timeline through disaster education topics at school. Teachers could learn how to create such a timeline through training such as that described above. The training introduced in this chapter currently involves examining scenarios based on actual school cases; however, in the future, it will be necessary to develop a comprehensive training programme that includes guidance on how to create a timeline. In the near term, given the limited time and opportunities that current teachers and school administrators have to participate in DRR training, in addition to these current efforts, it is critical to provide support to teachers as they tailor disaster education to children's developmental stages and learn how to deliver disaster education. Disaster education in schools is becoming increasingly important to improve the resilience of society as a whole and to enhance disaster prevention.

In Ishinomaki, an initiative has been launched to link disaster management training for teachers with disaster education for children (Sato et al. 2023). In the early stages of this initiative, trained teachers supported by experts asked children to think for themselves about potential evacuation sites in the event that there was a heightened risk that the nearby Kitakami River would flood while they were at school. Instead of the teachers thinking about such issues and unilaterally communicating information to the children through lectures and drills, in this initiative, two-way risk communication is realised—the child learners independently think about evacuation sites on their own. In addition, through the children's assigned homework, their parents became involved in considering emergency evacuation sites, which resulted in organic collaboration involving the schools and families on school disaster management and disaster education. These collaborations are still in their infancy, and it is hoped that they will expand.

4.7 Conclusion

This chapter provided an overview of how school disaster safety has evolved in Japan since the GEJE in 2011 with the increase in meteorological disasters and confirmed that progress is being made towards the implementation of more practical and effective education for mitigation and adaptation. A situation-based training for teachers developed in Ishinomaki City, Miyagi Prefecture was found as one of the efforts for realising practical and effective disaster and climate education. Disaster and climate education for children is a long-term investment in improving the capabilities of future citizens and society as a whole to realise sustainable development. Efforts to further deepen their integration between disaster and climate education should be made by involving not only schools but also parents and all school stakeholders in the community.

References

Fujioka T (2017) New school curriculum guidelines and environmental education: From the viewpoint of natural disasters and education for disaster risk reduction. Jpn J Environ Educ 27(1):6–11

Global Alliance for Disaster Risk Reduction & Resilience in the Education Sector (GADRRES) (2022) Comprehensive school safety framework 2022–2030 for child rights and resilience in the education sector (https://inee.org/sites/default/files/resources/The-Comprehensive-School-Safety-Framework-2022-2030-for-Child-Rights-and-Resilience-in-the-Education-Sector.pdf) (Accessed on 4 Jan 2024)

Geospatial Information Authority of Japan Homepage "GSI Map" (https://disaportal.gsi.go.jp/) (Accessed on 4 Jan 2024)

Ida Y (2019) Chiri-sogo toha nanika (What is "Geography" in the new Courses of the Study?) Trends Sci (November):10–14 (in Japanese only)

Japan Meteorological Agency (JMA) Homepage. "Bosai kisho joho to keikai reberu tono taiou ni tuite (Regarding correspondence between disaster prevention/weather information, and alert levels) https://www.jma.go.jp/jma/kishou/know/bosai/alertlevel.html (Accessed on 4 Jan 2024)

Ministry of Education, Culture, Sports, Science and Technology (MEXT) (2013) *Gakko Bosai no tameno Sanko Shiro, Ikiru Chikara wo Hagukumu Bosai kyoiku no Tenkai* (Revision to development of disaster prevention education to cultivate 'zest for living') (in Japanese only) (https://anzenkyouiku.mext.go.jp/mextshiryou/data/saigai03.pdf) (Accessed 20 Dec 2023)

Ministry of Education, Culture, Sports, Science and Technology (MEXT) (2017) *Shogakko gakushu shidou youryou kaisetu* (Elementary school curriculum guidelines commentary) (in Japanese only) https://www.mext.go.jp/content/20230308-mxt_kyoiku02-100002607_001.pdf (Accessed on 4 Jan 2024)

Ministry of Education, Culture, Sports, Science and Technology (MEXT) (2018) *Koto gakko gakushu shidou youryou kaisetu chiri kekishi hen* (High school curriculum guidelines commentary: Geography and history edition) (in Japanese only) https://www.mext.go.jp/component/a_menu/education/micro_detail/__icsFiles/afieldfile/2014/10/01/1282000_3.pdf (Accessed on 4 Jan 2024)

Ministry of Education, Culture, Sports, Science and Technology (MEXT) (2019) *Shizen saigai ni taisuru gakko bosai no kyoka oyobi jissentekina bosai kyoiku no shuishin ni tuite irai* (Request for strengthening school disaster safety against natural disasters and practical disaster education) (in

Japanese only) https://www.mext.go.jp/a_menu/kenko/anzen/1422067_00001.htm. (Accessed on 4 Jan 2024)
Ministry of Education, Culture, Sports, Science and Technology (MEXT) (2021b) *Gakko anzen suishin no tameno kyoshokuin muke kenshu kunren jizzen jireishuu* (A collection of practical examples of training and drills for teachers and staff to promote school safety) (in Japanese only) https://www.mext.go.jp/content/20210604-mxt_kyousei02-000015766_08.pdf. (Accessed on 4 Jan 2024)
Ministry of Education, Culture, Sports, Science and Technology (MEXT) (2022) Dai sanji gakkouanzen no suishin ni kansuru keikaku (The Third Plan on Promoting School Safety) (in Japanese only) https://anzenkyouiku.mext.go.jp/plan-gakkouanzen/data-3/3keikaku-zenbun.pdf. (Accessed on 4 Jan 2024)
Ministry of Education, Culture, Sports, Science and Technology (MEXT) (2021a) Gakko no "Kiki kanri manyuaru nado no hyoka minaoshi gaidorain (Guideline for evaluation and review of crisis management manual at school) (in Japanese only) https://anzenkyouiku.mext.go.jp/mextshiryou/data/kikikanri/kikikanri-all.pdf. (Accessed on 4 Jan 2024)
Ministry of Education, Culture, Sports, Science and Technology (MEXT) Website "MEXT X School Safety" (https://anzenkyouiku.mext.go.jp/) Accessed 20 Dec 2023 (in Japanese only)
Ikeda M, Nagata T, Kimura R, Yi T-Y, Nagamatsu S (2021) Analysis of current disaster prevention education materials developed in Japan-The future of disaster prevention education in relation to the course of study. J Soc Saf Sci 39:103–111
Disaster Management, Cabinet Office (2021) Homepage. "Hinan joho ni kansuru gaidorain no kaitei" (Revision of guidelines regarding evacuation information) https://www.bousai.go.jp/oukyu/hinanjouhou/r3_hinanjouhou_guideline/ (Accessed on 4 Jan 2024)
Ministry of Land, Infrastructure, Transport and Tourism (MLIT) Homepage Suibo ho no ichibu wo kaiseisuru horitsu ga sekou saremashita. (A law amending part of the Flood Control Act has been enforced) https://www.mlit.go.jp/mizukokudo/mizukokudo02_tk_000001.html (Accessed 20 Dec 2023) (in Japanese only)
Oda T, Ikeda M, Nagata T et al (2023) Development of an in-service teacher training program for disaster risk reduction in a new compulsory high school geography course in Japan. E-Journal GEO 18(2):199–213
Sakurai A (2016) Governance of the education sector and disaster risk reduction. In: Shiwaku K, Sakurai A, Shaw R (eds) Disaster resilience of education systems, experiences from Japan. Springer, pp 11–23
Sakurai A, Sato T, Murayama Y et al (2023) Development of an in-service teacher training program gor supporting an effective school evacuation plan: Examining emergency evacuation places during heavy rainfall. Jpn J Saf Educ 23(1):19–30
Sakurai A, Murayama Y, Sato T, Oda T (2022) Climate and disaster risk reduction education. In: Rajib S (ed) Handbook on climate change and disasters. Elgar handbooks in energy, the environment and climate change, pp 322–329. Edward Elgar, U.K
Sakurai A (2021) School safety management: International framework and Japanese practice. In: Jing Y, Han J-S, Ogawa K (eds) Risk management in East Asia: Systems and frontier issues, pp 167–204. Palgrave
Sato T, Hashimoto M, Sakurai A et al (2023) Improvement of school disaster management and implementation of disaster education for flooding risks: A case from Oyachi Elementary School in Ishinomaki city. Jpn J Saf Educ 22(2):37–48
Science Council of Japan (2016) Towards integrated promotion of environmental education (only in Japanese "*Kankyo-joho no tougou teki suishin ni mukete*" https://www.scj.go.jp/ja/info/kohyo/pdf/kohyo-23-t234.pdf (Accessed on 25 March 2024)
Tohoku University (2020) Online training: Understanding local disaster risk by utilizing maps of school districts, Collaborating Center of Disaster Education Research and Implementation. https://drredu-collabo.sakura.ne.jp/online. (Accessed on 4 Jan 2024)

Chapter 5
Mapping the Current State of Knowledge About Disaster Education Within Turkish National Curricula Through the Lens of Ecosystem Approach

Elanur Yilmaz-Na, Ozgur Erdur-Baker, and Gaye Tuncer-Teksoz

Abstract Every year, severe disasters impact millions of people. There is an ever-increasing need for more knowledgeable, ecologically mature, and accountable individuals to enact adaptation, mitigation, and preparedness behaviors against disasters. Schools are the best places to help individuals develop distinctive disaster-related knowledge and experiences. Community-based resiliency can also be increased as students can effectively communicate disaster risks and information that leads to behavior change. However, more integration of disaster-related issues into curriculum framework is needed to overcome the significant barriers to ineffective education practices within formal school settings. Therefore, this chapter's aim is twofold: (1) to analyze the Turkish national curriculum framework through a comprehensive and critical lens regarding its pre-defined expectations for different grades (1–12) and subject areas in mandatory courses and (2) to offer research-based suggestions on more promising disaster education. The ecosystem approach is employed as an analytic framework through which we explore how the national curricula incorporate different but interrelated disciplines of knowledge of disaster education toward a more inclusive education. The study reveals that disaster education does not encompass holistic and complementary perspectives for instilling pro-ecological behaviors and values, but rather disaster-related conceptions and discussions are separately included without illustrating the interaction with the concepts of ecology and biodiversity management and conservation, climate change, and sustainability lifestyles. These

E. Yilmaz-Na (✉)
Faculty of Education, Recep Tayyip Erdogan University, Rize, Türkiye
e-mail: eyilmaz1@lakeheadu.ca; elamsin@gmail.com

Faculty of Education, Lakehead University, Orillia, Canada

O. Erdur-Baker · G. Tuncer-Teksoz
Faculty of Education, Middle East Technical University, Ankara, Turkey
e-mail: erdur@metu.edu.tr

G. Tuncer-Teksoz
e-mail: gtuncer@metu.edu.tr

A. Yildiz and R. Shaw (eds.), *Disaster and Climate Risk Education*, Disaster Risk Reduction, https://doi.org/10.1007/978-981-97-5987-3_5

findings have important implications for revising the national curriculum framework on disaster education.

Keywords Disaster education · Ecosystem approach · Curriculum analysis · Türkiye

5.1 Introduction

The recent major earthquakes near the border of Türkiye-Syria have resulted in the death of tens of thousands of people, and many more have been injured. Thousands of buildings, schools, and hospitals have been destroyed, resulting in surviving without housing, clean water, and medical attention under harsh winter conditions (AFAD 2023). This devastating series of earthquakes in the region reminds us of three facts. We have learned that it has everything to do with (1) how we are aware of the risks surrounding the environment where we live, (2) how we prepare for such local risks, and (3) how we can recover afterward.

Similarly, with a worldwide increase in disasters, predominantly arising from climate change (UNDRR 2023), disaster risk reduction (DRR) is gaining importance in global agendas since the widespread adoption of the Sendai Framework for Disaster Risk Reduction 2015–2030 (Mizitori 2020). On the eighth anniversary of adopting the Sendai Framework (2015–2030), with its clear focus on the importance of reducing disaster risks, it is acknowledged that there is no such thing as a natural disaster. Instead, unplanned urbanization, incompetence of preparations, poverty, burning fossil fuels, and ecosystem destruction have exacerbated the frequency and intensity of disastrous events. Although there is no agreed definition of DRR, the UNDRR definition was widely cited as "preventing new and reducing existing disaster risk and managing residual risk, all of which contribute to strengthening resilience and therefore to the achievement of sustainable development" (UNDRR 2023). Aligned with the goal of DRR, the Hyogo Framework sets out five priorities for action, each elaborated into several specific areas of attention. Among these five priorities, the principle of raising awareness and education applies equally to DRR to build a culture of safety and resilience at all levels (UNDRR 2015). As concerns about disaster risks, climate change, and ecological problems have increased, education has been targeted as a keystone in managing the related issues (UNEP 2006).

In light of education's integral role in DRR, school education can lead to transforming societies and, ultimately, building the community's resilience. Recent studies show that only well-equipped individuals are more likely to deal with the uncertainty of the environmental, social, economic, and political future and can manage variability and adapt (UNICEF 2012). Aligned with this ultimate goal, the dramatic changes in knowledge, science, ecology, and critical perspectives in social contexts urge the formal education system to change its conventional approaches to curriculum policy and practices by empowering students in decision-making and planning for a more sustainable and resilient future (OECD 2018). In responding to the critical challenge of how formal education can empower students to survive and be resilient in the times of disasters, embedding disaster concepts and discussions in the curriculum have been suggested as an effective strategy (Yilmaz and Yildirim 2023). Accordingly, this cross-curricular approach helps the acquisition of knowledge, skills, and attitudes to reduce the risks of disasters and be prepared against them as well as proper response to them (Kawaga and Selby 2012).

Over the past decade, introducing disaster-related themes and discussions have evolved in the way of focusing surface-level knowledge acquisition with a lack of emphasis on developing behaviors, skills, and attitudes and reviewing the same concepts over and over without increasing complexity and following developmentally appropriate ways (e.g., Sozcu 2020). There is substantial research on textbook analysis (Degirmenci et al. 2019), examining disaster awareness (e.g., Dikmenli and Yakar 2019), disaster literacy (Sozcu 2019), disaster perceptions (e.g., Coban et al. 2017) and disaster preparedness (e.g., Sahin et al. 2018), on investigating teachers' understanding (Bozyigit and Kaya 2017) and implementation of extracurricular disaster education practices (Yilmaz and Yildirim 2023). However, research needs to be done on how disaster-related issues are conceptualized and represented in recently revised national curricular documents within the Turkish formal education system through a comprehensive and critical lens. While the previous studies were mainly focused on the explicit inclusion of nature-induced disaster concepts, our interest is to elicit and pinpoint both the explicitly and implicitly embedded concepts with a critical eye on the curricular manifestations. Specifically, this study attempts to explore to what extent human–ecosystem/environment–economic–society entanglement exists in the curricula in alignment with the understanding that human systems and the ecosystem are inevitably intertwined and cannot be viewed separately (Weldemariam et al. 2017). That is to say, this study seeks to broaden the scope of disaster education with the ecosystem approach by examining curriculum frameworks in terms of not only explicit mentions of disaster-related discussions and topics but also implicit discussions, incorporating the concepts related to ecology management and biodiversity degradation, climate change, and sustainability lifestyles.

Collectively, this curriculum analysis contributes to disaster-related research with a particular focus on exploring to what extent learning expectations within the Turkish national curricula can best fulfill the role of supporting elementary and secondary students to become personally responsible citizens in ecological matters. This approach is aligned with promising practices of disaster education not only for disaster preparedness, response, and recovery but also including reducing and adapting disaster risks before any disastrous events occur. Accordingly, this critical analysis attempts to answer the following research questions:

1. How is disaster education incorporated explicitly across the Turkish national curricula?
    a. What is the depth level of disaster education inclusion into curriculum expectations?

   b. What is the focus of disaster education across different grades and subject areas?

2. How is disaster education incorporated implicitly across the Turkish national curricula?
   a. What is the depth level of disaster education inclusion into curriculum expectations?
   b. What is the focus of disaster education across different grades and subject areas?

5.2 Theoretical Framework

This study employs the ecosystem approach to disaster risk reduction (Eco-DRR) as a theoretical framework. It is used as it allows us to engage in a comprehensive but relevant and context-related analysis procedure. Eco-DRR is the "sustainable management, conservation and restoration of ecosystems to reduce and adapt disaster risks, to achieve sustainable and resilient development" (Estrella and Saalismaa 2013, p. 30). There appears to be some agreement that Eco-DRR is a comprehensive and integrated approach with its commitment to sustainable development (Murti and Mathez-Stiefel 2019).

Since 2014 Eco-DRR has been recognized starting from the Convention of Biological Diversity. Up to now, significant progress has been made in coherence among the relevant policy mechanisms for Eco-DRR. According to the Eco-DRR approach, all ecosystem components are linked and connected (Tidball and Krasny 2011). While one component is disrupted, the ecological balance is affected adversely, possibly increasing the severity and frequency of disastrous events. Recently, the role of ecosystems is now embedded in the United Nations Sendai Framework for DRR for countries to take proper risk reduction and adaptation actions. Drawing on a review of the global agreement on Eco-DRR, several scholars have discussed that while individuals' vulnerability to disasters can be increased through the overuse of natural resources, resulting in ecological degradation (e.g., Renaud et al. 2013), individuals' commitment to using natural resources wisely and sustainably and to preserving the ecologies can achieve multiple benefits ranging from reducing their vulnerabilities to disasters to enhancing ecology and biodiversity conservation (e.g., Tora et al. 2022).

Disaster education lays a foundation for understanding how the human–ecosystem/environment–economic–society entanglement works (Rahma et al. 2023). Thus, it has been shown that disaster education is inevitably connected to environmental changes, and the role of human actions or inactions in the environment changes and, thereby, sustainable development (Ho and Seow 2017). However, a major criticism of the existing disaster education in Türkiye is related to its aim to transmit limited knowledge about the mechanism, natural causes, adverse impacts, and preparedness measures for disasters (Inal et al. 2018). Instead, recent research suggests that disaster education should focus on developing a sense of care for and responsibility toward the environment as well as developing pro-ecological behaviors of individuals (e.g., Apronti et al. 2015). Accordingly, the intersectional sustainability mindset further informs the current study, reducing risks and vulnerability, adapting pro-ecological measures and climate change, and developing environmental justice and scientific literacy (Anderson 2010). This is also exemplified in work undertaken by several scholars (e.g., Wahl 2017). In these studies, Bronfenbrenner's socio-ecological model (1979) was adopted to understand the complexities of relations between the environment and social system, including individuals and communities that build a base of human actions at multiple levels (Kilanowski 2017). Aligned with this understanding, in this study, disaster education refers to reorienting education practices linking environmental, physical, and social factors toward the ultimate goal of internalizing protective, sustainable, and pro-ecological literacy including knowledge, skills, attitudes, and behaviors at individual and collective levels (Rahma et al. 2023; Shaw et al. 2021).

5.2.1 *Urgency of Disaster Education*

Although the role of education in the fight against disasters has been of great importance (Torani et al. 2019), formal education needs to be utilized more for risk reduction purposes and adequately responding to disasters (Benadusi 2014). Thus, we no longer have time to rely on current formal education to develop attitudes and mindsets and to ensure that the next generation is more aware and accountable in their decision-making through traditional education approaches. According to the report of IPCC (2014), actions must be taken immediately. Therefore, children, the youth, stakeholders, actors, and policymakers are placed in a central position as each group makes decisions about necessary changes in the way they live, in which different lifestyles and standards of living to reduce risks should be adopted (Tanner et al. 2009). That is to say, a formal curriculum has a huge potential both educationally and conceptually by graduating young generations with the capacity for critical thinking; the conditions for a better life; and the knowledge, skills, and attitudes to participate in the creation of a sustainable and resilient future (Aghaei et al. 2018). However, education institutions are arguably falling short in their moral responsibility to produce graduates with critical consciousness of social problems to achieve real social change and justice so that students can develop better insides and behavior patterns encompassing sustainable practices (Apronti et al. 2015).

This situation leads to significant concerns that the critical function of educating students for a sustainable future must be addressed, or at least we claim it needs to be given a high priority in designing a promising formal curriculum. Thus, many research studies reveal that integrating disaster education into formal curricula is the best approach to educating DRR, environmentally sustainable development, and climate change adaptation and mitigation measures (e.g., Efthymis et al. 2014). Nevertheless, a comprehensive disaster education curriculum, which is required for raising future citizens, is generally not included in education systems worldwide or needs to be better addressed (Boon and Pagliano 2014). To date, disaster education often focuses on developing factual knowledge and gives little systematic attention to cognitive skills and affective components (Bulus-Kirikkaya et al. 2011). Furthermore, teachers only transmit factual knowledge in textbooks presenting fixed futures (Hosseini and Izadkhah 2007) which, in turn, students are more likely to predominantly perceive disaster risks as an abstract and a distant matter (Aydın and Çoşkun 2010), and students usually feel insecure about how to handle unpredictability (Mitchell et al. 2008). To address these challenges, a systematic review study on child-centered disaster education highlighted the importance of making all materials personally relevant to students (Amri et al. 2018). Thus, they could be more likely to take proper actions.

The existing body of research also suggests that the current disaster education is not effective in substantial behavioral changes (Johnson et al. 2014). This can be because the formal curriculum does not offer relevant and connected topics and discussions on relevant subjects (Kagawa and Selby 2012). Relatively little discussion, especially regarding the curriculum and how the process of disaster education is institutionalized in K-12 agendas has explored the underlying assumptions behind everyday disaster education practices or problematized pedagogy (Kawaga and Selby 2012; Sheehy et al. 2022). Based on the ecosystem approach, studies show the importance of integrating discussions about interconnected topics into formal education, such as emission reduction, healthy and sustainable lifestyles, environmental health, and energy conservation, as well as risk reduction and adaptation measures (Krasny and DuBois 2019). Hence, implementing an interdisciplinary and transdisciplinary disaster education curriculum needs to be more robust (Rahma et al. 2023).

5.2.2 Disaster Education Policy in Türkiye

In Türkiye, disaster education has been repeatedly called to be integrated into the formal education system. Accordingly, disaster education was initially embedded into school curricula under the "Disaster Protection and Safe Life" theme. This theme was mainly associated with knowledge of disasters, their negative impacts, and necessary safety measures. Furthermore, in parallel with this understanding, the Ministry of National Education (MONE) assumed teachers were the main actors in providing disaster education. Therefore, teachers were encouraged to take the responsibility of transferring the required knowledge to their students and families (Yilmaz and Yildirim 2023). With this in mind, supporting the curricular priority, non-formal and informal learning activities were developed to enrich the formal curriculum. Nonetheless, this initiative could not be carried out jointly and sustainably across the country (Yamazaki et al. 2017) unless disaster education is framed as an interdisciplinary subject to be part of national formal education (Yilmaz and Yildirim 2023).

Furthermore, the student learning goals of the national curricula have been changed to focus more on what is most important for students and national success in the twenty-first century. Accordingly, within the framework of the "2023 Education Vision'" published by the MONE, school curricula are identified as a keystone to raise qualified citizens equipped with the twenty-first-century competencies of the age and future with variations of lifelong learning skills, including environmental literacy. However, disaster education is not evident as an interdisciplinary, a multidisciplinary, or transdisciplinary theme; instead, it remains limited as it is only a part of the curriculum of life sciences in the years 1, 2, and 3, the curriculum of social studies in the years 4, 5, and 6, and in secondary geography curriculum (Basibuyuk and Pala 2023). In addition to the several forms of cross-curricular approach taken recently, as another progressive movement, the "Disaster Awareness" course is activated as an elective stand-alone course in elementary education for 6th graders to 8th graders (MONE 2023).

As another disaster education initiative at the national level, disaster education has been given weight by AFAD. Policymakers, government officers, non-governmental organizations, and other community members were recipients of raising awareness campaigns and training provided by AFAD (2022) through informative brochures and visual materials under the slogan "Disaster Ready Türkiye". Furthermore, there are a few research centers in disaster management at universities where training, consultation, and research activities are provided with a preliminary focus on first aid and risk awareness (e.g., Istanbul Technical University). However, these efforts are mostly limited to specific regions. Similarly, several municipalities located in the provinces vulnerable to disasters (e.g., Istanbul, Izmir, Kocaeli, Bolu) carry out disaster education, as listed in the study conducted by Gerdan (2019). Additionally, the Japan International Cooperation Agency (JICA) organizes various projects, including developing and implementing training activities and visual materials and organizing video conferences with different organizations in the field of disaster management to share Japanese experiences on DRR with governmental officers, technical staff, and emergency managers in Türkiye.

5.3 Method

This exploratory and interpretative research systematically reviews the Turkish national curricula, aiming to provide a significant opportunity to advance our knowledge of how disaster education is incorporated within the formal education system. This research mainly overviews the overt and explicit curriculum in which disaster education is deliberately taught as well as focusing on the implicit or covert curriculum, which does not describe clearly or directly but still whose forms and effects remain related (Eisler 2000).

Several scholars have highlighted the importance of managing, protecting, and improving healthy ecosystems to build resilience against disasters (e.g., Renaud et al. 2013; Triyanti et al. 2017). With this in mind, it has been well acknowledged that reducing disaster risks is closely conceptualized with a continuous cycle of mitigation and adaptation processes and activities that prevent hazardous events from turning into disasters through risk assessments, preparation, response, and recovery phases (e.g., Estrella et al. 2016).

Further, it has conclusively shown that there is an unambiguous relationship between climate change and the risks of disasters. Climate change increases the frequency and intensity of hazards, as well as individuals' vulnerability, and alters exposure patterns and trends of hazards (IPCC 2014). Still, there is an increasing concern over integrating this knowledge into educational policies (e.g., Rahma et al. 2023). Thus, this current research synthesizes this conventional understanding featured in the national curriculum by incorporating the interrelated concepts of ecology and biodiversity management and conservation, sustainable lifestyles, mitigation negative impacts of climate change as well as adaptation to climate change.

5.3.1 Data Collection Procedure

The research starts with collecting curriculum frameworks that are publicly accessible, specifically for the mandatory courses throughout all grades in elementary and secondary education. The underlying reason for selecting the curriculum frameworks of mandatory courses is that elective courses can be only offered depending on the course enrollment and teacher availability. That is why elective courses are not accessible by every student equally to address the urgent need of disaster education.

To review how disaster education is explicitly integrated into Turkish national curricula, a series of keyword searches is initially conducted looking for the terms: "disaster", "risk", "safety", and "emergency" across all curricula for different grades and subject areas to find where these terms are mentioned and then to quantify their mentions. Consequently, "climate change", "global warming", "recycling", "renewable energy", "sustainability", "protection", "environment", "ecosystem", "conservation", "biodiversity", "management", and "natural resources'" are also specified for further keyword searches to include the expectations implicitly related to disaster education. This maintains an additional but interrelated focus on ecosystem and biodiversity management and conservation (Kasada et al. 2022; Murti and Mathez-Stiefel 2019), sustainable lifestyles (Laurie et al. 2016), and climate change as a disaster risk driver (UNDRR 2015). This procedure also offers an effective way of providing a comprehensive picture of the curricula for disaster education. A systematic content analysis of curriculum documents for grades 1–12 in different subject areas is further employed for this specific purpose. A systematic curriculum review is conducted for all mandatory course curricula that incorporate related and identified keywords and terms: (1) life sciences, (2) social studies, (3) science, (4) English, (5) geography, (6), biology, (7) technology and design, and (8) visual arts. The curricula of interest are briefly described in the table presented in the appendix.

5.3.2 Data Analysis Procedure

In order to explore the depth of the integration of disaster education into the curricula, curriculum frameworks are analyzed by adopting a taxonomy developed by the corresponding researcher. This categorization is aligned with the four tenets of disaster education practices based on a consensus among scholars in the literature (e.g., Dufty 2014; Johnson et al. 2014; Kawaga and Selby 2012; Shaw et al. 2021). These four tenets are as follows: (1) knowledge-oriented learning for building factual and theoretical knowledge about and understanding of the disaster-related concepts, practices, science, and mechanism; (2) skills-oriented learning for developing necessary cognitive thinking and practicing skills; (3) affective-oriented learning for incorporating values, attitudes, and dispositions for promoting social capital and emotional resiliency to prepare for, respond to, and recover from; and (4) action-oriented learning for taking proper and required actions to be prepared for, resilient against, respond to, and recover from local and global disaster-related challenges at individual and collective levels. This taxonomy provides a holistic framework outlining four levels of learning depth in disaster education based on ecosystem approach, as summarized in Table 5.1.

As it is presented, learning starts by acquiring a basic grasp of the science and mechanics of how disaster occurs, how they impact human beings and other living things physically and socially, how frequently they occur, and how the trends and patterns in their occurrence change over time. Then, it moves progressively through the levels to push students' understanding deeper and their thinking higher to make

Table 5.1 Taxonomy for the level of depth of curriculum expectations regarding disaster education*

Knowledge oriented (1)	Skills oriented (2)	Affective oriented (3)	Action oriented (4)
Focusing on factual and theoretical knowledge about concepts/practices/measures/science/physical mechanism	Focusing on the application and analysis of factual information as well as the development of higher order thinking skills (e.g., critical thinking, problem-solving, systematic thinking) and generic skills for coping, self-protection, self-management, interpersonal, and communication	Including values, emotions, attitudes, and dispositions such as hope, self-efficacy, social consciousness, empathy, and justice.	Taking proper and required actions to be prepared for, resilient against, respond to, and recover from local and global challenges at individual and collective levels

* *The taxonomy is developed by the corresponding researcher aligned with the related research* (e.g., Dufty 2014; Johnson et al. 2014; Kawaga and Selby 2012; Shaw et al. 2021)

their knowledge more relevant and meaningful, ultimately to build personal, home-based, and community-based preparedness as well as social capital and emotional resilience. Notably, each level is valuable for the enhancement of learning and for building students' capacities to be proactive to deal with local and global disaster-related challenges. For coding reliability, a test–retest approach is applied (Belur et al. 2021). Four curriculum expectations identified after initial screening are randomly re-analyzed and re-coded to check the consistency in coding the depth of curriculum expectations among the researchers.

5.4 Results

This section portrays the state of knowledge about how disaster education is incorporated within Turkish national formal education in a systematic and detailed way by presenting and describing the following figures. The section is divided into two subsections to address two main research questions.

R.Q. 1: How is disaster education incorporated explicitly across the Turkish national curricula?

To address the first research question, the curricular frameworks are analyzed by starting with keyword search, including the terms used explicitly in disaster-related discussions such as "disaster", "risk", "safety", and "emergency". Figure 5.1 reveals the total number of times "disaster" is found in science, social studies, geography, English, and other courses throughout each grade in the Turkish national curricula. Specifically, there are far more explicit mentions of disaster-related conceptions and discussions in geography ($n = 10$), social studies ($n = 11$), life sciences ($n = 9$), and English ($n = 6$) than in science ($n = 3$) and biology ($n = 2$) courses. This data is a strong example of the courses that engage deeply with disaster-related concepts and discussions predominantly from a social sciences perspective rather than a natural science perspective in elementary and secondary education.

Furthermore, Fig. 5.2 portrays an overview of curriculum expectations distributed across subject areas. It is apparent from this data that there needs to be more coherence and uneven inclusion of curriculum expectations on disaster education in different subject areas. Specifically, there are far more disaster education expectations in life sciences (n = 7, 14.00%), English (n = 5, 7.10%), geography (n = 4, 11.8%) in social studies (n = 4, 12.20%) than in biology (n = 1, 7.10%) and science (n = 3, 7.10%), visual arts (n = 1, 5%), and technology and design (n = 1, 2.4%). While a higher percentage indicates that most expectations are found in a specific subject area, it mainly depends on the number of total curriculum expectations of the course. For instance, considering the total number of curriculum expectations, geography and social studies courses have the same number of curriculum expectations focusing on disaster-related concepts. However, the social studies course has the highest percentage. Accordingly, data from this figure shows a need for more priority of curriculum expectations on disaster education in mandatory courses.

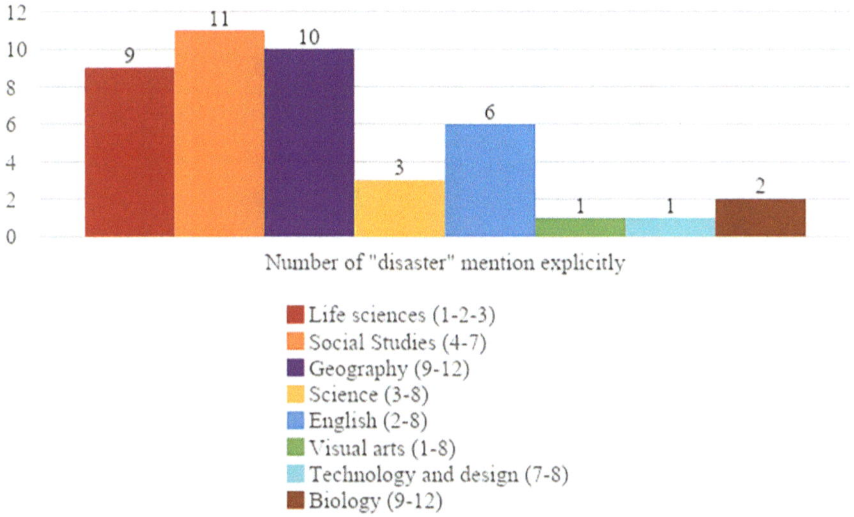

Fig. 5.1 Distribution of curriculum expectations explicitly mentioning "disaster" across subject areas

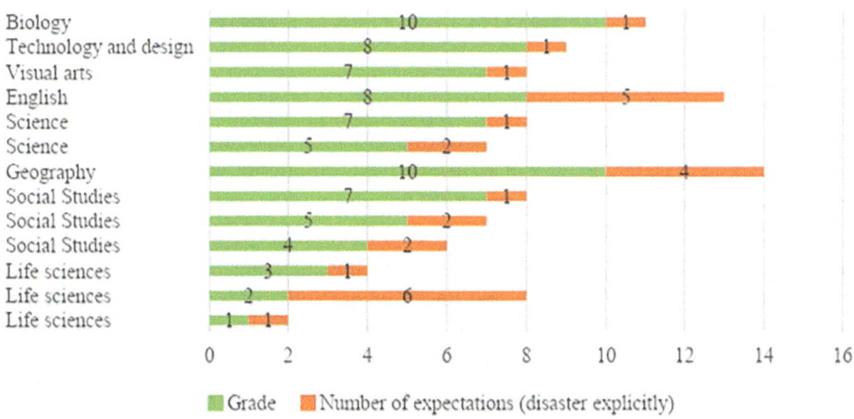

Fig. 5.2 Explicit disaster education curriculum expectations by grade across subject areas

To further investigate at which level of depth disaster education expectations are integrated within the curricula, we apply the taxonomy as presented in Table 5.1. Accordingly, Fig. 5.3 is quite revealing in several ways. It shows that most curriculum expectations engaged with disaster education are ranked at Level 1 across the curricula in different subject areas, which mostly focus on developing fundamental knowledge about disaster-related concepts, safety practices and measures, and scientific mechanisms of their occurrence. The taxonomy for the depth of curriculum also reveals that only two courses are ranked at high levels of depth to include

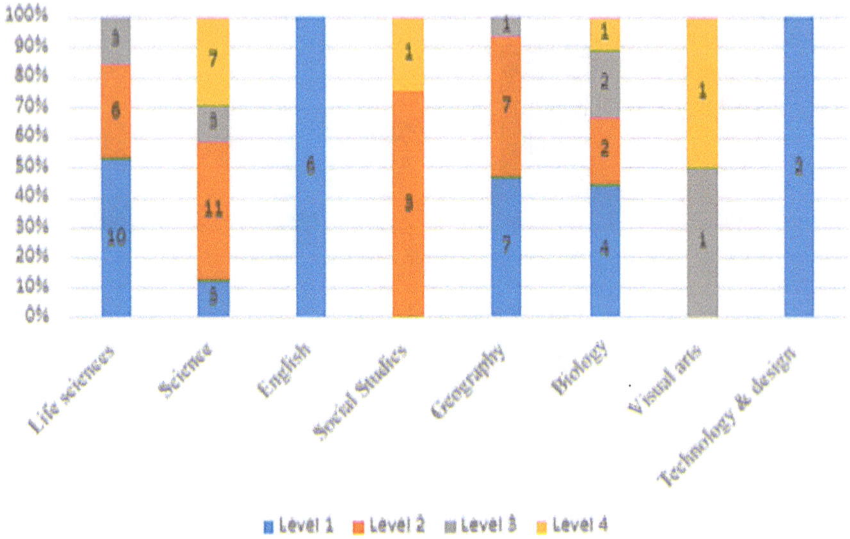

Fig. 5.3 Depth of explicit disaster education curriculum expectations across subject areas. The number of expectations can differ from those presented in Fig. 5.2 because some expectations have related but different subcategories, and follow-up explanations provided in the curriculum are taken into account while analyzing the depth of inclusion of curriculum expectations

disaster education. Notably, the social studies course has only one expectation that operates at Level 4-action-oriented learning, in which students propose a course of action to solve disasters as global issues through collaborative work. However, no mentions are found of encouragement of student involvement in any initiatives for building community resilience and in any school/community partnerships for disaster education being formally integrated into any of the curricula.

Another anticipated finding is that only two curriculum expectations are ranked at Level 3-affective-oriented learning in social studies and geography courses, where students share cultural values and attitudes (e.g., a sense of responsibility and solidarity) that enhance their commitment and willingness to take steps and measures to build disaster resilience. Equally important, this analysis also displays that at least one curriculum expectation is identified as operating at Level 2-skills-oriented in most of the mandatory courses, primarily because of students' engagement in higher order thinking skills to analyze the local conditions predominantly causing risk to them and to their community members as well as in practicing safety measures and procedures in the fight against disasters.

Furthermore, the content of each curriculum expectation is identified and allocated to a specific focus theme or category as a result of an inductive content analysis. Figure 5.4 presents the seven themes recurring throughout the curriculum expectation across different subject areas.

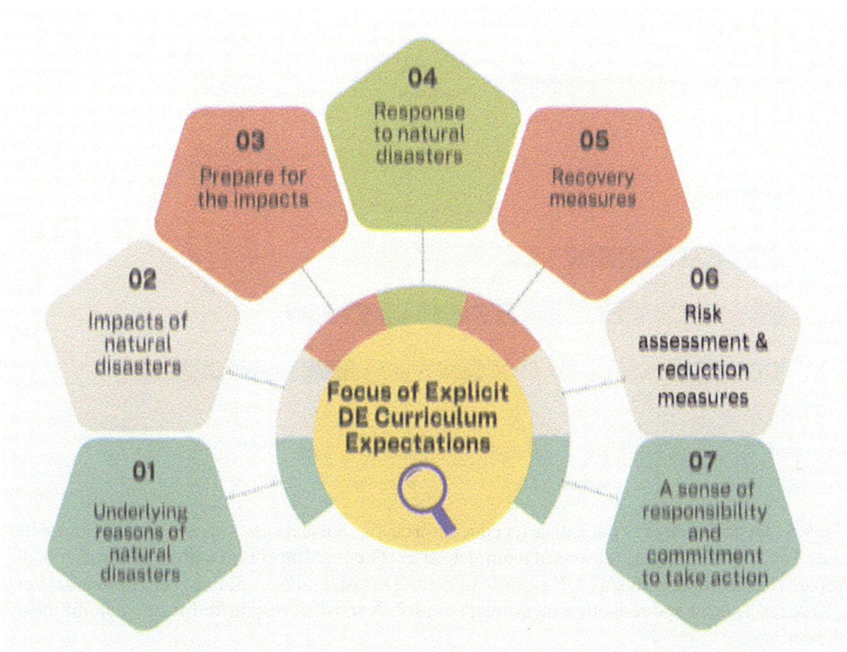

Fig. 5.4 Focus of explicit disaster education curriculum expectations

As presented in Fig. 5.5, almost all courses specifically focus on understanding natural disasters' underlying reasons and scientific mechanisms, except for the technology and design course. Under this theme, there is no mention of the impacts of human action/inaction on the occurrence of disasters. Along with this understanding, the common term "natural disasters" is commonly used in curricular materials. More importantly, the themes of recovery measures (in social studies courses), response to natural disasters (in life sciences and social studies courses), and embracing a sense of responsibility and commitment to take actions (in social studies and geography courses) are not particularly prominent in the data. When we take a look at these expectations closely, we find out that they are categorized as the higher level of learning (affective and action-oriented learnings).

R.Q. 2: How is disaster education incorporated implicitly across the Turkish national curricula?

To address the second research question, the curricular frameworks are also analyzed through a keyword search, in which the terms that are implicitly associated with disaster education are used ("climate change", "global warming", "recycling", "renewable energy", "sustainability", "protection", "environment", "ecosystem", "conservation", "biodiversity", "management", and "natural resources""). As can be seen in Fig. 5.6, the science course (3–8) has the best coverage of climate change and other environmental- and ecosystem-related concepts and discussions with 19

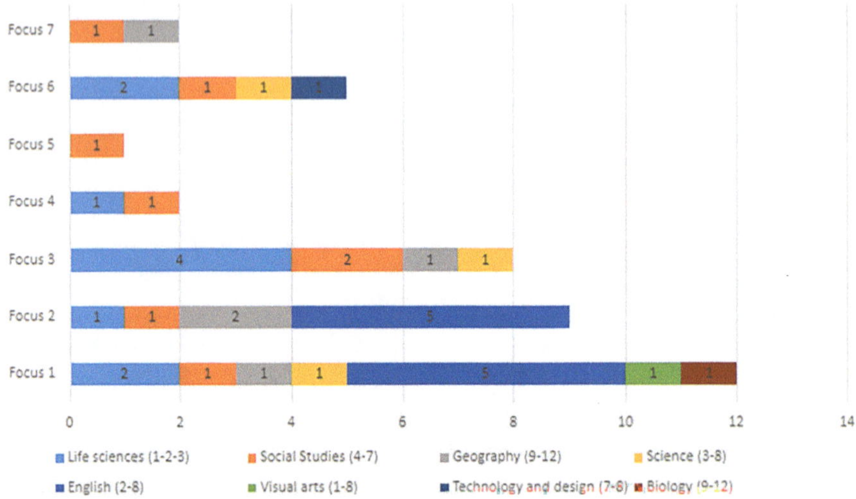

Fig. 5.5 Distribution of explicit disaster education curriculum expectations across focus and subject areas. Focus 1: Underlying reasons of natural disaster; Focus 2: Impacts of natural disasters; Focus 3: Prepare for the impacts; Focus 4: Response to natural disasters; Focus 5: Recovery measures; Focus 6: Risk assessment and reduction measures; Focus 7: A sense of responsibility and commitment to take actions

curriculum expectations compared to the other courses, followed by geography grade 9–12 ($n = 17$), life sciences grade 1–3 ($n = 15$), English grade 6 ($n = 8$), biology grade 10 ($n = 6$) and social studies grade 4–5–7 ($n = 4$), and technology and design grade 7 ($n = 4$). The present results are significant in at least one paramount respect. The emphasis on perspectives from natural sciences and social sciences across different grade levels can lead to deeper learning about concepts and discussions that are implicitly related to disaster education.

Accordingly, it is encouraging to compare this figure with the following figure (see Fig. 5.7), at which the level of depth of implicit curriculum expectations of disaster education integrated into subject areas is presented. As a result, a very important finding is that most of the curriculum expectations in science-related courses are classified as belonging to the upper categories of the taxonomy. In contrast, all other courses have curriculum expectations categorized as lower taxonomy categories.

Moreover, regarding the content of curriculum expectations covering related but implicit disaster-related concepts and discussions, seven broad themes emerged from an inductive content analysis, as illustrated in Fig. 5.8.

Furthermore, when we examine the weight of the focus of curriculum expectations across subject areas, these findings reveal inconsistent and uneven coverage as illustrated in Fig. 5.9. The most striking result from the data is that the minimal coverage of fundamental knowledge about ecosystems and biodiversity at both elementary and secondary education levels limits the focus on ecological management and climate change.

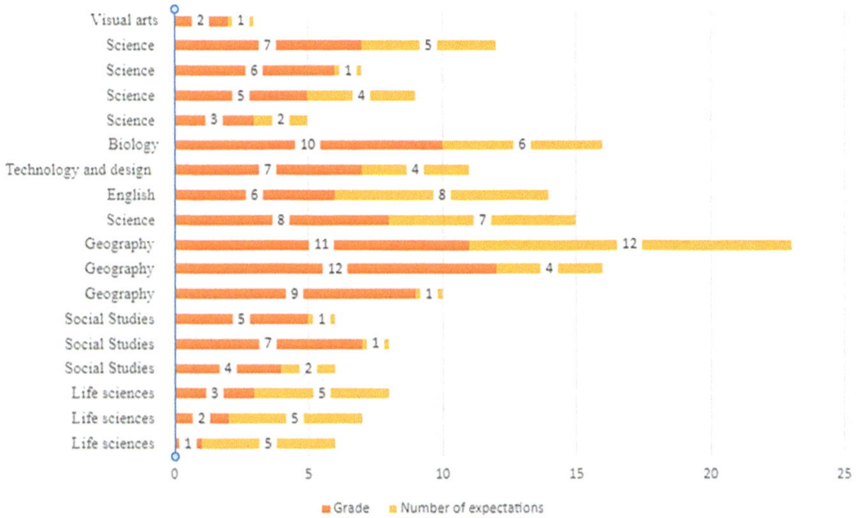

Fig. 5.6 Implicit disaster education curriculum expectations by grade across subject areas

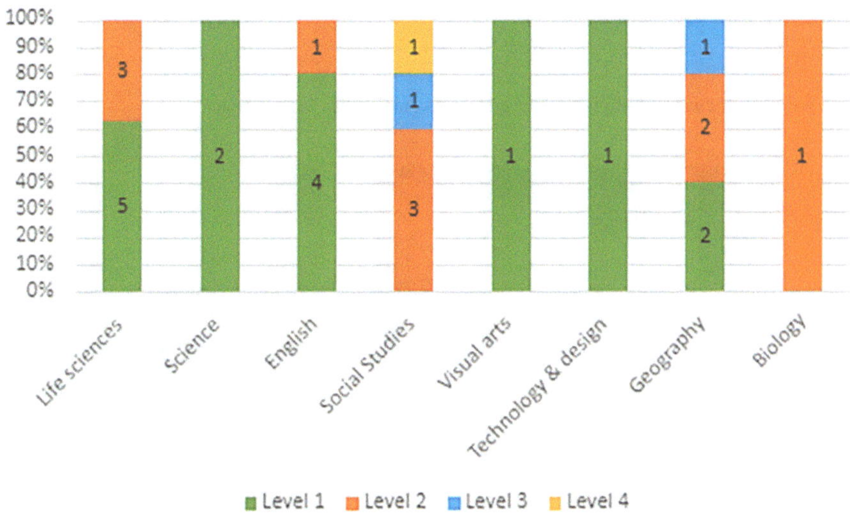

Fig. 5.7 Depth of implicit disaster education curriculum expectations across subject areas. The number of expectations can differ from those presented in Fig. 5.5 because some expectations have related but different subcategories and follow-up explanations provided in the curriculum frameworks are considered while analyzing the depth of inclusion of curriculum expectations

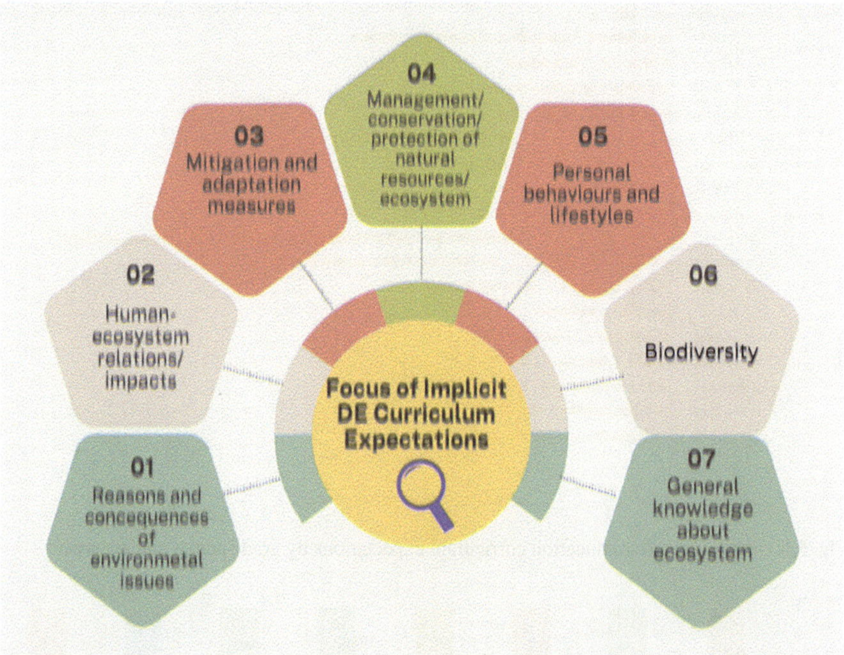

Fig. 5.8 The focus of implicit disaster education curriculum expectations

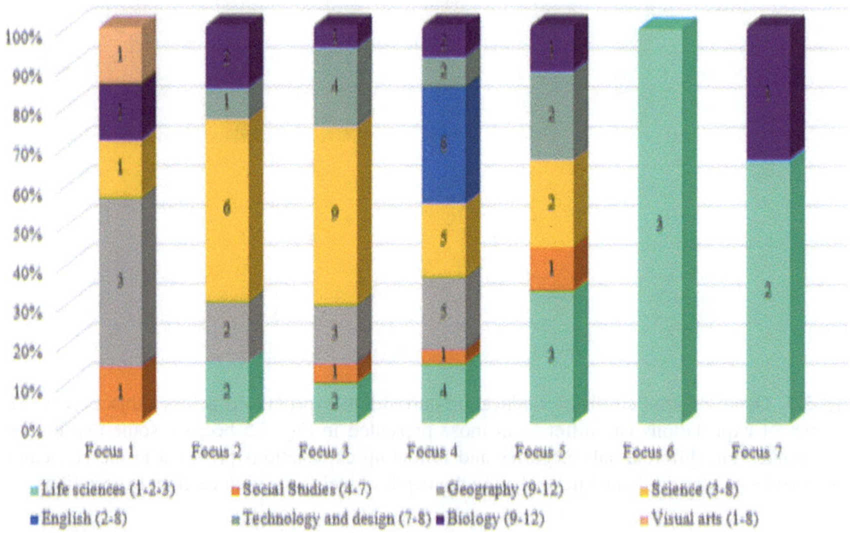

Fig. 5.9 Distribution of implicit disaster education curriculum expectations across focus and subject areas. Focus 1: Reasons and consequences of environmental issues; Focus 2: Human/ecosystem relations and impacts; Focus 3: Mitigation and adaptation measures; Focus 4: Management/protection/conservation of natural resources/ecosystem; Focus 5: Personal behaviors and lifestyles; Focus 6: Biodiversity; Focus 7: General knowledge about ecosystem

Another important finding is that there needs to be more discussion about reducing air, water, and soil pollution, and there is no focus on justice-oriented concepts, which are recently placed at the center of climate change and environmental education (IPCC 2018). Still, a critical finding reveals that the focus of mitigation and adaptation measures and management of natural resources are commonly used in natural sciences (life science, science, and biology), social sciences courses (geography and social science), and in the technology and design course. This is a promising finding because mitigation and adaptation measures are both necessary for addressing climate change and its impacts on increasing the severity of and occurrence of disasters (UNDRR 2015).

Additionally, when we look at these curricular expectations closely, all the implicit curricular expectations at the higher level of learning (affective and action-oriented learnings) are categorized under the focus of mitigation and adaptation measures.

5.5 Discussion

This study aims to shed light on the debates about the current status of disaster education through a systematic and critical analysis of the mandatory national curriculum frameworks accessible to the public. Key findings of the study indicate that the national curricula need to include a clear frame of disaster education, with consistent and even coverage across subject domains at different grade levels in elementary and secondary education. Expectedly, the current study presents that disaster education is covered from a social sciences perspective in the social studies course in elementary education and the geography course in secondary education. However, curriculum frameworks that are still in use are more dated because they generally discuss disaster concepts less often with a focus on disaster science and underlying reasons and scientific mechanisms. Rather, we all know that disaster is not a term describing a natural event, rather it is used for its negative impacts on and consequences for individuals, their properties, and a wider society. This limited conceptualization of disaster education can be related to the factors that influence how much space is explicitly or implicitly allocated to disaster education in the curriculum and the lack of national education policy on disaster education (Inal et al. 2018).

Although the shift to the competency-driven curriculum approach initiated in MONE (2023) is considered a promising step for facilitating experiential-based learning (McPherson et al. 2016), the integration of disaster education is still insufficient in terms of encompassing holistic learning about disaster and the environment and the acquisition of knowledge, skills, and values. We can imply that the Turkish formal curricula have a limited emphasis on multidisciplinary/interdisciplinary/transdisciplinary learning acquisition that addresses environmental, social, cultural, economic, ethical, and justice-oriented discourses.

Still, curriculum frameworks can change in actual classroom practices. Regarding disaster education, when the curricula are scrutinized and interpreted concerning the curriculum expectations, we can claim that teachers are provided with some guidance on the scope and the depth of disaster education. The explanations provided in the curricula can be utilized instrumentally in identifying space and opportunities and their interpretations. However, this capacity is only likely to be used if teachers are interested in and even actively looking for opportunities to teach about disaster education through a more integrative and transdisciplinary approach and if they feel confident that they can do so (Dwiningrum et al. 2022). Yet, the literature acknowledges that teachers do not show high commitment to preparing students against disasters (Wolff et al. 2017), and there is lack of adequate and qualified teachers who require professional development for adopting their teaching practices based on competency-driven approaches and for connecting relevant issues (Ho and Seow 2017).

As mentioned in the literature review, cross-curricular approach to disaster education also requires an action-orientation discourse (Apronti et al. 2015). Contrary to expectations, overall, the curriculum outcomes are predominantly ranked at the lower levels of depth across different subject domains rather than actively engaging students in taking actions and developing responsive attitudes and behaviors. From the general tone of the curriculum frameworks, student agency seems limited to proposing suggestions instead of creating action plans or making them happen. In addition, the curriculum frameworks do not clearly and explicitly state key values, such as responsibility, solidarity, and respect for nature. Instead, the curriculum frameworks expect teachers to read behind the lines in ambiguous discourse. Thus, regarding the expectations laid out in the curricular documents, the scarcity of explicit action- and affective-oriented considerations need to provide more guidelines and flexibility for teachers to promote and prepare students to be ready for unprecedented disasters as well as to tackle ecological challenges. This result may be explained by the fact that the curricula often focus solely on increasing knowledge about natural causes of disasters and preparedness measures. However, rather, they ignore the understanding of humans as part of nature and ecosystem processes, and taking actions can make a substantial difference toward a more resilient and sustainable future. Notably, it is better to mention the success of behavior-oriented learning in promoting resilient behaviors among disaster survivors (Benadusi 2013).

This current investigation, however, has a limited scope, as it solely focuses on the analysis of existing curriculum frameworks from grade 1 to grade 12 as of the end of November 2023 and does not include disaster education curricular policies in higher education. This research has thrown up many questions that need further investigation and will undoubtedly shed new light on disaster education in different education contexts. The evidence from this study further supports the curriculum reform with a strong engagement of key stakeholders in its development process (Mutasa and Coetzee 2019); otherwise, the actual implementation of disaster education will remain weak.

5.6 Conclusions and Recommendations

It is widely acknowledged that disaster education should focus on establishing a "culture of safety and resilience" (UNICEF 2014, p. 132). Thus, it is evident that there needs to be more understanding of how disaster education can be operationalized into the Turkish national curricula, and disaster-related discussions do not hold a key position in the curricular documents. At the general level, the curriculum framework lists a range of skills (e.g., agency, critical thinking, entrepreneurship, etc.) that are important for students in the twenty-first century. The curricular frameworks call for student-centered pedagogical approaches to disaster education (e.g., experiential, inquiry-based, problem-based, and project-based learning) to promote these skills. However, it fails to contextualize them more coherently regarding disaster-related discussions and learning outcomes. Contrary to expectations, we find out that curricular frameworks usually consider students as recipients of factual knowledge transmitted by teachers or textbooks, rather than students actively generating their own understanding. Thus, student agency is not promoted at all. As recommended in the literature, students should be seen as critical agents with a great deal to bring positive changes to the community and our future world. These findings suggest that student agencies and how to promote their agency should be recognized explicitly and clearly in the nationwide curricular frameworks.

Moreover, the curricular materials do not highlight ecological and planetary perspectives within the scope of formal disaster education. Accordingly, related learning expectations present ambiguity in connecting social, environmental, economic, and cultural systems. These findings urge authorities to develop curriculum policies and school practices to prepare students to survive and fight against unprecedented disasters. From this perspective, this study has important implications for facilitating a reorientation of the existing curricula toward relying on interdisciplinary/multidisciplinary/transdisciplinary approaches with a holistic view of the ecosystem. We believe that in formulating formal disaster education, it is critical to embrace a more holistic and cross-curricular perspective through which ecological, social, cultural, and economic factors are not viewed as separate systems. The curricular frameworks should be able to present this interconnection of human-related and ecological systems. It can thus be suggested that disaster education should not focus solely on natural disaster events. Instead, it must be designed systemically to involve all ecocentrism and pro-ecological behaviors to embrace the big ideas toward a safer and resilient future.

This shift also requires offering continuous opportunities for professional development of teachers to improve their teaching practices in classrooms. The current policies and initiatives for teachers' professional development do not lead to effective integration of disaster education into classroom practices (Yilmaz and Yildirim 2023). This implies collective efforts among policymakers, scholars, educators, and teacher trainers to establish support systems that prepare teachers to develop promising pedagogical approaches, and innovative assessment methods.

To accomplish this, the curriculum development process should be aligned with the scientific knowledge obtained from curriculum analysis research as well as widely recognized international policy agreements (e.g., Hyogo framework, Sendai framework) and global initiatives (e.g., greening education, whole school approach). Furthermore, effective implementation of disaster education necessitates a collaborative effort from all stakeholders in the local community. Thus, localized disaster resilience should be valued and represented while revising the current curricular materials or developing new ones. Accordingly, it is essential to develop a consultation mechanism in which all local stakeholders (e.g., the youth, scholars, educators, and community members) should be encouraged to actively participate in the decision-making process for disaster education. For that purpose, open dialog among partners should be built with a great deal of collective understanding and a sense of responsibility toward disaster education practices.

Let us now turn to the taxonomy of the depth of integration of disaster education. To the best of our knowledge, there is no taxonomy from respective scholars for ranking disaster education integration in the curriculum as of the time of this analysis. The taxonomy recognizes scaffolding learning through four levels. While the lower level of taxonomy focuses on the knowledge about disaster-related concepts and measures that we want students to acquire, the middle levels focus on the application and analysis of information as well as the development of skills for personal, home-based, and community-based preparedness and resilience. Finally, the upper levels of taxonomy highlight the equal importance of learning that involves internalizing values, emotions, attitudes, and dispositions as drivers of necessary individual and collective actions. The taxonomy could be widely adopted to develop curriculum expectations in disaster education. Considering this advantage, taxonomy would be a valuable instrument for studies of curriculum development, curriculum analysis, and assessment of students' learning on disaster education.

Appendix: Bright Spots of Turkish Curriculum Focused on Disaster Education

Subject	Goals/Teaching principles	Themes/units	Scope
Life sciences (1–3)	The main purpose of the program is to raise citizens who have basic life skills, know themselves, live a healthy and safe life, internalize the values of the society they live in, are sensitive to nature and the environment, research, produce and love their country	1.1. Life in our school	The subject is explained based on the necessity of rules such as using classroom equipment carefully, keeping the classroom clean, paying attention to timing in communication (listening, speaking, speaking appropriately, etc.), not damaging your friends and your friends' belongings, and keeping yourself and your surroundings clean. Care is taken to determine classroom rules with the participation of students
		1.4. Safe life	Students know that in case of emergency, the Emergency Call Center should be called by dialing the numbers 1–1–2. The correct actions to be taken during disasters and emergencies (earthquake, flood, fire) are emphasized
		1.6. Life in nature	The focus is on the animals in the immediate environment (fish, birds, reptiles, insects, domestic animals, etc.), what these animals feed on and where they shelter. If it is necessary, when observation is to be made, security measures are taken Students observe the garden plants, wild plants, and trees in the immediate environment and how the plants change over time (plants growing, shedding their leaves, blooming, etc.). These issues are emphasized by making use of their ponds. Students take care to protect the animals and plants in their immediate surroundings. Students become sensitive about keeping nature and the environment clean. The emphasis is on doing what is necessary to keep nature and the environment clean and warning the people around about this within the framework of courtesy rules. Students distinguish items that can be recycled. Focus is on materials such as plastic, paper, batteries, vegetable oil, and glass
		2.3. Healthy life	Emphasis is placed on personal cleanliness and environmental cleanliness

(continued)

(continued)

Subject	Goals/Teaching principles	Themes/units	Scope
		2.4. Safe life	Students learn to call the 112 Emergency Call Center; it is ensured that the information to be given is expressed correctly. Information is given about the Alo 183 Anti-Violence Line. Practices are made regarding the correct behaviors to be followed during disasters and emergencies (earthquake, flood, fire)
		2.6. Life in nature	The positive and negative effects of natural elements in the immediate environment (climatic conditions, landforms, soil fertility, water resources, etc.) on people are discussed. Students explain how natural disasters such as earthquakes, fires, and floods occur, focusing on cause-and-effect relationships. Emphasis is placed on the human factor in natural disasters. The importance of danger hunting against disasters and emergencies is discussed, and danger hunting is conducted in the classroom, corridors, and garden. Students are also encouraged to do the danger hunt at home. Students explain how to act before, during, and after an earthquake. The importance of keeping natural resources such as water, air, and soil clean; using them appropriately; and planting trees is emphasized for a livable world. Additionally, non-governmental organizations interested in the subject are introduced at a basic level. Emphasis is placed on the positive and negative effects of humans on the natural environment. Care is taken to give examples of its positive effects. Examples of endangered creatures are given. How materials such as plastic, paper, batteries, and glass are collected, and their areas of reuse are exemplified. The contributions of this process to the environment and sustainability are emphasized. In an emergency, evacuation route, exit doors, and how to use emergency exits are highlighted. The location of the meeting areas closest to where they live is indicated. An Emergency Information Card is prepared

(continued)

(continued)

Subject	Goals/Teaching principles	Themes/units	Scope
Social studies (4–7)	1. Students explain the interaction between humans and the environment by recognizing the general geographical features of the environment they live in and the world and to develop their spatial perception skills 2. Students raise awareness of the limitations of the natural environment and resources, trying to protect natural resources with environmental awareness and having a sustainable environmental understanding	4.3. People, places and environment	While teaching this learning area, students acquire skills such as perceiving space, using maps, location analysis, drawing and interpreting tables, graphs, and diagrams, with the value of sensitivity to the natural environment
		5.3. People, places and environment	While teaching this learning area, students develop a sense of sensitivity to the natural environment and solidarity, as well as skills such as map literacy, environmental literacy, and observation
		7.7. Global connections	While teaching this learning area, students develop specific values such as peace and respect, as well as skills such as cooperation, problem-solving, and recognizing stereotypes and prejudice
Science (3–8)	1. In the process of exploring nature and understanding the relationship between human beings and the environment, students adopt scientific process skills and scientific research approaches and produce solutions to the problems encountered in these areas 2. Students raise awareness of the mutual interaction between individuals, the environment, and society 3. Students raise awareness of sustainable development regarding society, economy, and natural resources 4. Students raise interest and curiosity about the events occurring in nature and the immediate environment and develop attitudes 5. Students develop fundamental knowledge of astronomy, biology, physics, chemistry, earth and environmental sciences, and science and engineering applications	F.3.6. Journey to the world of living things/ living things and life	In this unit, students can distinguish between living and non-living beings, recognize the environment they live in, keep it clean, protect and care for it; observe the natural and artificial environment, and explain it with examples to gain a consciousness of savings in the use of resources, to take individual responsibility, and also to gain awareness of healthy living and well-being
		F.3.6.2. Me and my environment	In this unit, the following subjects and concepts are covered: school and its environment, environmental cleanliness, nature, forests, parks, gardens, buildings, national parks, natural monuments, etc.
		F.3.7. Electric vehicles/ physical phenomena	In this unit, students observe the electric vehicles around them and distinguish the purpose of use of these vehicles and the electricity sources they use; know what to pay attention to when using vehicles; get to know batteries, discover how batteries are used to operate various devices, and realize the negative effects of battery waste on the environment
		F.4.5.5. Sound pollution	Students explain the negative effects of sound pollution on human health and the environment
		F.4.6. Journey to the world of living things/ living things and life	This unit is aimed at helping students acquire the resources necessary for life and the knowledge and skills for sustainable use of resources

(continued)

(continued)

Subject	Goals/Teaching principles	Themes/units	Scope
		F.5.6. Journey to the world of living things/ living things and life	In this unit, students question the causes and consequences of environmental problems, raise awareness about biodiversity, endangered species, and what needs to be done to protect these species, awareness of environmental problems caused by human activities, and knowledge and skills to solve these problems
		F.6.4. Matter and heat/ matter and its nature	In this unit, students understand that matter consists of moving particles, the changes that occur in matter, the density of matter by using mass and volume, and the importance of density for living things. By examining heat conduction and insulation, students understand the contribution of thermal insulation technology to the family and country economy, fuel types, and the effects of fuels used for heat purposes on the environment
		F.7.4.5. Domestic Waste and Recycling	In this unit, the following concepts are covered: domestic solid waste materials, domestic liquid waste materials, recycling, and reuse
		F.8.6. Energy conversions and environmental science/ living things and life	In this unit, students understand photosynthesis, respiration, and energy transformations; explain the food chain and the elements that make up this chain; and discover the relationship between the elements, recognize the cycles of matter in life-related to environmental science, know the environmental problems, and offer solutions to environmental problems and gain knowledge and skills regarding these issues

(continued)

(continued)

Subject	Goals/Teaching principles	Themes/units	Scope
English (2–8)	Not stated	6.9. Saving the planet	In this unit, students respond to the following questions and provide simple suggestions: —What should we do to save our world? —We should save energy —We can use less water and electricity. —We should recycle the batteries —We should not harm animals —Turn off the lights —Don't waste water —Unplug the T.V. The following concepts are covered: air/water/noise pollution, cut down damage, garbage, electrical device, harm, litter, plug, unplug, recycle, rubbish, reduce, save, trash, and waste
		7.9. Environment	In this unit, students describe simple processes such as getting the seeds and planting and watering them Students express their obligation: —What should we do for our environment? —We have to start using public transportation —Stop polluting the rivers—We must stop destroying forests Students give explanations/reasons, as in —Rain forests are important because they are necessary for oxygen —We should protect wild animals because they are important for the balance of nature The following concepts are covered: balance, climate, eco-friendly, efficient, global warming, greenhouse effect, increase nature protect, pollute/pollution, renewable, responsible, solar/wind energy, take action, temperature, threaten, waste

(continued)

(continued)

Subject	Goals/Teaching principles	Themes/units	Scope
Geography (9–12)	1. Students develop geographical skills within the framework of human-nature relationship 2. Students make the connection between the basic elements of the universe and life 3. Students understand the functioning and change of natural and human systems 4. Students raise awareness of understanding and protecting the spatial values of the country and the world, starting from their surroundings 5. Students raise a sense of responsibility for the functioning of the ecosystem 6. Students understand the importance of spatial planning for the harmonious unity and continuity of nature and human beings 7. Students develop a sense of commitment to saving in the use of natural and human resources 8. Students make sense of the functioning of natural and human systems in local and global interaction 9. Students understand the importance of harmonizing economic developments with nature 10. Students evaluate natural disasters and environmental problems and develop practices for protection and precautions against them 11. Students understand the role of environmental, cultural, political, and economic organizations active in international relations at regional and global levels	9-4– Environment and society	This unit covers topics related to how people use the natural environment and the changes that occur under human influence in the natural environment, respectively
		10-4– Environment and society	This unit covers the causes and characteristics of natural disasters, their distribution and effects in the world and in Türkiye, and measures of protection from disasters, respectively
		11-4– Environment and society	In this unit, types of environmental problems, the effects of the use of mineral and energy resources on the environment, the use of non-renewable resources, environmental consequences of natural resource use in countries, the impact of different land use practices on the environment, mechanism, frequency, intense and global effects of environmental problems, sustainable use of natural resources are included
		12-4– Environment and society	This unit covers the limitations of the natural environment, policies and practices to prevent environmental problems in countries, the effects of environmental organizations and agreements on environmental management and protection, and threats to the common natural and cultural heritage

(continued)

(continued)

Subject	Goals/Teaching principles	Themes/units	Scope
Biology (9–12)	In the units "Ecosystaem Ecology and Current Environmental Problems" and "Community and Population Ecology", students should be directed to research, observe, and examine the living things around them and their interactions with the environment in which they live, and students should be engaged in practices. In addition, within the scope of school and environmental opportunities, field trips, botanical garden trips, natural history museum trips, protected areas trips, etc. should be organized	10.3. Ecosystem ecology and current environmental problems	This unit covers the key concepts of decomposer, food web, food pyramid, food chain, bioaccumulation, ecosystem, energy pyramid, environmental problems, ecological footprint, carbon footprint, water footprint, and matter cycles
Technology and design (7–8)	Students are expected to be able to observe, examine, be sensitive to their environment, feel responsible for the problems that affect human life, be able to propose innovative and original solutions to these problems by using analytical thought systematics, have self-confidence, and have the ability to work together	7. A. 1. I am learning technology and design	A two- or three-dimensional design product can be examined in this unit. For example, when examining poster designs, the value of responsibility in matters such as environmental awareness and savings is emphasized
		7. C. 1. Energy conversion and design	In this unit, students learn technologies to obtain clean and sustainable energy by using natural resources such as water, wind, and sun and designing a product
		8. A. 1. Development of Innovative Thinking and Protection of Ideas	Emphasis is placed on how to protect inventions, inventions, and designs. It is emphasized that the protection of intellectual and industrial property rights is a legal and social responsibility, the contribution of social progress and environmental awareness to civilization, and the need to act in accordance with ethical rules
Visual arts (1–8)	Lessons are associated with current events that happen in the local and surrounding environments and that are of interest to the students (economic, environmental, and cultural sustainability, natural disasters, environmental awareness, occupational health, safety, scientific developments, technological developments, global warming, healthy nutrition, harmful habits and ways of protection, saving energy resources, etc.)	1.1. Visual communication and formatting	It is emphasized that artists can observe what is happening around them more carefully. It is also stated that this is their responsibility and requires sensitivity
		2.1. Visual communication and formatting	Phenomena that disrupt and destroy the natural environment can be emphasized. It can be explained what a poster is and why it is made. They may be asked to give examples of these phenomena from their environment and then make posters about environmental awareness
		8.1. Visual communication and formatting	Economy, natural disasters, environmental awareness, occupational health and safety, scientific developments, technological developments, etc., topics can be used

References

AFAD (2023) Tukiye's disaster management and emergency management presidency. 06 Şubat 2023 Kahramanmaraş (Pazarcık ve Elbistan) depremleri saha çalişmalari ön değerlendirme raporu [06 February 2023 Kahramanmaraş (Pazarcık and Elbistan) earthquakes field studies preliminary assessment report] Accessed from https://deprem.afad.gov.tr/assets/pdf/Arazi_Onrapor_28022023_surum1_revize.pdf

Aghaei N, Seyedin H, Sanaeinasab H (2018) Strategies for disaster risk reduction education: A systematic review. J Educ Health Promot 7(98). https://0.4103/jehp.jehp_31_18

Amri A, Haynes K, Bird DK, Ronan K (2018) Bridging the divide between studies on disaster risk reduction education and child-centred disaster risk reduction: A critical review. Children's Geogr 16(3):239–251. https://doi.org/10.1080/14733285.2017.1358448

Anderson A (2010) Combating climate change through quality education. Brookings Global Economy and Development

Apronti PT, Osamu S, Otsuki K, Kranjac-Berisavljevic G (2015) Education for disaster risk reduction (DRR): Linking theory with practice in Ghana's basic schools. Sustain 7(7):9160–9186. https://doi.org/10.3390/su7079160

Aydın F, Coşkun M (2010) Observation of the students' "earthquake" perceptions using phenomenographic analysis (primary education 7th grade–Turkey). Int J Phys Sci 5(8):1324–1330

Basibuyuk A, Pala SM (2023) Hayat bilgisi, sosyal bilgiler ve coğrafya dersi öğretim programlarının afet eğitimi açısından incelenmesi. Erzincan Univ J Fac Educ 25(2):184–197. https://doi.org/10.17556/erziefd.1063242

Belur J, Tompson L, Thornton A, Simon M (2021) Interrater reliability in systematic review methodology: exploring variation in coder decision-making. Sociol Methods & Res 50(2):837–865. https://doi.org/10.1177/0049124118799372

Benadus M (2014) Pedagogies of the unknown: Unpacking 'culture' in disaster risk reduction education. J Contingencies Cris Manag 22(3):174–183. https://doi.org/10.1111/1468-5973.12050

Benadusi M (2013) The two-faced Janus of disaster management: still vulnerable, yet already resilient. Southeast Asia Res 21(3):419–438. https://doi.org/10.5367/sear.2013.0164

Boon HJ, Pagliano PJ (2014) Disaster education in Australian schools. Aust J Environ Educ 30(2):187–197. https://doi.org/10.1017/aee.2015.8

Bozyiğit R, Kaya B (2017) Determining the cognitive structures of geography pre-service teachers on some concepts related to nature-induced disasters. Marmara Geogr J 35:55–67

Bulus-Kirikkaya E, Cakin O, Imali B, Bozkurt E (2011) Earthquake training is gaining importance: the views of 4th and 5th year students on Earthquake. Procedia-Soc Behav Sci 15:2305–2313

Çoban M, Sözbilir M, Göktaş Y (2017) Determining preparation perceptions before the earthquake of individuals experienced earthquake: A case study. East Geogr J 22(37):113–134

Değirmenci Y, Kuzey M, Yetişensoy O (2019) Disaster awareness and education in social studies textbooks. E-Kafkas J Educ Res 6(2):33–46

Dewey J (1986) Experience and education. Educ Forum 50(3):241–252

Dikmenli Y, Yakar H (2019) Examining of preservice teachers' disaster awareness levels. YüzüncüYıl Univ J Educ 16(1):385–416. https://doi.org/10.23891/efdyyu.2019.130

Dufty N (2014) What is disaster resilience education? In: Paper presented at the Australian and New Zealand Disaster and emergency management conference surfers Paradise, Gold Coast (QLD)

Dwiningrum SIA, Pratiwi PH, Sumunar DRS (2022) Integrated the Green School in Ecosystem-Based Disaster Risk Reduction (Eco-DRR): Lesson learned from educational practices in the schooling (No. 8853). EasyChair

Efthymis L (2014) Disaster Data Centre—an innovative educational tool for disaster reduction through education in schools. J Power Energy Eng 2:25–40

Eisler R (2000) Tomorrow's children: A blueprint for partnership education in the 21st century. Adolescence 35(137):221

Estrella M, Renaud FG, Sudmeier-Rieux K, Nehren U (2016) Defining new pathways for ecosystem-based disaster risk reduction and adaptation in the post-2015 sustainable development agenda. Ecosystem-based disaster risk reduction and adaptation in practice, pp. 553–591
Estrella M, Saalismaa N (2013) Ecosystem-based DRR: An overview. In: Renaud F, Sudmeier-Rieux K, Estrella M (eds) The role of ecosystems in disaster risk reduction. University Press, United Nations, pp 26–47
Gerdan S (2019) Disaster trainings as a social responsibility field. Int J Manag Adm 3(5):101–110
Ho LC, Seow T (2017) Disciplinary boundaries and climate change education: Teachers' conceptions of climate change education in the Philippines and Singapore. Int Res Geogr Environ Educ 26(3):240–252. https://doi.org/10.1080/10382046.2017.1330038
Inal E, Kaya E, Altıntaş KH (2018) Evaluating the formal education in terms of sufficiency of disaster education in Turkey. Atatürk Univ J Kazım Karabekir Educ Fac 37:114–127
Intergovernmental Panel on Climate Change (IPCC) (2018) Special report on global warming of 1.5 °C. https://www.ipcc.ch/sr15/ Accessed 29 Nov 2023
IPCC (2014) Climate change 2014: Synthesis report. Contribution of working groups I, II and III to the Fifth assessment report of the intergovernmental panel on climate change [Core Writing Team, R.K. Pachauri and L.A. Meyer (eds.)]. IPCC, Geneva, Switzerland, p. 151
Izadkhah YO, Hosseini M (2007) Disaster preparedness strategy through earthquake education and training of classified target groups. In: Proceedings of the 2nd international conference on integrated natural disaster management (INDM), Tehran
Johnson VA, Ronan KR, Johnston DM, Peace R (2014) Evaluations of disaster education programs for children: A methodological review. Int J Disaster Risk Reduct 9:107–123. https://doi.org/10.1016/j.ijdrr.2014.04.001
Kagawa F, Selby D (2012) Ready for the storm: Education for disaster risk reduction and climate change adaptation and mitigation1. J Educ Sustain Dev 6(2):207–217. https://doi.org/10.1177/0973408212475200
Kasada M, Uchida K, Shinohara N, Yoshida T (2022) Ecosystem-based disaster risk reduction can benefit biodiversity conservation in a Japanese agricultural landscape. Front Ecol Evol. https://doi.org/10.3389/fevo.2022.699201
Kilanowski JF (2017) Breadth of the socio-ecological model. J Agromedicine 22(4):295–297. https://doi.org/10.1080/1059924X.2017.1358971
Kırıkkaya EB, Ünver AO, Çakın O (2011) Teachers views on the topic of disaster education at the field on elementary science and technology curriculum. Necatibey Eğitim Fakültesi Elektron Fen Ve Mat Eğitimi Derg 5(1):24–42
Krasny ME, DuBois B (2019) Climate adaptation education: Embracing reality or abandoning environmental values. Environ Educ Res 25(6):883–894. https://doi.org/10.1080/13504622.2016.1196345
Laurie R, Nonoyama-Tarumi Y, Mckeown R, Hopkins C (2016) Contributions of education for sustainable development (ESD) to quality education: A synthesis of research. J Educ Sustain Dev 10(2):226–242. https://doi.org/10.1177/097340821666144
McPherson S, Anid N M, Ashton WS, Hurtado-Martín M, Khalili N, Panero M (2016) Pathways to cleaner production in the Americas ii: application of a competency model to experiential learning for sustainability education. J Clean Prod 135:907–918. https://doi.org/10.1016/j.jclepro.2016.06.138
Mitchell T, Haynes K, Hall N, Choong W, Oven K (2008) The roles of children and youth in communicating disaster risk. Children Youth Environ 18(1):254-279. https://doi.org/10.1353/cye.2008.0036
Mizutori M (2020) Reflections on the Sendai framework for disaster risk reduction: Five years since its adoption. Int J Disaster Risk Sci 11:147–151. https://doi.org/10.1007/s13753-020-00261-2
Murti R, Mathez-Stiefel SL (2019) Social learning approaches for ecosystem-based disaster risk reduction. Int J Disaster Risk Reduct 33:433–440. https://doi.org/10.1016/j.ijdrr.2018.09.018

Mutasa S, Coetzee C (2019) Exploring the use of experiential learning in promoting the integration of disaster risk reduction into primary school curriculum: A case of Botswana. J Disaster Risk Stud 11(1):1–8. https://hdl.handle.net/10520/EJC-143f4ba5c4
OECD (2018) Preparing our youth for an inclusive and sustainable world the OECD PISA global competence framework. Accessed from https://www.oecd.org/education/Global-competency-for-an-inclusive-world.pdf
Rahma A, Mardiatna D, Hizbaron DR (2023) Developing a theoretical framework: School ecosystem-based disaster risk education. Int Res Geogr Environ Educ:1–18. https://doi.org/10.1080/10382046.2023.2214041
Şahin Y, Lamba M, Öztop S (2018) Determination of disaster awareness and preparedness level of university students. J CivIzation Stud 3(6):149–159
Selby D, Kagawa F (2018) Archipelagos of learning: Environmental education on islands. Environ Conserv 45(2):137–146. https://doi.org/10.1017/S0376892918000097
Shaw R, Sakurai A, Oikawa Y (2021) New realization of disaster risk reduction education in the context of a global pandemic: Lessons from Japan. Int J Disaster Risk Sci 12:568–580. https://doi.org/10.1007/s13753-021-00337-7
Sheehy K, Vackova P, van Manen S, Saragih Turnip S, Rofiah K, Twiner A (2022) Inclusive disaster risk reduction education for Indonesian children. Int J Incl Educ:1–17. https://doi.org/10.1080/13603116.2022.2115156
Sozcu U (2019) Determining natural disaster literacy levels of preservice teachers. Unpublished PhD Thesis, Kastamonu University Institute of Social Sciences, Kastamonu
Sozcu U (2020) Disaster education in Turkey: Trends in theses and articles between 2003 and 2020. J Pedagog Res 4(3): 418–441. https://doi.org/10.33902/JPR.2020465083
Tanner T, Garcia M, Lazcano J, Molina F, Molina G, Rodriguez G, ... Seballos F (2009) Children's participation in community-based disaster risk reduction and adaptation to climate change. Particip Learn Action 60:54–64. https://eprints.soas.ac.uk/id/eprint/31355
Renaud FG, Sudmeier-Rieux K, Estrella M (eds) (2013) The role of ecosystems in disaster risk reduction. United Nations University Press
Tidball KG, Krasny ME (2011) Toward an ecology of environmental education and learning. Ecosphere 2(2):1–17. https://doi.org/10.1890/ES10-00153.1
Tora TT, Degaga DT, Utallo AU (2022) Vulnerability management practices for sustainable livelihood security in drought-prone Gamo lowlands: An empirical insight from southwest Ethiopia. Heliyon 8(12). https://doi.org/10.1016/j.heliyon.2022.e12055
Torani S, Majd PM, Maroufi SS, Dowlati M, Sheikhi RA (2019) The importance of education on disasters and emergencies: A review article. J Educ Health Promot 8(1):85. https://doi.org/10.4103/jehp.jehp_262_18
Triyanti A, Walz Y, Marfai MA, Renaud F, Djalante R (2017) Ecosystem-based disaster risk reduction in Indonesia: Unfolding challenges and opportunities. In: Djalante R, Garschagen M, Thomalla F, Shaw R (eds) Disaster Risk Reduction in Indonesia: Progress, Challenges, and Issues. Springer, pp 445–467
UNDRR (2015) A-Sendai Framework for Disaster Risk Reduction 2015–2030. https://www.undrr.org/publication/sendai-framework-disaster-risk-reduction-2015-2030
UNICEF and UN/ISDR (2011) Children and disasters: Building resilience through education. https://www.unisdr.org/files/24583_childrenanddisastersbuildingresilie.pdf Accessed 29 Nov 2023
UNICEF (2014) Towards a learning culture of safety and resilience: Technical guidance for integrating disaster risk reduction in the school curriculum. UNESCO
United Nations Children's Fund (UNICEF) (2009) Disaster risk reduction in school curricula: Case studies from thirty countries. https://unesdoc.unesco.org/ark:/48223/pf0000217036 Accessed 29 Nov 2023
United Nations Disaster Risk Reduction (UNDRR) (2023) Sendai framework terminology on disaster risk reduction. Accessed from https://www.undrr.org/terminology/disaster-risk-reduct

ion#:~:text=Disaster%20risk%20reduction%20is%20aimed,the%20achievement%20of%20sustainable%20development Accessed 18 Jan 2024

United Nations Environment Program (UNEP) (2006) Annual report. https://www.unep.org/resources/annual-report/unep-2006-annual-report Accessed 29 Nov 2023

Wahl DC (2017) Ecoliteracy: Learning from living systems. Designing Regenerative Cultures. https://medium.com/age-of-awareness/ecoliteracy-learning-from-living-systems-e162df608981 Accessed 29 Nov 2023

Weldemariam K, Boyd D, Hirst N, Sageidet BM, Browder JK, Grogan L, Hughes F (2017) A critical analysis of concepts associated with sustainability in early childhood curriculum frameworks across five national contexts. Int J Early Child 49(3):333–351. https://doi.org/10.1007/s13158-017-0202-8

Wolff LA, Sjöblom P, Hofman-Bergholm M, Palmberg I (2017) High performance education fails in sustainability?—A reflection on Finnish primary teacher education. Educ Sci 7(1):32. https://doi.org/10.3390/educsci7010032

Yamazaki Y, Özmen B, Ergünay O, Kaneko F, Miyazaki R (2017) Development and its dissemination of school disaster and emergency management plan in the republic of turkey. In: 16th World conference on earthquake, 16WCEE 2017 Santiago Chile, January 9th to 13th 2017 Paper N° 276 Registration Code: S-276

Yilmaz E, Yildirim A (2023) Too little, too late: Addressing the roles of a school safety programme in earthquake preparedness of elementary students in Turkey. Asia Pac J Educ 43(4):1049–1070. https://doi.org/10.1080/02188791.2021.1965956

Chapter 6
Managing Risk in the April 2015 Earthquake: Autoethnographic Narratives of a Teacher and a Student

Tikaram Poudel and Hiralal Kapar

Abstract In these autoethnographic narratives of a university teacher and a student, we reflect on the intricate dynamics of managing educational activities in an emergency. We locate our experiences focusing on the aftermath of the devastating April 2015 earthquake with a magnitude of 7.8Mw in Kathmandu. We explore the struggles we encountered and cope with the challenges in the educational context with special reference to the Department of Language Education at Kathmandu University during the uncertainty caused by the April 2015 earthquake. Drawing on the autoethnographic approach, we bring our anecdotal narratives to understand the complex interplay of risk, resilience, and educational activities. Our experiences of our efforts provide insight into re-establishing a sense of normalcy facing the disruption caused by the earthquake disaster. We analyze these narratives thematically within the theoretical frameworks of risk society theory and the theory of resilience. This paper contributes to the discourse of emergency in education with specific reference to Nepal.

Keywords Autoethnographic narratives · Education in an emergency · Risk management · Risk society theory · Nepal 2015 Earthquake

6.1 Introduction

We begin this chapter by understanding risk management in educational settings following the April 2015 earthquake in Nepal. This understanding enables us to tell readers why this area of research is interesting and needs an academic discussion. After making our positionality clear on our argument, we bring anecdotes that provide

T. Poudel (✉)
Department of Language Education, Kathmandu University, Lalitpur, Nepal
e-mail: tikaram.poudel@kusoed.edu.np

H. Kapar
Kathmandu University, Lalitpur, Nepal
e-mail: hiralalkapar@kusoed.edu.np

A. Yildiz and R. Shaw (eds.), *Disaster and Climate Risk Education*, Disaster Risk Reduction, https://doi.org/10.1007/978-981-97-5987-3_6

our perspectives on the impact of the earthquakes, particularly in educational space, the challenges we faced, and how we coped with these uncertainties.

The April 2015 earthquake disrupted the daily lives of the people in affected areas (Harris 2015), resulting in approximately 9000 fatalities and more than 200,000 people homeless. The earthquake generated fear, confusion, and uncertainty. The impact of the disaster on educational settings was immense. Many children lost their lives; several of them lost their residences and parents. Sapkota and Neupane (2021) observed that the loss and injury of family members and economic loss, such as damage to their houses, caused a significant decline in the annual test scores in several earthquake-affected rural schools. The Department of Language Education at Kathmandu University also had a similar fate in this disaster. As university professors, we promptly and swiftly adapted to the new reality in order to safeguard the well-being of our family members and students. Since most of the university infrastructure that housed our department was almost damaged, we could not think of continuing our classes as we used to do before the disaster. In this new context, our role extended to ensure the safety of students in the dilapidated buildings. Amid the chaos, we shifted to an online mode of learning, though we were not much prepared for it. We gradually picked up the pace and shifted to online space from traditional classroom activities, offering support and guidance to students grappling with personal loss and displacement.

This duoethnographic study delves into the personal narratives of a university teacher (myself) and one of my students (the co-author of this paper) to gain deeper insights into how we navigated the tremors of the earthquake and the subsequent challenges it posed to higher education in Nepali context. In 56 sec, the earthquake demolished some 50,000 classrooms (Awale 2021). Thank God. The quake hit on Saturday, a weekly public holiday in Nepal. Otherwise, the death toll would be beyond our imagination (Basnet 2020).

6.1.1 Earthquake and Education in an Emergency

Education in an emergency comprises learning opportunities for learners of all ages during and after crises or disasters. It incorporates early childhood development, primary, secondary, non-formal, technical, vocational, higher, and adult education. It ensures learning opportunities for affected individuals, children, and women despite adverse situations, supporting them and addressing their educational needs during emergencies like internal conflicts, wars, natural disasters, and public health emergencies (Inter-Agency Network for Education in Emergencies (INEE) 2010). The United Nations General Assembly (UNGA) 2010 resolution on the Right to Education in Emergency Situations ensures the right to education for all affected by conflicts, natural calamities, or health emergencies (Nicolai et al. 2015). The phase of a crisis determines the response from concerned agencies, including preparedness, acute emergencies, and protracted crises (Nicolai et al. 2015). Education in an emergency underscores the importance of maintaining educational continuity by addressing the holistic needs of affected individuals, ensuring their safety and resilience to challenging circumstances.

According to the estimation of the Government of Nepal, the April 2015 earthquake destroyed nearly 8,000 schools, including public and private (USAID 2023). This destruction put over two million students out of schools and universities. Educational institutions that were safe sheltered the displaced people, leading to delayed reopening of these institutions. The disaster forced students to several vulnerabilities like child labor, human trafficking, especially for girls, and loss of jobs. These vulnerabilities impacted them psychologically, leading to a high rate of dropouts. Basnet (2020) argues that the earthquake and its aftershocks affected people physically and psychologically. Several of them lost their homes, and others had to demolish theirs because they were dangerous. They did not have enough food, water, or sanitation; they constantly feared epidemics like diarrhea would spread. Students, both at schools and universities, could not ensure when they could continue their studies; the anxiety of loss of academic year caused them mental stress, for some leading to post-traumatic stress disorder (PTSD).

The National Education Policy 2019 (NEP) of Nepal recognizes the crucial role of education during crises and emergencies (Ministry of Education 2019). Policy 16 of NEP 2019 underscores that in case of emergency, a coordination committee, including the provincial education ministers under the leadership of the federal education minister, will be formed. The policy document emphasizes the need to address risks proactively and build resilience to ensure the continuity of education during emergencies. The NEP advocates for adaptive responses and resilience-building strategies in the face of adversity (Ungar 2011).

6.1.2 Theoretical Underpinning

Ungar's (2011) resilience theory provides a theoretical framework for this study. This theory addresses how individuals and communities respond to adversity and traumatic events. In this context, the traumatic event of the earthquake disrupted the physical environment and the emotional and psychological well-being of teachers, students, and their families. Using Ungar's resilience theory, we explore how we and our institution exhibited resilience to bounce back from adversity caused by the earthquake, adapt positively, and grow in the face of challenges. The concept of resilience is evident in how we coped with these traumatic events, exhibited adaptability, and found ways to navigate the immediate crisis. Another aspect of resilience theory is social support networks in coping with adversity. In an emergency, teachers and their family members' emotional and mental well-being are prioritized for effective academic functioning.

Besides Ungar's theory of resilience, we relied on Beck's theoretical perspective to explain the issues we raised on Beck's risk society elaborated in his seminal work *Risk Society: Towards a New Modernity* (1992). He internalized that contemporary societies live in uncertainty and fear because of modernization and technological advancements. He argued that these risks are unprecedented, systemic, and transnational. These risks, such as nuclear disasters and biological weapons or pandemics like COVID-19, caused by the interplay between science and technology, threatened the very existence of humanity, unlike traditional risks of premodernity, such as famine or local wars. This phenomenon of uncertainty and fear caused by reflexive modernization is systematic and has global implications. Although Beck is more concerned with human-induced risk, we found his theory to explain the issues posed by the April 2015 earthquake on the educational system of Nepal, in general, and Kathmandu University, in particular, such as being socialized, morally stronger, and cooperative at the same time. During the disaster, people were morally weak, and they were searching for social and emotional counsel to bear the grief of losing their family and relatives. We understand that the risk we faced was caused by a natural hazard, not a predominantly manufactured risk, to use Beck's term. To a certain extent, human-induced risks were responsible for the complete halt of educational activities because of the damage to university infrastructure and unpreparedness despite living in a high seismic zone.

In this regard, Beck's theory is a discourse on societal factors, such as the university authority and government actions to mitigate in educational settings. Societal factors, such as the structure of governance, socioeconomic disparities, and cultural norms, also contribute to the vulnerability of risks in educational systems. Beck argues that societal inequalities and power dynamics, such as the marginalization of certain communities and heightened vulnerabilities during disasters like the April 2015 earthquake, intensify the safety issues of citizens in Nepal. In our context, safety issues refer to the awareness of mitigating the risks during disasters like the April 2015 earthquake. For example, schoolchildren were taught to hide under a table during an earthquake as a protective measure. When the earth began to shake, children outside the house ran to their houses and hid under tables. Unfortunately, the house collapsed, and many of them died. We failed to protect the lives of these children because of the improper instructions.

We also understand that societal factors can mitigate risks through collaborative efforts, effective educational institutions, and the governance of social networks. Beck's emphasis on reflexive decision-making processes within society provides a framework to analyze how societal factors shape risk management strategies in educational settings, offering insights into measures to enhance preparedness and resilience in the face of future challenges.

6.2 Methodology

We used an autoethnographic research perspective to explore the risk management and coping strategies we adopted in the aftermath of the April 2015 earthquake. This reflexive research perspective combines researchers' anecdotes and ethnography to understand a phenomenon like education in emergencies like the April 2015 earthquake through personal experiences (Ellis and Bochner 2000). We found this approach suitable for our study because it allows us to reflect deeply on geographic context. Autoethnography is a unique research method because it positions us as members of a university professor and student in which we engage ourselves as a part of the story we tell, unlike traditional ethnography (Anderson 2006).

As autoethnographic researchers, we are conscious that we treat our reflections and narratives on the aftermath of the April 2015 earthquake in accordance with Kathmandu University policies and procedures. We present our anecdotes and narratives and their interpretations and analyses connecting the cultural undertones of our community. We ensured the trustworthiness of the information generated and analytical rigor through a collaborative critique and comprehensive evaluation of our narratives (Stevens et al. 2021).

6.2.1 Autoethnography #1. University Professor

The April 2015 earthquake in Kathmandu reminds me of the opening line of Eliot's *The Waste Land:* 'April is the cruellest month of the year'. It was a Saturday, a weekly holiday in Nepal. My wife and I had a morning walk longer than usual and had a leisurely lunch at home. After lunch, we were having a nap. Then, suddenly, the entire building began to shake, producing unusual sounds beyond our comprehension. There were unusual sounds from the neighborhood, shouting and crying. My daughter was in her room and could not come out because the shaking caused the door's latch to get stuck. This situation intensified the panic. In the meantime, in the bathroom, the gas cylinder that we used to heat water fell and hit the bathroom tap, which broke, causing a flood. My family's experiences during the earthquake, with the daughter trapped in her room and the household facing additional challenges such as a gas cylinder falling and causing a flood, intensified the terror. My family's response to these adversities in the aftermath of the earthquake was crucial for coping and moving to a new normal. The availability of external resources and support systems fosters resilience (Ungar 2011). The shaking continued for 56 seconds, and we felt everything came to a standstill.

6.2.2 Autoethnography #2. University Student

The unforgettable afternoon shrinks Nepali's heart when I recall the black day of the last earthquake. Yes, it was Saturday, April 25, 2015. All the schools in Nepal were closed as it was a weekly holiday. I was a graduate student at the Department of Language Education at Kathmandu University. After cleaning our clothes and room, everything began to shake when I was playing a computer game with my friend, Raju. Suddenly, we heard a group screaming outside. "*Bhag Bhukamph!*" (Run! Earthquake!). We came out of our room and saw the crowd on the road, and everyone was rushing to save their lives. We also rushed to be down on the road.

I saw people crying and lying on the ground as they were scared by the hits of the earthquake. The frequent hits of the earthquake made us mentally weak as only negative news was coming from out. On the day, we felt the frequency of the hits every 5–10 mins. Everybody prayed to overcome, and there was no way out. As everyone tried to reach their family and relatives, there were huge traffic jams everywhere: the network, roads, and airways. No telephone call was available. It was another big issue of the day.

I tried a lot to talk to my younger brother, who was in the other part of the city. My mind was totally occupied with taking my brother back from Kathmandu. After 3 days, I contacted him and found he was safe, and I felt relieved. It was the story of everyone in Kathmandu. It was also everyone's case, as the news of the deaths made people scared.

The terrible news further caused panic; we were very tense listening to the news of the destruction. The news of death and physical damage were major headlines of almost every news page. Newspapers reported that the death toll reached 9,000, and over 1,900 were injured (Rafferty 2023). This catastrophe impacted every student like me. Following the news and its impacts, the Government of Nepal announced a month of school holidays. Almost every radio and television channel shared procurement and risk management strategies targeting school students during the earthquake.

Many students lost their family, relatives, and classmates, too. These human casualties badly influenced students' mental states. Students stopped reading and attending classes for several months (Save the Children 2015). The Government planned psychological counseling for school-level students. I take the decision as part of coping with the situation in compliance with the resilience theory (Ungar 2011).

As a university student, I found the decisions of the Kathmandu University management committee supporting students in recapturing the mental state of their regular university routine activities. The strategies applied in managing the aftermath of the 2015 earthquake at the School of Education Kathmandu University align closely with the principles of resilience theory in disaster risk management. Resilience theory emphasizes the capacity of individuals, communities, and systems to anticipate, respond to, recover from, and adapt to adverse situations (Paton 2003).

Kathmandu University decided to opt for an online learning mode by creating self-learning materials (SLMs). For this, teachers uploaded these learning materials to the Moodle platform for students' easy access. In the context of the earthquake response, the establishment of temporary learning spaces and the distribution of educational materials reflect adaptive strategies to ensure the continuity of education despite the immediate challenges (Berkes 2007). The focus on psychosocial support and capacity building for teachers aligns with enhancing the social and human capital within a community to better cope with and recover from disasters (Manyena 2006). Additionally, the emphasis on community engagement and collaboration in rebuilding efforts resonates with the collaborative and interconnected nature of resilience, where communities actively participate in decision-making and contribute to their recovery (Paton 2006). By integrating these strategies, the response to the Nepal earthquake embodies the resilience theory's principles, highlighting the importance of adaptability, social cohesion, and community empowerment in the face of disaster.

6.3 Managing Risk at the University

In this section, we delve into the multifaceted significance of risk management in educational settings, as illuminated by the experiences of a university professor and a student personally touched by this catastrophe. Risk management covers the potential pitfalls before facing threats (Oznacar 2018). Our stories remind readers that risk management is not merely a bureaucratic procedure; it is an intrinsic commitment to the safety and growth of students, as well as the perpetuity of knowledge.

Ensuring safety and continuing unforeseen challenges posed by the April 2015 earthquake was an enormous challenge for all of us. We felt this responsibility more acutely when the earthquake hit Nepal in April 2015 and disrupted the educational activities at Kathmandu University. The calamity shook not only our homes and university infrastructure but also the foundation of our preparedness for risk management in an academic setting. It also taught us how to face adversities and develop resilience.

The earthquake almost damaged Kathmandu University buildings, causing multiple cracks in the walls and ceilings. The continuous aftershocks further deteriorated the situation. One day, about a week after the main Earthquake, we were listening to the harrowing stories from our colleagues in the courtyard of the Kathmandu University building; the entire earth began to shake, and the nearby multistoried housing building also began to shake as if the big building is dancing in the air. We were confronted with increasing hazards that transcend borders irrespective of individuals and institutions. The intensity of the shaking of the nearby multistoried housing building indicates the uncontrollability and unpredictability of risks intensified by modernity (Beck 1992).

The April 2015 earthquake created uncertainties in the educational space of Nepal. Beck's risk society theory (1992) highlights those uncertainties are central to contemporary life. The earthquake disaster intensified the uncertainties, leading to perennial fear. Although Beck's theory primarily emphasizes the impact of modernization on societies, we extend this theory to explain the risks and uncertainties posed to our academic life during the phenomenal disaster of the April 2015 earthquake. Damage to the university buildings and continuous aftershocks heightened the uncertainties in our academic activities. In the meantime, the dilapidated buildings further posed a physical risk to the lives of students, staff, and teachers, particularly in unpredictable natural disasters like the April 2015 earthquake. Beck argues that risks are not confined to national or local experiences like the April 2015 earthquake. Such risks and uncertainties could be global, as we experienced in the COVID-19 pandemic affecting entire humanity and we experienced collectively. This situation arose because the precariousness of the university buildings indicates the unpreparedness of university authority and government regulations to the vulnerability and the unpredictability of risks. The earthquake's impact extended beyond physical damage to educational infrastructure; it disrupted the functioning of higher education institutions, affecting student–teacher dynamics, curricular activities, and institutional resilience. The interplay of social, economic, and institutional factors contributed to the vulnerability of higher education systems in the face of the earthquake, emphasizing the systemic nature of risks and their far-reaching consequences on educational landscapes.

Our resilient tendency enabled us to adapt to the unpredicted phenomenon of adversity. Resilience refers to individuals' or communities' ability to adapt positively to adversity (Ungar 2011). We felt insecure; we were afraid, and uncertainty prevailed. However, we had to act. Ungar (2011) explains that during adversity, we feel insecurity, fear, and uncertainty when we encounter challenges. About a month later, we decided to go for online classes. We did not have adequate infrastructure for online classes. We were not well prepared and trained. However, we had little experience working online on a collaborative project with a Norwegian University and Kathmandu University. In our context, we adapted and found alternative ways of handling the situation despite a lack of preparedness and adequate infrastructure.

We decided to go for blended mode and shifted some of our classes to a private school to run our classes in the evening. Finally, we moved unanimously to an online mode of teaching and learning, though the transition was not easy. However, our fate of struggle was waiting ahead. When we started evening classes in a private school building, we had an economic blockade and ran out of petroleum products, leading to a complete halt in public transport. We were forced to go online without a second thought because we had no other options. The experience of some of our colleagues with the collaborative project with the Norwegian colleagues encouraged us to use the positive adaptation of the online learning mode. The experience that we gained in the project became a valuable resource for handling new challenges (Ungar 2011). The apt decision of the University to shift classes to online mode indicates flexibility, resourcefulness, and trust in the available human resources to handle emergencies. In the situations we face, individuals and institutions find ways to overcome the adverse situations and continue functioning in the face of adversity, which is the adaptive nature of resilience (Ungar 2011). Following Beck, we argue that the novel dimension of online classes was itself revealing because we understood that the shift to an online learning mode was instrumental in reducing the prevailing uncertainties. We realized that institutional procedures of online learning mode enabled us to repair our credibility as an institution of higher learning (Beck 1992). This collective reflexivity and form of discourse enabled us to present a new model of future diffusion of teaching and learning activities.

6.4 Resilience in Managing the Risk

As a university professor and a student, we experienced the seismic disruption that the April 2015 earthquake shattered the historic architectural monuments that we boast of being one of the oldest Asian civilizations. It also deeply caused a collective psychological disruption with hopelessness and depression among the youngsters.

The April 2015 earthquake shattered the entire educational system at Kathmandu University as well as the entire country. We understood the significance of risk management in educational settings. This natural calamity underscored my role as an educator, i.e., an educator's responsibility extends beyond the classroom to the safety and well-being of students and the continuity of knowledge generation in the face of unpredicted challenges. The earthquake shattered physical infrastructure and created chaos in our physical and mental well-being. We understood that risk management was not merely an administrative concern but a commitment to safeguarding the future of our students and our society. This fateful catastrophe reminded us of the resilience from the disaster and the unwavering determination to continue educating. This seismic tremor taught us a profound lesson: we could continue the pursuit of knowledge despite such calamities by effective risk management strategies.

As a university professor, the seismic disruption caused by the April 2015 earthquake in Nepal shattered physical structures and deeply embedded paradigms within the educational landscape. This calamity brought to the fore the stark reality delineated by Beck's theory of risk society, illuminating the vulnerabilities inherent in modern educational systems when faced with unpredictable disasters (Beck 1992). The earthquake acted as a catalyst, compelling a profound re-evaluation of risk management within the educational context and unveiling its significance beyond administrative protocols. It delineated a paradigm shift, encapsulating the educator's expanded role from the confines of a mere purveyor of knowledge to a guardian of safety, well-being, and the continuity of learning amid unforeseen challenges.

Ungar's resilience theory resonated profoundly amid the chaos unleashed by the earthquake, offering a lens to understand the educator's response in navigating through the tumultuous aftermath (Ungar 2011). Amid the chaos and uncertainty that engulfed both physical and mental realms, resilience emerged as a beacon. It encompassed the ability to bounce back and, more importantly, the unwavering commitment to adapt and grow in the face of adversity. This theory aptly depicted the determined spirit of educators, who, despite the upheavals, remained resolute in their pursuit of continuing education, showcasing an unyielding dedication that transcended the calamity's disruptions.

Kathmandu University understood that we had to act despite all adversities. We decided to go online and continue the classes. Some of our colleagues were working on their doctoral research on information and communication technology (ICT) issues funded by a Norwegian project. Initially, they trained us on how to make online classes effective. We began to digitalize our curriculum as a first response to disaster management.

6.5 Conclusion

Reflecting on our anecdotes, we elucidate the intricate challenges that the Department of Language Education at Kathmandu University faced in the aftermath of the catastrophic April 2015 earthquake in Kathmandu. We document the profound impact of the earthquake on managing risk, resilience, and the emotional well-being of students and teachers despite heightened psychological stress, vulnerability, and disrupted teaching and learning opportunities. Theoretical frameworks of risk society theory (Beck 1992) and resilience theory (Ungar 2011) enabled us to explain the complexities of managing risk and building resilience strategies. The risk society theory provided insight into the systematic nature of risk in modern society, and the resilience theory helped us explain how individuals and their institutions respond to natural disasters like earthquakes in an educational setting. Our autoethnographic accounts recognize the role of teaching and student communities in responding to disasters, emphasizing psychosocial support for affected individuals and highlighting adaptive strategies to mitigate the risk in continuing educational activities. These reflections underscore the role of agency in mitigating risk by supporting the initiatives taken by the university authorities, such as shifting to online learning mode and collective engagements to address the challenges. The disruption that the earthquake caused reminds us of the fact that the educational setting serves students and their welfare, not just a bureaucratic procedure. The resilience, adaptability, and determination of professors and students in the face of unforeseen challenges reaffirm the value of preparedness and innovative approaches in managing risks during emergencies. Finally, these narratives contribute to the discourse on education in emergencies, highlighting the need for proactive risk management strategies, community involvement, and resilience-building measures to ensure educational continuity during crises.

References

Anderson L (2006) Analytic autoethnography. J Contemp Ethnogr:373–395. https://doi.org/10.1177/0891241605280449

Awale S (2021) Lessons not learned from 2015. Retrieved September 23, 2023, from https://nepalitimes.com/: https://nepalitimes.com/here-now/lessons-not-learned-from-2015-o2etq6zu

Basnet BK (2020) Earthquake and its impacts on education: Aftermath Nepal quake 2015. Eur Educ Res III(3):101–118. https://doi.org/10.31757/euer.332

Beck U (1992) Risk society: Towards a new modernity. Sage

Berkes F (2007) Understanding uncertainty and reducing vulnerability: Lessons from resilience thinking. Natural Hazards 2:283–295

Ellis C, Bochner AP (2000) Autoethnography, personal narrative, reflexivity. In: Denzin NK, Lincoln YS, Handbook of qualitative research. Sage, pp. 733–768

Harris G (2015) Nepal school system left shattered in the aftermath of the quake. Retrieved September 23, 2023, from https://www.nytimes.com/2015/05/14/world/asia/nepal-earthquake-shatters-school-system.html: https://www.nytimes.com/

Inter-Agency Network for Education in Emergencies (INEE) (2010) Minimum standards for education: Preparedness, response, recovery. Inter-Agency Network for Education in Emergencies
Manyena SB (2006) The concept of resilience revisited. Disasters 30(4):434–450
Ministry of Education (2019) National education policy. Government of Nepal
Nicolai S, Hine S, Wales J (2015) Education in emergencies and protracted crises. Overseas Development Institute
Oznacar B (2018) Risk management strategies in school development and the effect of policies on tolerance education. In: Altınay F (ed) Open and equal access for learning in school management. IntechOpen. Retrieved from https://www.intechopen.com/books/open-and-equal-access-for-learning-inschool-management
Paton D (2003) Disaster and crisis: Understanding the complexity. J Contingencies Cris Manag 11(1):3–6
Paton D (2006) Disaster resilience: Integrating individual, community, institutional, and environment perspectives. In: Paton D, Johnstone D (ed) Disaster resilience: An integrated approach. University of Tasmania, pp 306–319
Rafferty JP (2023) Nepal earthquake of 2015. Retrieved from https://www.britannica.com/topic/Nepal-earthquake-of-2015: https://www.britannica.com/
Sapkota JB, Neupane P (2021) The academic impacts of 2015 Nepal earthquake: Evidence from two secondary schools in Sindhupalchowk district. Educ Sci, 11(8). https://doi.org/10.3390/educsci11080371
Save the Children (2015) Nepal Earthquake: Education cluster situation report. https://reliefweb.int/report/nepal/nepal-earthquake-education-cluster-situation-report-8
Stevens DD, Chetty R, Bertrand Jones T, Yallew A, Butler-Henderson K (2021) Doctoral supervision and COVID-19: Autoethnographies from four faculty across three continents. J Univ Teach Learn Pract, 18(5). https://doi.org/10.53761/1.18.5.6
Ungar M (2011) The social ecology of resilience: Addressing contextual and cultural ambiguity of a nascent construct. Am J Orthopsychaitry 81(1):1–17
USAID (2023) https://2017-2020.usaid.gov/nepal/fact-sheets/emergency-education-response. Retrieved from https://2017-2020.usaid.gov/nepal/: https://2017-2020.usaid.gov/

Chapter 7
Incorporating Climate Change and Disaster Education into Accounting Curriculums

Lutfa Tilat Ferdous, Fathima Roshan Rakeeb, and Lisa Wakefield

Abstract There is a growing trend to incorporate climate reporting into accounting education due to concerns and pressures arising from global climate change, natural disasters, and changing stakeholder preferences and expectations. The purpose of this book chapter is to assess the inclusion of climate change and disaster-related topics in accounting curricula, which is in response to the increasing demand for accountants to take on new roles that involve supporting organisations in disaster-related impact management, reporting, and disclosure. Accounting curricula need to be transformed to produce accounting professionals who can help organisations deal with the risks associated with climate change and disaster-related risks. This chapter integrates climate change and natural disaster curricula into accounting education, emphasising professional skill development, industry, and community engagement. The findings of this chapter can be helpful for universities and professional accounting bodies to incorporate climate change and disaster education topics into their accounting curriculum for students. The inclusion of issues related to climate change and disaster education will (i) equip students with the necessary knowledge and skills to navigate climate change and disaster risks, as well as to conceptualise adopting conservative accounting practices and disclosures, (iii) encourage professional engagement with operational practices relating to climate change and disaster risk management, and (iv) raise students' interest in climate and natural disaster risk accounting, as well as sustainability, environmental, social, and carbon emissions issues.

Keywords Climate change and disaster education · Accounting curriculum · Reconceptualisation of accounting education · Education support

L. T. Ferdous (✉) · F. R. Rakeeb · L. Wakefield
Business School, University of Leicester, University of Leicester, Leicester LE1 1RQ, UK
e-mail: ltf4@leicester.ac.uk

F. R. Rakeeb
e-mail: frr4@leicester.ac.uk

L. Wakefield
e-mail: ljw57@leicester.ac.uk

A. Yildiz and R. Shaw (eds.), *Disaster and Climate Risk Education*, Disaster Risk Reduction, https://doi.org/10.1007/978-981-97-5987-3_7

7.1 Introduction

> To get all businesses involved in solving the world's toughest problems, we must change the accounting rules (Bakker 2013).

This chapter presents a reflective discussion of reforming accounting education to correspond with the growing urgency for engagement from both the industry and professionals. Increasingly, organisations recognise the importance of environmental and social impacts, both adverse and beneficial, on their businesses and broader society as a result of the climate crisis (Lovell and Mckenzie 2011). In this chapter, we discuss the vital role accounting education can play in addressing climate change and disasters to ensure that it is not only relevant but thrives in this age of climate change and disasters. This paper focuses on improving accounting education, highlighting the incorporation of climate change education, environmental awareness, the cultivation of professional skills and active engagement with industries as pivotal components within teaching and learning methodologies. Moreover, it explains how the essential support structure can enable such a profound and impactful transformation.

Businesses play a crucial and urgent role in achieving decarbonisation (Zhang et al. 2023; de Sousa Jabbour et al. 2019). Anthropogenic greenhouse gas (GHG) emissions are causing significant climate change, posing one of the biggest challenges of our time. In the UK, recent research (Department for Energy Security and Net Zero 2023) shows that the average resident emits 11.7 tonnes of CO_2-equivalent emissions each year, five times higher than the Paris Agreement recommendations. It is crucial for corporations and governments to work together and reach a mutual understanding in order to reduce carbon emissions in crucial industries. Globally Responsible Leadership Initiative and Principles of Responsible Management Education are initiatives established by the UN Global Compact in response to a call for more responsible business leadership through education (Godemann et al. 2014). Accounting education should equip future leaders with knowledge of climate change, disaster risk, and carbon emissions to ensure sustainability. Investors are increasingly interested in understanding how climate change can affect a company's accounting practices. For instance, the presentation of financial statements, inventory reporting, property plant and equipment reporting, assets impairment, and the recognition, measurement, and disclosure of liabilities may all be impacted (ACCA 2022). CIMA (2010) report documents a deficiency in the involvement of accounting and finance professionals in managing climate change. Therefore, accounting education still has a long way to go before it equips upcoming accountants with the knowledge and skills to comprehend current environmental challenges (Cho et al. 2022).

Academia has an opportunity to re-evaluate organisational methods and expand the scope of accounting education beyond technical elements toward a more comprehensive and conceptual comprehension (McGuigan 2021). This transition is essential to equip students with the essential skills and knowledge to navigate challenges and orient toward the future, aiding organisations, governments, and society in constructing and preserving an ethical and sustainable future (Tharpos 2022).

7.1.1 Contribution of Chapter

Business education providers should address global problems like climate change and environmental sustainability in their curricula as a means of integrating climate change into business education (Molthan-Hill et al. 2019; Linnenluecke et al. 2015; De Aguiar and Fearfull 2010). Molthan-Hill et al. (2019) conducted a survey of 45 countries and concluded that higher education institutions lack a structured approach to integrating climate change education into their curricula. There has been relatively little scholarly research on the role of managerial and financial accounting in identifying and evaluating climate risks. This gap in research is notable, particularly considering the rising demand from investors for information on investment risks associated with climate change (Linnenluecke et al. 2015). Researchers have consistently criticised accounting education for its limited scope (Boyce et al. 2019; Khan and Gray 2016; Botes et al. 2014; De Aguiar and Fearfull 2010). This criticism has been consistent over time and is something that needs to be addressed.

To address this gap, our chapter makes a contribution to accounting curricula and education scholarship. Initially, this chapter addresses the persistent demands for the revitalisation and enhancement of accounting education, amplified by the challenges posed by climate change and the management of climate-related risks (Boyce et al. 2019; Botes et al. 2014; De Aguiar and Fearfull 2010). Second, the chapter adds to the discourse concerning the significance of accounting education and offers multiple recommendations aimed at ensuring its alignment with organisational, societal, and environmental concerns to make it more resilient for the future. Third, this chapter highlights the importance of having a comprehensive set of professional skills, as well as active involvement in the industry, in order to address climate change. These skills and active involvement are pivotal for adeptly handling amplified accountability, ethical deliberations, governance, and sustainability challenges currently encountered by both organisations and society (Linnenluecke et al. 2015; Lovell and MacKenzie 2011).

Finally, the book chapter highlights the challenges and opportunities encountered by accounting educators in preparing students, with a focus on climate change, for the accounting field in the future. The findings are valuable for accounting educators, accounting departments, program administrators, professional accounting bodies, and universities to design accounting programs.

This book chapter is structured as follows. Firstly, it discusses the integration of climate education within accounting and the awareness of accountants regarding climate issues within both organisational and societal contexts. Next, it suggests ways to reconceptualise accounting education to address the challenges posed by climate change. Following this discussion, the chapter proposes a structure for empowering accounting educators to initiate and implement meaningful change in the context of climate change and disaster. Finally, the paper concludes with some final remarks on accounting education in the future.

7.1.2 What Climate Education in Accounting

Integrating climate change education systematically into formal education systems is a critical and potent approach to combat the challenges posed by the climate crisis. Stevenson et al. (2017) present a compelling argument for implementing climate education through cooperative problem-solving and inventive approaches involving various environmental, societal, economic, political, and educational establishments and frameworks. With a multidimensional approach, the education system can empower students with the expertise and abilities to effectively undertake the intricate challenges of climate change and natural disasters.

Education on climate change explores local and global mitigation and adaptation, as well as their interactions (Reid 2019). Furthermore, climate change education encompasses natural disasters and climate risk awareness, such as 'bushfires, floods, droughts, prolonged heatwaves, cyclones, tsunamis, storms', and other types of natural hazards (Stevenson et al. 2017, p. 69). Accounting education has a crucial role to play in addressing the global challenge of climate change. Integrating climate change into the curriculum allows students to gain a deeper understanding of the effects climate change has on businesses, economies, and society as a whole. Accounting education must emphasise climate and disaster risk reduction in order to build community resilience by systematically identifying, assessing, and reducing risk. This knowledge can assist students in developing effective strategies to mitigate the impacts of climate change and promote sustainable economies while advocating for responsible practices in the public interest.

Accounting education is the fundamental building block for future accountants, which embraces changes and provides a directional focus for students. Linnenluecke et al. (2015) demonstrated the significant impact that accounting roles can have in facilitating organisational climate change adaptation. Accounting professionals can assist organisations in preparing for and mitigating climate change risks by performing three critical functions. Firstly, through risk assessment, they can evaluate the organisation's vulnerability and adaptive capacity. Secondly, by valuing the costs and benefits of adaptation, they can help in making informed decisions. Lastly, through disclosure functions, they can provide valuable information on risks associated with climate change impacts. With the help of accountants, organisations can be better prepared to tackle the challenges of climate change and safeguard their future.

De Aguiar and Fearfull (2010, p. 65) suggest integrating global climate change into accounting education curricula and leveraging climate change disclosure as a potent educational instrument by:

(i) Encouraging students to understand the fundamentals of climate change.
(ii) Contributing to the scrutiny of conservative accounting practices.
(iii) Critically analysing organisational responses to climate change.
(iv) Assisting students in identifying and accounting for carbon emissions from various sources to foster climate consciousness.

7.2 Accountants' Role in Climate Awareness for Organisations and Society

Climate change poses the most significant threat to humanity, and the effects of this phenomenon on lives, livelihoods, and ecosystems are widely documented in the literature (Nyberg and Wright 2022). In the relationship between accounting and society, it is intended to broaden its scope beyond narrow conceptions of accounting. Accounting plays a constitutive role not only within the boundaries of a particular firm but also across social systems (Lovell and MacKenzie 2011). Organisations are increasingly being required to provide 'accounts' of not only their financial performance but also their social and environmental performance (Deegan 2013). Organisations need to develop policies on climate change issues as the effects of climate change are rapidly increasing and impacting all aspects of society, including organisations that can be held accountable for their actions (De Aguiar and Fearfull 2010; MacKenzie 2009).

The COVID-19 pandemic has provided a unique opportunity for accountants to take on a leadership role in addressing environmental, social, and governance issues. For instance, the objective of achieving net-zero emissions by 2050 is compelling organisations and governments to create sustainable business practices, with a specific emphasis on climate reporting and carbon accounting. Currently, policymakers are concentrating on learning from COVID-19's impacts rather than understanding the economic, financial, and fiscal consequences of compounded risks. A robust economic, fiscal, and financial resilience is imperative in the face of these complicated and cumulative shocks, particularly due to the escalating effects of climate change (Ranger et al. 2021; Leoni et al. 2021).

As organisations shift towards the retrieval phase, accounting professionals have a unique opportunity to take charge and lead the way in crafting a short-term recovery plan while also laying the foundation for long-term organisational resilience and risk management (Tharapos 2022). Accountants can contribute to building a more sustainable future for organisations and societies by focusing on issues such as climate change, disaster risk, and environmental, social, and governance.

7.2.1 Reconceptualising Accounting Education to Address Climate Change and Challenges

Accounting, accountability, and management practices have played a significant role during the global COVID-19 pandemic. The study of their role can foster an understanding of the role they play during emergencies, and rethinking these lessons is vital to improving the readiness of both organisations and society, aiming to prevent future crises (Leoni et al. 2021). As a result of the COVID-19 pandemic, higher education has undergone profound and lasting changes, signalling that the former way of operating will not be reinstated (Tharapos 2022).

The inability of accounting education to contribute to the intellectual development of students has been associated with a "superficial learning approach" and "ethically underdeveloped moral positions" imparted by accounting educators (De Aguiar and Fearfull 2010, p. 66; Gray and Collision 2002). To address this issue, it is necessary to re-evaluate existing accounting curricula and teaching methods and determine how they can be improved to promote intellectual and ethical development among students. Sikka et al. (2007) suggest that accounting textbooks should engage with the social responsibilities of organisations. Such approaches that redefine the accounting curriculum to address climate change may be implemented. This includes adopting pedagogical approaches that motivate and encourage development and utilising interactive and collaborative learning methodologies to analyse corporate disclosures related to climate change (De Aguiar and Fearfull 2010). In addition to learning accounting principles and practices, it is also important to include philosophical, ethical, and critical thinking skills in accounting education; such measures can lead students towards a deeper understanding of the implications of climate change and climate-related disasters. The inclusion of climate accounting in accounting textbooks, along with an emphasis on environmental and social company performance, serves to motivate students to assume greater responsibility for minimising their personal emissions. This integration enables accounting education to effectively prepare students for a career as skilled and ethical professionals with a strong intellectual foundation. Students may also seek information from a range of corporate disclosures and reporting practices by using climate change disclosure and climate accounting as educational tools. This approach can encourage students to consider how these sources might influence or initiate positive changes in the organisation over time.

Accounting students should have the capability to initiate significant transformations and pave the way for an organisation's and society's sustainable future. This can be achieved by actively embracing climate consciousness, integrating disaster risk management into financial reporting, prioritising transparency, adopting new accounting standards that account for environmental and social impacts, and ensuring accountability of influential financial entities for their actions.

As evidenced by Lovell and MacKenzie (2011), the accounting profession has taken a more proactive approach towards the role of accounting and accountants in assisting organisations to tackle and adjust to climate change, compared to the academic literature. This highlights the profession's commitment towards addressing the pressing matter of climate change. Accounting education should undergo an improvisation that incorporates climate change disclosure and reporting practices across various levels. Integration of such approaches within accounting education may inspire students to explore innovative approaches in accounting for climate change and disaster risk issues. This broader perspective could encompass the wider impact of organisational actions, ultimately aiming to present a more comprehensive and democratic portrayal of an organisation.

7.2.2 *Incorporating Climate Change Accounting Curriculum and Implications*

Companies are experiencing tremendous pressure from external forces to accomplish climate challenges and maintain a sustainable business environment. Based on the survey results from 1200 financial executives across Europe, a more proactive approach is essential to climate-related liabilities. In regard to climate-related liabilities, 70% of respondents felt pressure from clients and 61% from three or more stakeholders (Coppola et al. 2019).

The level of pressure experienced by various stakeholders tends to differ depending on the industry. In sectors such as tourism, automotive, consumer goods, and energy and utilities, executives report high pressure to act across all stakeholder groups (Coppola et al. 2019). Coppola et al. (2019, p. 3) explore the potential benefits that climate change presents to businesses: First, companies can enhance their resource efficiency, such as by boosting their energy efficiency, and thereby reduce their expenses. Second, climate change may lead to the creation of new products and services that are less carbon-intensive or facilitate carbon reduction for others. Finally, companies can ensure the durability of their supply chains by reducing their reliance on volatile fossil fuel prices and shifting to renewable energy sources. The combination of these measures can cultivate competitiveness and open up new market opportunities.

Climate change is a growing concern for investors, who want to know how companies are preparing for its impact on their business models, risk management strategies, and financial statements. Climate change might pose challenges to a company's operations, and investors are interested in what the company plans to do to deal with these challenges (Pankratz et al. 2023). In order to maintain investor confidence, companies must take a proactive stance on climate change and demonstrate a commitment to sustainability. Industry requires climate-ready graduates who can meet stakeholder and investor demands at various levels. Gray and Collison (2002) emphasised the importance of accounting education in equipping accountants with the necessary environmental knowledge and skills to enable them to serve the public interest through an accounting curriculum.

Climate-related matters that affect financial performance and position should be disclosed by companies to ensure transparency and accountability. Following the Presentation of financial statements (IAS1), companies are mandated to estimate future cash flows for impairment testing and calculate decommissioning obligations. In another case, in climate-related matters, material uncertainties can undermine a company's ability to remain a going concern (IFRS 2023). This demonstrates a company's commitment to addressing climate change and facilitates informed decision-making while fostering a more sustainable future.

The impact of climate-related factors on inventory valuation (IAS 2) could make a company's inventory obsolete, reduce the value of its inventory, or increase the cost of completing a project. Companies must carefully assess the impact of climate-related issues on their fixed assets. This includes property, plant, and equipment (IAS

16), as changes to estimated residual value and expected useful lives can result from obsolescence or legal restrictions on asset use. Sustainable asset management enables companies to mitigate risks and promote long-term success through informed decisions. Climate-related risks can have a significant impact on a company's ability to respond to asset impairment (IAS 36). For instance, a company may experience a decrease in revenue or an increase in costs due to the effects of climate change on its operations. This could lead to the impairment of assets, including property, plant, and equipment (PPE) or intangible assets. Furthermore, climate-related risks also affect the recognition, measurement, and disclosure of liabilities (IAS 37) related to things such as government penalties imposed for not meeting climate-related targets. Climate-related events, such as wildfires, floods, and policy changes, can negatively impact a borrower's ability to repay a lender, thereby affecting the lender's credit losses. This can also impact a company's financial instruments (IFRS 9) (IFRS 2023). Disclosures related to climate change affect the fair value measurement of assets and liabilities in financial statements (IFRS 13). For instance, potential climate-related legislation and its impact on market participants' views can influence the fair value of an asset or liability (IFRS 2023, p. 5). In light of climate-related issues, companies have an opportunity to proactively assess the potential impact on their future taxable profits (IAS 12). This can lead to a better understanding of deferred tax asset recognition and derecognition (IFRS 2023) and enable businesses to make informed decisions that contribute to long-term financial sustainability. Thomson and Bebbington (2005) argued that social and environmental reporting can transform accounting education by illustrating how organisations should align their operations with society and the environment. Teaching accounting from a critical perspective can be enhanced by incorporating climate change education (de Aguiar and Fearfull 2010). In addition to training students to understand different accounting applications, users, and requirements, analysing comparative information can motivate them to develop better disclosure mechanisms and behaviour.

Bebbington and Larrinaga-Gonzalez (2021) stated that environmental policies on climate change could impact accounting in three ways: by incorporating emissions allowances, providing additional information on climate change risks, and reporting uncertainty. Accounting educators should incorporate the above-mentioned climate-related disclosure, asset life cycle, and supply chain interruptions into their curriculum (Khosa et al. 2023). With this integration, students can demonstrate the tangible effects of climate change on the assets, liabilities, and costs companies report on their financial statements. As an example, suppose a student analyses a beverage company that relies heavily on water resources for production. They investigate the impact of climate change-induced water scarcity on the company's financial statements. The student collects data on changing weather patterns, drought frequency, and water stress indices in the regions where the company sources water. They predict potential future water scarcity scenarios due to climate change. The student then assesses how these changes might impact the company's financial statements. They predict increased costs associated with water conservation measures or potential relocation, potential liabilities, and increased operating expenses. Finally, the student prepares a report or presentation demonstrating how these climate change-induced impacts translate into potential adjustments in the company's financial statements. This approach would also aid students in comprehending the potential risks (physical, transitional, and liability risks) posed by climate change and sustainability for businesses. While advising businesses on risk assessment and mitigation strategies, students would be encouraged to adopt a holistic viewpoint when assessing and mitigating risks related to societal and environmental factors (Khosa et al. 2023).

Addressing climate change has emerged as an urgent priority on the global agenda. Treaties such as the 2015 Paris Agreement are examples of international collective efforts that have brought together 196 ratifying countries around the world to limit global warming through binding emission targets (UN 2023). An expanding coalition of countries, including the largest polluters—China, United States, India and the European Union, are committing to achieving net-zero emissions (UN Emissions Gap Report 2023). This contemporary global movement demands for substantial policy restructuring within firms, challenging their role and contribution in tackling this critical issue. Recent studies have shown that capital market participants incorporate firm's environmental footprint into their valuations, underscoring the benefits for firms in managing their emissions. Firms with lower environmental footprint are associated with high profitability (Homroy 2023), better performance (Misani and Pogutz 2015; Busch and Hoffman 2011; Al-Tuwaijri et al. 2004), lower cost of capital (Sharfman and Fernando 2008; Homroy 2023) and ultimately higher firm value (Matsumara et al. 2014; Benkraiem et al. 2022).

Integrating climate change discourses within the realm of management accounting education is imperative to equip future business leaders with the necessary skills, knowledge and mindset to combat the global environmental crisis. Unlike financial accounting, which focuses on producing financial statements for external users, management accounting[1] concentrates on providing strategic information to senior management to facilitate informed decision-making. Therefore, management accounting information should consider strategically relevant information that enables sustainable value creation (AICPA & CIMA). Key functions of management accounting include product and process costing (including relevant costing), pricing, budgeting and variance analysis, managing resources in the short-term (cost-volume-profit relationships) and long-term (capital investment decisions) and performance management and control (ICAEW 2021). While most of the techniques used in the management accounting functions are based on a calculative and positivistic logic, on the other hand, discourses of sustainability and climate change extend to applied ethics (Shearman 1990; Berg and Lyngstadås 2023). Therefore, the challenge here is to have the right pedagogic approach to help students understand and develop critical and analytical thinking to evaluate these competing institutional logics (Besharov and Smith 2014).

One of the main topics covered in a typical management accounting course is performance management. The Balanced Scorecard (BSC) framework developed by

[1] Management accounting is defined as "... the sourcing, analysis, communication, and use of decision-relevant financial and nonfinancial information to generate and preserve value for organisations"—AICPA and CIMA (2023, p. 3).

Kaplan and Norton (1992) provides a comprehensive view of organisational performance from four distinct perspectives—financial, customer, internal process, and learning and growth. BSC, therefore, is a powerful tool that can aid strategy mapping goal congruence if used correctly (Kaplan and Norton 2000). While the BSC is well known for recognising the need for balancing short and long-term performance for organisational success, it has been criticised for its failure to incorporate environmental performance (Kang et al. 2015; Hahn and Figge 2018). Indeed, performance management systems reflect organisational priorities. Therefore, the first step in the right direction is recognising environmental performance as a strategic prioritya

Stakeholder theory is one of the theoretical frameworks that has underpinned the development in sustainability accounting, which emphasises the significance of considering and addressing the interests, needs, and concerns of various individuals or groups, known as stakeholders, who are affected by or can affect the actions and decisions of an organisation (Freeman 1984). Accordingly, a widely accepted rationale is that managers seek to garner stakeholder support and secure the "social license to operate" (Deegan and Unerman 2011). Similarly, the 'Triple Bottom Line (TBL)' framework by Elkington (1997) has also advocated for the consideration of an organisation's environmental, social and economic impact when assessing sustainability. While the TBL framework has been criticised for oversimplification of these distinct domains and their intricate inter-relationship, it still remains a key development in the field of sustainability (Elkington 2018). In addition, extant literature has also documented evidence of a significant relationship between environmental performance and financial performance (de Burgos-Jimenez et al. 2013; Iwata and Okada 2011; Chen and Xie 2022).

Combining theoretical underpinnings of sustainability and empirical evidence in performance management provides a great starting point to open discussions in classrooms to elaborate on the interconnectedness of business decisions to various stakeholder interests and how organisations can supplement standard performance measurement (related to profitability, liquidity etc.) with non-financial performance indicators relating to environmental sustainability, such as the UN SDGs (Berg and Lyngstadås 2023).

Developing key performance indicators (KPI) that are aligned with the organisation's goals is another key responsibility of a management accountant. There are several indicators a company could use to track its environmental performance, such as measures related to environmental costs and revenue. For example, Jasch (2003) categorises environmental costs into,

(i) waste disposal and emission treatment costs—costs related to treatment, disposal and clean-up costs of existing waste and emissions.
(ii) prevention and environmental management costs—costs related to prevention of waste and emissions
(iii) wasted material—costs of wasted materials, including energy and water
(iv) production costs of non-product output—costs of residual material to cost object
(v) environmental revenue—sales derived from waste or grants of subsidies.

Note that the framework above captures internal costs that are incurred within an organisation and assumes (and expects) the external costs to be borne by the general public, consumers and suppliers, raising the question of whether stakeholders are willing to accept such costs.

While cost-based models prioritise the monetary impact of firms' operations, organisations aiming to improve their stakeholder relationships and accountability should employ more robust measures that encompass their wider impact on the environment. One such potential measure is carbon performance. Carbon performance captures firms' environmental impact by measuring how well an organisation manages its Greenhouse Gas (GHG) emissions compared to its industry peers (Hoffman and Busch 2008; Velte et al. 2020). The GHG protocol has pioneered the establishment of global standards for accounting and measuring emissions. Under the protocol, emissions are measured under three different scopes. While Scope 1 measures direct emissions arising from core business operations, Scopes 2 and Scope 3 measure indirect emissions resulting from the consumption of purchased energy and activities throughout the company's entire value chain, respectively (GHG Protocol 2023). Measuring emissions under these three scopes will not only allow companies to quantify and manage their direct carbon footprint but also propel them to assess their indirect impact on the environment. Real and simulated case studies can be used in the classroom to elaborate on the importance of incorporating such key performance indicators into performance management systems within organisations. A similar approach can also be utilised in teaching other areas, such as incorporating environmental considerations within costing concepts, recognising environmental impact in standard Net Present Value (NPV) valuations and capital budgeting decisions (Berg and Lyngstadås 2023), and hence embedding climate change impact into management accounting curriculum.

7.2.3 The Role of Information and Communication Technology (ICT) in Promoting Climate Change Literacy

Education systems should promote resilience capabilities among students and other communities to face the challenges of natural disasters and climate change (Benavot et al. 2024). Information and Communication technology (ICT) is a powerful medium that can be helpful in illustrating the consequences of individual and collective decisions on climate change and the environment (Fauville et al. 2020). The Internet, along with the progressive development of geospatial data browsers, virtual globes, immersive visualisation displays, interactive applications and technologies such as big data, blockchain, and artificial intelligence, has notably expanded the possibilities for a more climate literate public (Niepold et al. 2007; Appelbaum et al. 2017; Rauter, 2021; Tharapos, 2022). Therefore, embedding digital competencies—which is regarded as an essential skill for a 21st-century living—into climate change literacy is fundamental to building an information and knowledge society that empowers people's lives (Pérez-Escoda et al. 2021; Facer and Selwyn 2021).

Immersive technologies are powerful tools for creating subjective impressions of actively engaging in a realistic experience that is ideal for demonstrating the impact of climate change (Dede 2009). By considering/allowing multiple perspectives, immersive technologies encourage critical thinking that promotes situated learning and deep learning through active student engagement, creating an interesting and enjoyable experience for the learner (Fauville et al. 2020; Dede 2009; Biggs 2003). Such technologies can be incorporated into accounting pedagogy for effective learning and teaching. For example, blended case-based learning is an effective pedagogical approach in accounting education (Tran and Herzig 2023; Prado et al. 2020; Anderson and Schiano 2014). Therefore, embedding immersive technologies into simulated case studies, such as explaining how accounting standards and various accounting-related decisions (management of environmental costs, capital budgeting, etc.) ultimately impact the environment and society, will allow students to evaluate the inter-dependencies of multiple dimensions within organisations—economy, environment and society—in decision-making. However, this calls for investment in the appropriate technological infrastructure within universities and higher educational institutions.

7.3 Higher Education's Role in Climate Change and the Future of Accounting Education

With the rapid and continued development of artificial intelligence (AI), numerous articles in the business press have queried whether there is a future for the accounting profession (Griffin 2019; Mok and Zinkula 2023) and whether accounting education will be of use to the students of the future in its current format (Hopper 2013).

However, the roles "at risk" from the increasing automation of the profession are those repetitive process-driven data analysis tasks, and accounting and accounting education is more than this. AI can and should be used to undertake process tasks (Meiryani et al. 2022), thus allowing accountants to focus on providing thought leadership. It will allow accountants to support and influence organisations' critical strategic issues, including the sustainable future of the entity and the ecosystem in which it operates.

The dictionary definition (OED 2023) of "account" considers two clear and related concepts: to account for something or to hold individuals to account. These two differing purposes are reflected within the syllabus of accounting programmes both within higher and professional education as well as within the roles accounts undertake in industry, practice and the third sector. Accountants prepare reports showing economic performance and the stewardship of resources in accordance with prescribed guidelines and expectations and provide strategic guidance on the content of these reports. This guidance allows stakeholders to hold the organisation's leadership to account for the organisation's performance and its impact.

For many, the current perceived role of an accountant and the reports generated only reflect the quantifiable performance of an organisation, but this ignores the skills and behavioural expectations required of anyone calling themselves a professional (Beatton 2022). The integration of the fundamental principles of ethics (IFAC 2020) into accounting education is important for ensuring that accountants are trained to question decisions and provide clear guidance on strategic policies considering uncertainty, risk, ambiguity and suffering (Issa and Pick 2010) as well as the overall economic profit.

The further development of accounting education to allow students to develop professional scepticism and ethical mindsets required of accounting education is to move from facilitating an education focused on compliance-based transactional knowledge to a more critical consideration of systems and processes. This will allow accountants of the future to be in a position to provide guidance and support to organisations and governments to understand the impact on both global economies and the ecosystem in which organisations operate (Bebbington et al. 2021).

The need for accounting education to prepare bright, capable, questioning accounting students who are able to lead and support strategic decision-making is not new. The IASB Framework for the Preparation and Presentation of Financial Statements, issued in 1989, defined the role of financial accounting as the provision of information that is useful for aiding stakeholders in decision-making. In the current era, the Anthropocene, the decision-making processes supported by accountants need to move from focusing on the shareholder theory stance of profit maximisation to recognising that humanity's actions and decisions are responsible for the current and increasing climate crisis.

Accounting education needs to prepare accountants to provide guidance on more than just economic outcomes and needs to change the way in which reporting occurs radically. While academics and corporations have tried developing different reporting methods and frameworks for the inclusion of sustainable information (IIRC 2021; GRI, 2002; Puma, 2012; Gibson et al. 2001)[2] these methods of reporting focus on historical performance rather than allowing stakeholders to consider what impact the decisions will have in the future and how these will impact the planetary boundaries (Rockstom et al. 2023). Accounting education needs to provide opportunities for accounting students to develop a capacity for critical inquiry of the current reporting frameworks and to "radically re-evaluate their worldviews" (Boler and Zembylas 2003) Classrooms need to be a safe space to allow students to position themselves within the discomfort of trying to solve the wicked problems of "climate change, disturbed weather patterns, collapsing ecosystems, species extinction, pollution from industrial farming practices, deforestation, desertification, huge oceanic dead zones,

[2] https://www.integratedreporting.org/resource/international-ir-framework/.
https://www.globalreporting.org/.
https://annual-report.puma.com/2021/en/sustainability/environment.html.

and numerous other ecological issues emissions" (Waddock 2013) while ensuring that economic stability is retained.

The radical change required, therefore, falls into the remit of the changing accounting education both within Higher Education Institutes and the qualifications run by the Professional Bodies. Sustainability has started to be integrated into education, with universities integrating sustainability into their mainstream undergraduate programmes, developing specialist master programmes in sustainable accounting and the professional bodies integrating small sustainability elements into all qualification assessments and designing CPD programmes to support the change of mindset needed for accountants, both now and in the future, to be at the forefront of decision making. Although these developments are a step in the right direction, they are insufficient to bring about the radical change that is necessary.

7.3.1 Education Support Centres and Communities of Practice

The UN Global Sustainable Development Report (2023) states, "Education systems need to be transformed to ensure that everyone is equipped with the knowledge and skills that enable innovation, resilience and creativity" (UN GSDR 2023, p. 107). If education systems were to fulfil such expectations in a constantly evolving global environment, it is necessary to embed the right support systems and infrastructure within higher education institutions. Several studies show that 'culture' is an essential factor in promoting sustainable development (Levy and Marans 2012; Dzimińska et al. 2020). Therefore, combining relevant resources with a culture that endorses climate awareness could effectively contribute towards enhancing climate change literacy.

There are a number of ways university and support staff could facilitate—for example, specialist staff with knowledge in climate science can provide expert advice to staff and students on making climate-conscious choices related to both learning and personal life; universities can discourage printing and usage of paper by increasing access to digital devices (for example, laptops, learning pads, clickers etc.); expert ICT staff could help educators in creating digital content and provide training on using specialist software/hardware (Davey et al. 2019).; support centres can also provide additional training for developing the necessary skills for blended or online teaching environments (Tharapos 2022).

Educational communities offer more than just a platform to exchange best practices; they also cultivate a feeling of belonging and secure space to openly discuss various issues and challenges (Sadiq 2021). Communities of practice are 'groups of people who share a concern or a passion for something they do and learn how to do it better as they interact regularly' (Wenger 2011; p. 1). Effective collaboration with formal academic communities (such as British Accounting and Finance Association

(BAFA), American Accounting Association (AAA), Accounting and Finance Association of Australia and New Zealand (AFAANZ)), practitioner communities (such as Chartered Institute of Management Accountants (CIMA), Association of Chartered Certified Accountants (ACCA), Institute of Chartered Accountants in England and Wales (ICAEW), Institute of Chartered Accountants of Scotland (ICAS)) is therefore necessary to improve climate literacy in the overall quality of accounting education through dissemination of research and teaching achievements (Tharapos and Marriott, 2020; Tharapos and O'Connell 2022).

7.4 Conclusions and Recommendations

Human-induced climate change has been characterised as the most substantial and widespread threat of the present era due to its profound consequences on both the natural environment and societies (UN 2022). A growing focus on climate change and sustainability reporting has sparked a discussion about the necessity of redesigning the role of accountants. This shift aims to guide organisations to document their impacts on the environment and develop strategies to mitigate and address these effects. The traditional role of an accountant involves recording, analysing, and reporting financial metrics. However, it can be expanded to include climate and environmental reporting that contributes to society and the environment. The chapter aims to present a future-focused perspective on accounting education's ability to remain relevant in light of climate change. We propose improving accounting education by integrating climate change education, environmental knowledge, professional skill development, and industry involvement at the core of learning and teaching. This chapter has several contributions to the development of accounting education and pedagogy. First, in response to the ongoing demands for the reinvigoration and advancement of accounting education, this chapter highlights the significance of climate change and the management of climate-related risks, which have been emphasised in recent years (Boyce et al. 2019; Botes et al. 2014; De Aguiar and Fearfull 2010). Second, this chapter explicitly asserts that accounting education must be future-proofed to align with organisational, societal, and environmental concerns. This chapter also emphasises the importance of professional skills and active engagement in the industry for addressing climate change (Linnenluecke et al. 2015). Lastly, the book chapter discusses how accounting educators are adapting to climate change and preparing students for the future of accounting.

Future accounting curricula should incorporate training on recording, reporting, and disclosing climate and sustainability-related information material to an entity's stakeholders in accordance with IFRS standards.[3] This approach would also allow students to understand better business risks associated with climate change and

[3] The objective of IFRS S1 is to require entities to disclose information about sustainability risks and opportunities that are useful for decision-making. The objective of IFRS S2 is to require entities to disclose information about their climate-related risks and opportunities, which can be used by

sustainability. They would be motivated to embrace a comprehensive approach to risk assessment and mitigation while advising businesses about climate and environmental risks. Future 'accounting graduates will need to become 'comfortable with complexity, possess an ability to adapt to new situations and be agile enough to approach problems with an openness to the novel, hitherto unimaginable perspectives' (McGuigan 2021, p. 384). Accounting educators, accounting departments, program administrators, professional accounting bodies, and universities can use the research findings in designing accounting programs.

References

ACCA (2022) Climate-related disclosures and investor focus. https://www.accaglobal.com/ie/en/student/exam-support-resources/professional-examsstudy-resources/strategic-business-reporting/technical-articles/climate.html. Accessed 10 Nov 2022

AICPA and CIMA (2023) GMAP—Global Management Accounting Principles, 2nd edn. https://www.aicpa-cima.com/resources/download/gmap-global-accounting-principles. Accessed 25 Dec 2023

Al-Tuwaijri SA, Christensen TE, Hughes KE II (2004) The relations among environmental disclosure, environmental performance, and economic performance: a simultaneous equations approach. Acc Organ Soc 29(5–6):447–471

Anderson E, Schiano B (2014) Teaching with cases: A practical guide. Harvard Business Review Press

Appelbaum D, Kogan A, Vasarhelyi M, Yan Z (2017) Impact of business analytics and enterprise systems on managerial accounting. Int J Account Inf Syst 25:29–44

Bakker P (2013) Accountants will save the world. Harvard business review. https://hbr.org/2013/03/accountants-will-save-the-worl. Accessed 15 Nov 2023

Beaton GR (2022) Why professionalism matters more than ever. https://ssrn.com/abstract=4196212 or https://doi.org/10.2139/ssrn.4196212

Benavot A, Williams JH, Naidoo J (2024) Should we not be educating for resilience? Leveraging SDG4 in times of crisis. Int J Educ Dev 104:102929

Benkraiem R, Shuwaikh F, Lakhal F, Guizani A (2022) Carbon performance and firm value of the World's most sustainable companies. Econ Model 116:106002

Berg T, Lyngstadås H (2023) We're only in it for the money? Developing sustainable literacy through management accounting curriculum. Account Educ:1–20. https://doi.org/10.1080/09639284.2023.2270475

Besharov ML, Smith WK (2014) Multiple institutional logics in organizations: Explaining their varied nature and implications. Acad Manag Rev 39(3):364–381

Biggs JB (2003) Teaching for quality learning at university. Buckingham: Open University Press/Society for Research into Higher Education. 2nd edn

Blessinger P, Wankel C (2012) Innovative approaches in higher education: An introduction to using immersive interfaces. In: Increasing student engagement and retention using immersive interfaces: virtual worlds, gaming, and simulation. Emerald Group Publishing Limited, pp 3–14

Boler M, Zembylas M (2003) Discomforting truths: The emotional terrain of understanding difference. In: Pedagogies of difference. Routledge, pp 107–130

Botes V, Low M, Chapman J (2014) Is accounting education sufficiently sustainable? Sustain Account, Manag Policy J 5(1):95–124

---

users of general-purpose financial reports to make decisions concerning providing resources to the entity.

Boyce G, Narayanan V, Greer S, Blair B (2019) Taking the pulse of accounting education reform: Liberal education, sociological perspectives, and exploring ways forward. Acc Educ 28(3):274–303

Busch T, Hoffmann VH (2011) How hot is your bottom line? Linking carbon and financial performance. Bus Soc 50(2):233–265

Chang Y, Du X, Zeng Q (2021) Does environmental information disclosure mitigate corporate risk? Evidence from China. J Contemp Account Econ 17(1):100239

Chartered Institute of Management Accountants (CIMA) (2010) Accounting for climate change. How management accountants can help organisations mitigate and adapt to climate change

Chen Z, Xie G (2022) ESG disclosure and financial performance: Moderating role of ESG investors. Int Rev Financ Anal 83:102291

Cho CH, Senn J, Sobkowiak M (2022) Sustainability at stake during COVID-19: Exploring the role of accounting in addressing environmental crises. Crit Perspect Account 82:102327

Coppola M, Krick T, Blohmke J (2019) Feeling the heat? Companies are under pressure to act on climate change and need to do more—Deloitte Insights. https://www2.deloitte.com/content/dam/insights/us/articles/43123_feeling-the-heat/DI_Feeling-the-heat.pdf. Accessed 10 Nov 2023

Davey B, Elliott K, Bora M (2019) Negotiating pedagogical challenges in the shift from face-to-face to fully online learning: A case study of collaborative design solutions by learning designers and subject matter experts. J Univ Teach & Learn Pract 16(1). https://doi.org/10.53761/1.16.1.3

De Aguiar TR, Fearfull A (2010) Global climate change and corporate disclosure: Pedagogical tools for critical accounting? Soc Environ Account J 30(2):64–79

de Burgos-Jiménez J, Vázquez-Brust D, Plaza-Úbeda JA, Dijkshoorn J (2013) Environmental protection and financial performance: An empirical analysis in Wales. Int J Oper Prod Manag 33(8):981–1018

Dede C (2009) Immersive interfaces for engagement and learning. Science 323(5910):66–69

Deegan C (2013) The accountant will have a central role in saving the planet… really? A reflection on 'green accounting and green eyeshades twenty years later.' Crit Perspect Account 24(6):448–458

Deegan C, Unerman J (2011) Financial accounting theory, 2nd edn. McGraw-Hill, Berkshire

Department for Energy Security and Net Zero (2023) 2022 UK greenhouse gas emissions, provisional figures. https://assets.publishing.service.gov.uk/media/6424b8b83d885d000fdade9b/2022_Provisional_emissions_statistics_report.pdf. Accessed 28 Dec 2023

Dzimińska M, Fijałkowska J, Sułkowski Ł (2020) A conceptual model proposal: Universities as culture change agents for sustainable development. Sustainability 12(11):4635

Elkington J (1997) Cannibals with forks: the triple bottom line of 21st century business. Capstone, Oxford

Elkington J (2018) 25 years ago I coined the phrase 'triple bottom line.' here's why it's time to rethink it. Harvard Business Review. https://hbr.org/2018/06/25-years-ago-i-coined-the-phrase-triple-bottom-line-heres-why-im-giving-up-on-it

Facer K, Selwyn N (2021) Digital technology and the futures of education: Towards 'Non-Stupid' optimism. Paper commissioned for the UNESCO Futures of Education report

Fauville G, Queiroz ACM, Bailenson J N (2020) Virtual reality as a promising tool to promote climate change awareness. Technol Health:91–108

International Financial Reporting Standards (IFRS) (2023). Effects of climate-related matters on Financial statements, International Financial Reporting Standards Publication. https://www.ifrs.org/content/dam/ifrs/supporting-implementation/documents/effects-of-climate-related-matters-on-financial-statements.pdf

Gibson K, Gray R, Liang Y, Dey C (2001) The silent accounts project: Draft silent and shadow accounts 1999–2000. In: Paper presented at BAA Scottish Group Conference, Stirling, August

Godemann J, Haertle J, Herzig C, Moon J (2014) United Nations supported principles for responsible management education: Purpose, progress and prospects. J Clean Prod 62:16–23

Gray R, Collison D (2002) Can't see the wood for the trees, can't see the trees for the numbers? Accounting education, sustainability and the public interest. Crit Perspect Account 13(5–6):797–836

Griffin O (2019) How Artificial Intelligence will impact accounting, ICAEW. https://www.icaew.com/technical/technology/artificial-intelligence/artificial-intelligence-articles/how-artificial-intelligence-will-impact-accounting

GRI (Global Reporting Initiatives) (2002) Sustainability reporting guidelines. https://www.globalreporting.org/. Accessed 22 Auguest 2024

Hahn T, Figge F (2018) Why architecture does not matter: On the fallacy of sustainability balanced scorecards. J Bus Ethics 150:919–935

Hoffmann VH, Busch T (2008) Corporate carbon performance indicators: Carbon intensity, dependency, exposure, and risk. J Ind Ecol 12(4):505–520

Homroy S (2023) GHG emissions and firm performance: The role of CEO gender socialization. J Bank Finance 148:106721

Hopper T (2013) Making accounting degrees fit for a university. Crit Perspect Account 24(2):127–135

ICAEW (2021) Management accountant: What is the role and what are you expected to do? https://jobs.icaew.com/article/management-accountant-what-is-the-role-and-what-are-you-expected-to-do. Accessed 25 Dec 2023

IFAC (2020) Accountancy skills evolution: impact of COVID-19 & the path forward. Retrieved October 4, 2021, from https://www.ifac.org/knowledge-gateway/preparing-future-ready-professionals/discussion/accountancy-skillsevolution-impact-covid-19-path-forward

IIRC (2021) International Framework. International Integrated Reporting Council. https://integratedreporting.ifrs.org/wp-content/uploads/2022/08/IntegratedReportingFramework_081922.pdf. Accessed 22 August 2024

Issa T, Pick D (2010) Ethical mindsets: an Australian study. J Bus Ethics 96:613–629

Iwata H, Okada K (2011) How does environmental performance affect financial performance? Evidence from Japanese manufacturing firms. Ecol Econ 70(9):1691–1700

Jabbour ABLS, Luiz JVR, Luiz OR, Jabbour CJC, Ndubisi NO, Oliveira JHC, Hourneaux Junior F (2019) Circular economy business models and operations management. J Clean Prod 235:1525–1539

Jasch C (2003) The use of Environmental Management Accounting (EMA) for identifying environmental costs. J Clean Prod 11(6):667–676

Kang JS, Chiang CF, Huangthanapan K, Downing S (2015) Corporate social responsibility and sustainability balanced scorecard: The case study of family-owned hotels. Int J Hosp Manag 48:124–134

Kaplan RS, Norton DP (2000) Having trouble with your strategy? Then map it. Focusing your organization on strategy with the balanced scorecard. Harv Bus Rev 49(5):167–176

Kaplan RS, Norton DP (1992) The balanced scorecard. Harvard Business School press, Boston

Khan T, Gray R (2016) Accounting, identity, autopoiesis+ sustainability: A comment, development and expansion on Lawrence, Botes, Collins and Roper (2013). Meditari Account Res 24(1):36–55

Khosa A, Pandey R, Wilkin C (2023) Accounting curricula and climate crisis: Evidence from Australia and New Zealand. University Of Tasmania. Conference contribution. https://hdl.handle.net/102.100.100/599580

Leoni G, Lai A, Stacchezzini R, Steccolini I, Brammer S, Linnenluecke M, Demirag I (2021) Accounting, management and accountability in times of crisis: lessons from the COVID-19 pandemic. Account, Audit Account J 34(6):1305–1319

Levy BL, Marans RW (2012) Towards a campus culture of environmental sustainability: Recommendations for a large university. Int J Sustain High Educ 13(4):365–377

Linnenluecke MK, Birt J, Griffiths A (2015) The role of accounting in supporting adaptation to climate change. Accounting & Finance 55(3):607–625

Lovell H, MacKenzie D (2011) Accounting for carbon: the role of accounting professional organisations in governing climate change. Antipode 43(3):704–730
MacKenzie D (2009) Making things the same: Gases, emission rights and the politics of carbon markets. Acc Organ Soc 34(3–4):440–455
Matsumura EM, Prakash R, Vera-Munoz SC (2014) Firm-value effects of carbon emissions and carbon disclosures. Account Rev 89(2):695–724
McGuigan N (2021) Future-proofing accounting education: educating for complexity, ambiguity and uncertainty. Revista Contabilidade Finanças 32(87):383–389
Meiryani M, Aprilia KR, Warganegara DL, Yanti Y (2022) Challenges of the accounting profession in the era of the industrial revolution 4.0. In: Proceedings of the 2022 international conference on e-business and mobile commerce, pp 39–46
Misani N, Pogutz S (2015) Unraveling the effects of environmental outcomes and processes on financial performance: A non-linear approach. Ecol Econ 109:150–160
Moka A, Zinkula J (2023) ChatGPT may be coming for our jobs. Here are the 10 roles that AI is most likely to replace. Business Insider https://www.businessinsider.com/chatgpt-jobs-at-risk-replacement-artificial-intelligence-ai-labor-trends-2023-02?r=US&IR=T. Accessed 20 Nov 2023
Molthan-Hill P, Worsfold N, Nagy GJ, Leal Filho W, Mifsud M (2019) Climate change education for universities: A conceptual framework from an international study. J Clean Prod 226:1092–1101
Niepold F, Herring D, McConville D (2007) The case for climate literacy in the 21st century. In: Fifth international symposium on digital earth. https://earthtosky.org/content/climate/PDF_Resources/niepold%20et%20al%20case%20for%20climate%20literacy.pdf. Accessed 29 Nov 2023
Nyberg D, Wright C (2022) Climate-proofing management research. Acad Manag Perspect 36(2):713–728
OED (2023) Account, in Oxford English dictionary. Oxford University Press, Oxford
Pankratz N, Bauer R, Derwall J (2023) Climate change, firm performance, and investor surprises. Manage Sci 69(12):1–48
Pérez-Escoda A, Lena-Acebo FJ, García-Ruiz R (2021) Digital competences for smart learning during COVID-19 in higher education students from Spain and Latin America. Digit Educ Rev 40:122–140
Prado AM, Arce R, Lopez LE, García J, Pearson AA (2020) Simulations versus case studies: Effectively teaching the premises of sustainable development in the classroom. J Bus Ethics 161:303–327
GHG Protocol (2023) About GHG protocol. https://ghgprotocol.org/about-us. Accessed 25 Nov 2023
Ranger N, Mahul O, Monasterolo I (2021) Managing the financial risks of climate change and pandemics: What we know (and don't know). One Earth 4(10):1375–1385
Rauter H (2021) The role of blockchain, big data, internet of things and artificial intelligence for transformative climate action. Planning climate smart and wise cities: a multidisciplinary approach. Springer International Publishing, Cham, pp 99–137
Reid A (2019) Climate change education and research: possibilities and potentials versus problems and perils? Environ Educ Res 25(6):767–790
Rockström J, Gupta J, Qin D, Lade SJ, Abrams JF, Andersen LS, Zhang X (2023) Safe and just earth system boundaries. Nature:1–10
Bebbington J, Larrinaga C, O'Dwyer B, Thomson I (eds) (2021) Routledge handbook of environmental accounting, 1st edn. Routledge. https://doi.org/10.4324/9780367152369
Sadiq K (2021) Communities of practice as a multidisciplinary response in times of crisis: adapting to successful online learning practices. Account Res J 34(2):134–145
Sharfman MP, Fernando CS (2008) Environmental risk management and the cost of capital. Strateg Manag J 29(6):569–592
Shearman R (1990) The meaning and ethics of sustainability. Environ Manage 14:1–8
Sikka P, Haslam C, Kyriacou O, Agrizzi D (2007) Professionalizing claims and the state of UK professional accounting education: some evidence. Account Educ: Int J 16(1):3–21

Stevenson RB, Nicholls J, Whitehouse H (2017) What is climate change education? Curric Perspect 37:67–71

Tharapos M (2022) Opportunity in an uncertain future: reconceptualising accounting education for the post-COVID-19 world. Acc Educ 31(6):640–651

Tharapos M, Marriott N (2020) Beauty is in the eye of the beholder: Research quality in accounting education. Br Account Rev 52(5):100934

Tharapos M, O'Connell BT (2022) Transnational communities of practice: Their development, operation, and contribution. J Stud Int Educ 26(3):299–317

Thomson I, Bebbington J (2005) Social and environmental reporting in the UK: a pedagogic evaluation. Crit Perspect Account 16(5):507–533

Tran TT, Herzig C (2023) Blended case-based learning in a sustainability accounting course: An analysis of student perspectives. J Account Educ 63:100842

United Nations Environment Programme (2023) Emissions Gap Report 2023. Nairobi. https://www.unep.org/resources/emissions-gap-report-2023

UN Global Sustainable Development Report (2023) Times of crisis, times of change: Science for accelerating transformations to sustainable development. United Nations, Independent Group of Scientists appointed by the Secretary-General, New York

UN (2022) United Nations Meeting Coverage and Press Release. https://press.un.org/en/2021/sc14445.doc.htm. Accessed 23 Nov 2022

Velte P, Stawinoga M, Lueg R (2020) Carbon performance and disclosure: A systematic review of governance-related determinants and financial consequences. J Clean Prod 254:120063

Waddock S (2013) The wicked problems of global sustainability need wicked (good) leaders and wicked (good) collaborative solutions. J Manag Glob Sustain 1(1):7

Wenger E (2011) Communities of practice: A brief introduction. Retrieved September 30, from https://scholarsbank.uoregon.edu/xmlui/bitstream/handle/1794/11736/A%20brief%20introduction%20to%20CoP.pdf?sequence=1

Zhang A, Tay HL, Alvi MF, Wang JX, Gong Y (2023) Carbon neutrality drivers and implications for firm performance and supply chain management. Bus Strateg Environ 32(4):1966–1980

Chapter 8
Schools as Agents of Social Transformation in the Socio-Environmental Disasters Prevention in Brazil

Patricia Mie Matsuo and Rosana Louro Ferreira Silva

Abstract Education plays a fundamental role in creating a culture of prevention and reduction of socio-environmental disaster risks, especially in a continental country like Brazil, which faces a diversity of disasters. This chapter presents the first Brazilian diagnosis of how schools address disasters in their educational spaces. The analysis was based on school practices registered in the four annual editions (from 2016 to 2019) of the #LearnToPrevent Campaign, promoted by Cemaden. Hundreds of public schools from all regions of the country developed several educational activities. Rain gauges emerged as one of the most common practices, offering pathways for the adoption of more active and contextualized perspectives through multiple didactic approaches. However, there are gaps to be overcome to make these practices more critical and emancipatory, promoting questioning of the root causes of disaster risk scenarios and the co-creation of measures to transform the economic and political systems that perpetuate social asymmetries in the country. Despite the absence of formalization of disaster risk reduction in the national curriculum, schools have played a crucial role as agents of social transformation, breaking certain barriers and establishing bridges, especially with Civil Defense, in the process of co-constructing knowledge and pioneering school practices for disaster prevention based on their socio-environmental reality.

Keywords Formal education · Environmental education · Campaign · Didactic approach · Rain gauges

P. M. Matsuo (✉)
Institute of Advanced Studies, Universidade de São Paulo, Sao Paulo, Brazil
e-mail: patricia.matsuo@alumni.usp.br

R. L. F. Silva
Biosciences Institute, Universidade de São Paulo, Sao Paulo, Brazil
e-mail: rosanas@usp.br

A. Yildiz and R. Shaw (eds.), *Disaster and Climate Risk Education*, Disaster Risk Reduction, https://doi.org/10.1007/978-981-97-5987-3_8

8.1 Introduction

Disasters are becoming increasingly frequent in the reality of a large part of the world's population and have become a frequent topic in the media and in family conversations. Several societies face countless impacts generated by the occurrence of multiple disasters. In this context, education represents a strategic component to contribute to the creation of cultures of prevention and disaster risk reduction, as well as the strengthening of sustainable and resilient societies.

Reference documents in Disaster Risk Reduction (DRR) such as the Hyogo Framework for Action, the Sendai Framework for Disaster Risk Reduction, and the National Policy for Civil Defense in Brazil prioritize education and recommend the integration of knowledge about DRR at all levels of formal education (UNISDR 2006; UNISDR 2015; Brazil 2012a).

In this context, the objective of this chapter is to present the Brazilian panorama of how disasters are addressed in the formal educational system, examine the didactic approach adopted in those school initiatives, and identify potential and gaps for strengthening Disaster Risk Reduction Education (DRRE) in Brazil.

To begin, we provide an overview of disaster scenarios in our country and delineate the interconnections between environmental education and DRRE. Next, we delve into the strategy of mobilizing society through campaigns, with a particular emphasis on the Centro Nacional de Monitoramento e Alertas de Desastres Naturais (National Center for Monitoring and Early Warning of Natural Disaster—Cemaden) Campaign #AprenderParaPrevenir (#LearnToPrevent). Following this, we present the results of utilizing rain gauge activities as a core component to discuss multidimensional approaches to disaster risk reduction. Finally, we offer reflections and recommendations for strengthening DRRE efforts in Brazil.

8.2 Brazil: A Tropical Country with Risk Scenarios

The Amazon, soccer, indigenous people, carnival, biodiversity and the hospitality of the people are some of the images commonly associated with Brazil. However, this idealized vision often creates a false sense of security in relation to "traditional" disasters such as volcanoes, earthquakes and tsunamis, among both the global population and Brazilians themselves.

Despite the absence of these events, the country has increasingly faced other types of disaster. Between 2013 and 2022, events such as floods, flash floods, landslides, and droughts affected over 4.2 million people across 93% of Brazilian municipalities (CNM, 2023). In 2023, the country recorded its highest number of disasters, 1,341 events occurred in over half of Brazil's 5,568 municipalities, affecting 14.5 million Brazilians in 2,797 municipalities (Agência Câmara de Notícias 2023; Cemaden 2023; MCTI 2024).

In recent years, Brazil has experienced a series of unprecedented disasters, including forest fires in the Amazon, Pantanal and Cerrado biomes (Barbosa et al., 2022; Mataveli et al., 2022), drought in the Amazon (Clarke et al. 2024; Jimenez et al. 2019), heatwaves in several regions of the country (Calim Costa et al. 2024; Libonati et al. 2022) and the resurgence of water-borne diseases epidemics such as dengue. This disease is transmitted by the *Aedes aegypti* mosquito, which reproduces in environments facilitated by climate change, high temperatures and abundant breeding sites due to high rainfall (Barcellos et al. 2024).

It is crucial to highlight that disaster risk management in Brazil confronts the challenge of shifting two important paradigms: changing the focus of action from post-disaster response to prevention and recognizing that these disasters are not natural. Instead, they are considered socio-environmental disasters, constructed from complex processes resulting from the interaction of natural elements such as rainfall, with climate change, as well as multiple social factors. These factors include population growth, social inequalities, poverty, structural racism, and deficiencies in housing, educational, and sanitation policies (Oliver-Smith et al. 2017; Maskrey 1993).

An example of a disaster that exposes these social inequalities occurred during the 2023 carnival in the municipality of São Sebastião, situated on the northern coast of the state of São Paulo (Soares 2023). The neighborhood of Barra do Sahy experienced the most significant recorded rainfall in the country, with 626 mm of precipitation in just 6 h (CEPDEC 2023). Consequently, it resulted in several landslides, 64 deaths, and 3 thousand people displaced (PMSS 2023a). This volume of rain fell on a landscape divided by the highway: on one side, closer to the beach and with a flatter terrain, were hotels, restaurants, and vacation residences of the wealthy class. On the opposite side, situated on steeper slopes,[1] lived lower-income residents.

Most of these victims belonged to the most vulnerable groups of society: the poor, black people, immigrants, residents of suburbs and non-whites, who were forced to live in high-risk areas due to intense real estate speculation driven by tourism in the region. Unfortunately, this event serves as further evidence of the numerous "racist disasters"[2] in Brazil.

Among those affected, students were directly impacted, as 14 schools had to suspend classes, jeopardizing the fundamental right to education of children and youth. This occurred because the schools were directly affected by the water and mud from the landslides, turned into shelters for displaced families, or because people's mobility was interrupted by the obstruction of access roads (PMSS 2023b).

Analyses of the impact of disasters on the teaching–learning process remain scarce in Brazil, but one study provided a preliminary diagnosis of the physical vulnerability of schools located in the 957 municipalities monitored by the Cemaden. This study

---

[1] The occupation of hillsides in Brazil is directly related to the process of colonization and slavery of the black, marginalized and invisible population, built dwellings in illegal and undervalued areas of hillsides, perpetuating social structures in these risk-prone spaces (Gomes 2022).

[2] In his latest book, Metamorphosis of the World, Beck (2018) portrays the "racist flood" that occurred in New Orleans after Hurricane Katrina in 2005, as it mainly affected poor, black and peripheral communities.

revealed the presence of 2,443 schools located in risk areas; 1,798 are public and 645 are private. Among these, 729 are situated in hydrological risk areas, while 1,714 are in geological risk areas (Marchezini et al. 2018). Considering that the survey only covered only municipalities with official mapped risk areas at the time and represented only 17% of Brazilian municipalities, this number is likely to be even higher.

These school risk scenarios reinforce the need and urgency to address socio-environmental disasters from a preventive perspective aimed at creating a culture of protection and disaster risk reduction.

8.3 Dialogues Between Environmental Education and Disasters

In times when disasters reach unprecedented levels, as seen in the recent landslides, floods, and forest fires in Brazil, we must ask ourselves, what is the "role" of education to face this complex problem? The Brazilian educator and philosopher Paulo Freire once said, "if education alone does not transform society, without it, society also does not change" (Freire 2000, p. 77).

In this context, environmental education plays an important role in contributing to disaster risk management, particularly in the formative process of society, especially in densely populated urban areas with less security infrastructure. So this affected group will have the ability to intervene and participate in the formulation of public policies that prevent, mitigate, and prepare people for the impacts of disasters (Carniatto et al. 2018).

In Brazil, environmental education is considered a citizen's right and a state duty by the 1988 Federal Constitution and the Environmental Education National Policy (Brasil 1999), which presents a set of objectives and principles for Brazilian environmental education.

Although there is a plurality of conceptions of environmental education (Sauvé 2010; Silva and Campina 2011), we have defended the urgency of Critical Environmental Education, which finds support in the perspective of popular education, with significant contributions from Paulo Freire. This approach understands education as a social, dialogical, collective, and emancipatory construction. It presents the complexity of the human-nature relationship and emphasises the need for strengthening civil society in the collective pursuit of social transformations (Silva and Campina 2011). In this critical conception, discussions of conflicts, power relations, and participation are fundamental to constructing meanings and bringing complexity to the understanding and intervention in the socio-environmental reality.

Considering the role of schools in the construction of sustainable and resilient societies (Trajber 2019), we argue the need for environmental education that promotes the encouragement of articulation of scientific, artistic, cultural, and traditional knowledge. This emphasizes environmental education as a right to life, as well as "the interrelationships between the environmental dimension and social justice, human rights, health, work, consumption, ethnic, racial, gender, sexual diversity, overcoming racism and all forms of discrimination and social injustice" (Portugal and Sorrentino 2018, p. 11), in other words, a curriculum that moves towards equity, sustainability, and inclusion.

Environmental education can contribute to the processes of understanding threats and engaging the community in defining and implementing strategies that effectively contribute to the mitigation of their vulnerability. This means strengthening the individual and collective capacities of all social actors, allowing the mobilization of local knowledge for empowerment, promoting participation in land use and occupation decision-making processes, and building changes in how risks are perceived and dealt with in these territories (Da-Silva-Rosa et al. 2015; González-Gaudiano and Maldonado-González 2017; Sulaiman et al. 2021).

Indeed, these elements highlight a productive dialogue between the principles of Critical Environmental Education and DRRE, which is complex and committed to processes of transformation and emancipation.

The legal recognition of DRRE in Brazil occurred in 2012 through the Protection and Civil Defense National Policy, which established the need to create a national culture of disaster prevention and to offer support for teachers through the development of educational materials related to disasters (Brasil 2012a).

However, the incorporation of DRRE into Brazil's main curricular reference document, the National Common Curricular Base (Base Nacional Comum Curricular—BNCC) (Brasil 2018), appears disjointed and superficial. Additionally the emphasis on disaster prevention remains an optional subject at Brazilian schools. The National Curricular Guidelines for Environmental Education remain one of the few normative instruments that contribute to the development of DRRE. This document considers that curriculum planning should contribute to "establishing relationships between climate change and the current model of production, consumption, social organization, aiming at the prevention of environmental disasters and the protection of communities" (Brasil 2012b).

This highlights a significant gap in the Brazilian education system's approach to addressing the increasingly urgent need for disaster preparedness and resilience-building among teachers and students. However, several projects and initiatives are attempting to include DRRE in different educational spaces, especially schools. Through interinstitutional and intergenerational partnerships, they have progressed slowly but resisted the political setbacks experienced in recent years in the country, as the Campaign #AprenderParaPrevenir (#LearnToPrevent).

8.4 Campaign as Mobilization Strategy

Campaigns represent temporary strategies adopted and recommended in various official documents. The United Nations Office for Disaster Risk Reduction (UNDRR) promotes annual thematic campaigns focused on mobilizing, sensitizing, and educating society, sharing best practices, supporting collaborative learning, and fostering a culture of disaster prevention, resilience, and citizenship (UNISDR 2015).

The school context began to receive attention in the 2006–2007 Campaign—"Disaster Risk Reduction Begins at School"—with the aim of integrating Disaster Risk Reduction (DRR) into the education system in risk areas and promoting society's participation in educational processes (UNISDR 2006). In 2009, the Campaign "One Million Safe Schools and Hospitals" was launched to raise awareness to modernize school buildings within safety standards in order to create resistance to natural threats (UNISDR AP 2010).

In Brazil, the Cemaden Education Program, inspired by the International Day for Disaster Reduction, created in 2016 the national Campaign #AprenderParaPrevenir (#LearnToPrevent) with the aim of placing the topic of disasters on the agenda of Brazilian society, especially within the school environment. In addition, it seeks to mobilize the creation of spaces for dialogue and knowledge construction, as well as to map and value the DRRE practices developed in all regions (Panzeri et al. 2020).

The Campaign represents the mobilization component of the Cemaden Education Program, which works with the idea of the school as the central core of disaster prevention actions. This involves knowledge construction, scientific research on local risks, and projects for intervention and local transformation within their communities (Cemaden 2023).

Initially created exclusively for schools, the Campaign began to include in subsequent editions, other active audiences in protection and prevention actions in Brazil, such as Civil Defenses, Universities, NGOs, Scout Groups (Matsuo et al. 2021), and more recently, churches, social movements, and other collectives.

Participation in the Campaign occurs from a voluntary perspective through an invitation to share narratives describing the DRRE actions and projects developed. All submitted practices are published and are available for open access on the Cemaden Education webpage, serving as a database for dissemination and consultation of inspiring practices (https://educacao.cemaden.gov.br/).

For each annual edition, aspects such as the theme, audience, participation format, and awards are defined based on the experience of previous campaigns. To date, Cemaden Education coordinated seven editions (2016, 2017, 2018, 2019, 2020, 2021 and 2023). However, in this chapter, we will present the results of a doctoral research,[3] which aimed to analyze school practices from the first four editions held in the pre-pandemic period between 2016 and 2019 (Table 8.1).

[3] Best 2023 Ph.D. research in Brazil, awarded by Fundação Coordenação de Aperfeiçoamento de Pessoal de Nível Superior—Prêmio CAPES de Tese, Teaching category.

Table 8.1 Characterization of the analyzed editions of the Campaign #AprenderParaPrevenir

<table>
<tr><th>Edition year</th><th>Theme</th><th>Audiences</th><th>N° Participation</th></tr>
<tr><td>1st–2016</td><td>Education in disaster risk prevention</td><td>- Final Years of Elementary School
- High schools</td><td>53</td></tr>
<tr><td>2nd–2017</td><td>Education in disaster risk prevention</td><td>- Schools (from all levels)
- Civil defenses</td><td>67</td></tr>
<tr><td>3rd–2018</td><td>Water [too much or too less] = disasters?</td><td rowspan="2">- Schools (from all levels)
- Civil defenses
- Universities</td><td>86</td></tr>
<tr><td>4th–2019</td><td>Reducing disaster risk: educational actions in climate change times</td><td>103</td></tr>
</table>

8.5 Overview of School Practices in DRRE

The first four editions of the Campaign #AprenderParaPrevenir received a total of 309 DRRE practices. Below, we highlight the general aspects of 238 practices directly related to the school audience, the focus of this study:

8.5.1 Public Schools as Active Agents

The Brazilian basic education system comprises three levels: (a) Preschool Education, for children up to 5 years old[4]; (b) Elementary Education, divided into Early Years (1st–5th grades) starting at age 6, and Final Years (6th–9th grades) for students from age 11; and (c) High School (1st–3rd grades) starting at age 15.

Initially, the campaign was aimed at the young students of Final Years of Elementary School and High School due to the skills and competencies required for the development of monitoring activities and scientific research on disasters (Trajber et al. 2017). In this context, most of the DRRE school practices submitted to the Campaign were developed with the involvement of this juvenile audience, as expected. However, the younger audience of the early stages of the education system (Preschool Education and Early Years of Elementary School) was integrated into the second edition of the campaign due to the interest and curiosity shown regarding the theme. It is important to note that the Early Years of Elementary School participation was the most significant in the 2018 and 2019 editions.

This engagement in the development of activities on risks and disasters was encouraged by the mobilization of local actors, especially the Civil Defenses, who

[4] Ages are just a reference. In Brazil there is a rate of age-grade distortion, that means, in each grade we have older students than recommended (Brazil 2020).

strategically articulated the municipal education networks responsible for the early stages of the education system. Some of these Civil Defenses are considered references in the national scenario, either for their performance in disasters, in the educational field, or for their research production in the area of DRR and environmental education. Examples include the Civil Defenses of Jaboatão dos Guararapes in the state of Pernambuco (Northeast region), Blumenau in Santa Catarina (South region), and Maricá in the state of Rio de Janeiro (Southeast region).

This interest and involvement of teachers from the early stages in the theme were also evidenced by the composition of almost 70% of the participants in a teacher training course for disaster reduction in Angra dos Reis, a municipality in the state of Rio de Janeiro with a historical occurrence of floods and landslides (Sato et al. 2017). In this context, while there is a potential opportunity to develop actions for this children's group, there is also the need to create or adapt strategies and didactic methodologies with appropriate languages to their specificities.

Considering that children are the most vulnerable to disasters (UNISDR 2006), but also the most receptive to learning about protection and prevention (Petal 2008). This openness to including activities related to DRR with this child audience allows for the integration of concepts, methodologies, and practices from the early years of literacy, contributing to the formation of a new generation that is safer, more critical, and resilient.

Most of these educational practices were carried out in public schools (97%). This result raises important reflections for advancing the process of DRRE integration. The first relates to the socio-environmental inequality scenario since public schools can be located in areas at risk of flooding, flash floods, and landslides. This raises the question of whether these schools addressed the disaster theme because, unfortunately, it's part of the student's life, and also they are more exposed to risk than students from private schools. The second presents a distorted view of the potential of public schools in Brazil, which are often unjustly accused of being stagnant and not bringing innovations like private schools. However, even with structural deficiencies and the lack of DRRE formalization, 238 schools created disaster prevention practices in their educational spaces.

The representation of public schools in the campaign editions reflects the commitment and creativity of teachers in creating and adapting an emerging and essential theme in today's context. This set the need to incorporate school knowledge and DRRE didactic methodologies into disaster risk management processes.

8.5.2 *Diversity of Practices and Approaches*

This constructing process of school knowledge also was identified in a diversity of DRRE practices, represented in the 85 different didactic activities, ranging from lectures, drills, experiments, science fairs, monitoring, risk mapping, field trips, to the creation of warning systems.

Activities involving rain gauges, whether in automatic, semi-automatic, or especially in handmade systems (made with PET plastic bottles), were one of the most developed at schools. Probably because it is a practical action, it is easy to obtain materials, has a simple system, and has a low cost for installation and monitoring.

The rain gauges were used in different formats, perspectives, and objectives, a condition that motivated a multidimensional analysis of the adopted didactic approaches. The classification categories were inspired by a study that mapped the inclusion of DRR in the education systems of 30 countries (UNESCO; UNICEF 2012). Brazil was not included in it, so we revisited the categories and adapted to the Brazilian socio-educational reality. We thus established five approaches: Expository, Communicative, Experiential, Investigative, and Citizen. These approaches were also associated with a gradient ranging from less active to more active and reflect the diversity and dynamism of the educational strategies (Fig. 8.1).

The activity with rain gauges falls under an *Expository* approach when, for example, the teacher presents the equipment to the students in a very illustrative way, without involving them in its construction. In this traditional teaching conception, characterized by a vertical and unilateral transfer of knowledge, students play a passive role as receivers of information. This educational model, referred to as "banking education," sees students as recipients of deposits of knowledge. They have the task of memorizing, repeating, storing and taking them in exams to prove that they are good recorders (Freire 1987). This vision of education denies dialogicity and needs to be reviewed and reconfigured towards a more autonomous, critical, and emancipatory approach. Here, we are referring not only to DRRE practices but also to the entire Brazilian educational system.

Fig. 8.1 Dimensions of didactic approaches

On the other hand, when activities with rain gauges allowed students to have direct contact with these rainfall measurement devices in a practical and interactive way, these practices adopted characteristics of the *Communicative* approach, which allowed an active student participation in the process of constructing handmade rain gauges using recyclable materials.

Activities with rain gauges can be seen as 'end' activities when their main objective is to facilitate direct contact with automatic and semi-automatic equipment and/or to construct handmade rain gauges. However, they can also be considered 'middle' activities when they serve as a preliminary step for the development of other practices, such as rainfall monitoring at the school or sharing data with institutions involved in risk and disaster management in the municipality.

Rainfall monitoring, followed by the construction of rain gauges, was a commonly adopted practice in schools, involving students at different levels of participation. They could choose the installation location, collect daily information, organize data, interpret and analyze data, and compare it with data from available automatic equipment on research institute websites.

We also observed several adaptations and strategies to obtain more accurate records: (i) to read, using colored water, a ruler, or adhesive attached to the external area; (ii) to level the base of the PET bottle with stones, resin, cement, and marbles; (iii) to prevent the proliferation of insects (vectors of diseases such as dengue) in the water stored in the rain gauge by using protective screens (Fig. 8.2).

High school students from a public school carried out a series of experiments as part of a project involving rain gauges, exploring different models of PET bottles, different types of materials such as soil and cement to level the base, and different methods to fix the rain gauges and facilitate reading (Trajber et al. 2017; Trajber 2018).

All these means of rainfall monitoring and their adaptations were part of a process of testing, observation, and comparison of results. They are closely linked to the

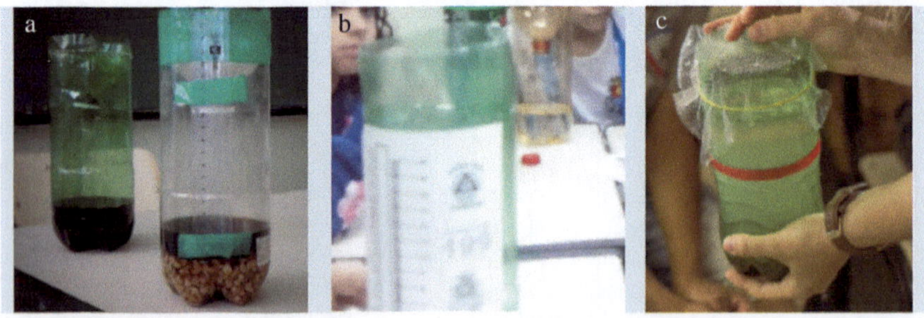

Fig. 8.2 Details of adaptations in the construction of handmade rain gauges with PET bottles. **a**: Use of stones for leveling the base and ruler for reading on the external part of the rain gauge. **b**: Rain gauge with adhesive scale in millimeters and phone number of the Civil Defense. **c**: Use of screen to prevent the proliferation of insects in the water stored in the rain gauge. *Source* (Schools from São Paulo and Minas Gerais States)

integrative and multidimensional process of experiencing in the *Experiential* and *Investigative* didactic approaches.

It is important to highlight the importance of revisiting and reflecting on the role of these students within academic research projects, seeking to go beyond the collection of pluviometric data. It is important to recognize them as active agents in the analysis and interpretation of rainfall information. Offer them the opportunity to understand the conditions and to transform the local socio-environmental processes that directly impact their living spaces.

An outcome of monitoring with handmade or semi-automatic rain gauges installed at the schools is the use of this information beyond the simple production of tables and graphs. It concerns the socio-environmental value added to this information collected by school communities.

Some schools proposed sharing this monitoring data from the *Citizen* perspective. The first proposal was related to using this rainfall information to create disaster risk warning systems for their communities.

> […] to help the community predict the risks of flooding, because our municipality has already had difficult times during periods of heavy rainfall That's why the project's title 'It rains, rain without stopping[5]' is an apology for the amount of rain and how to use the handmade rain gauge to help and alert the community before a disaster occurs (School from Acre State).

The second integration proposal identified occurred through sharing the monitoring data from the schools' rain gauges with institutions such as Civil Defense or research institutions that are part of the Protection and Defense Civil System that develop projects in schools and maintain a historical record of rainfall and weather events.

However, of all 49 initiatives that reported the construction and monitoring of rain gauges, only 12 mentioned the sharing of this data. This information highlights the gap between the production of meteorological equipment and the sharing of the collected data in a rainfall monitoring network, whether at a neighborhood, watershed, or municipal level. Incorporating this type of information into monitoring systems could both enhance local data in community management and enrich and recognize the school as a social agent in the process of constructing this knowledge.

Despite the necessary advances in DRRE practices, the rain gauge showed to be an example of an integrative central theme for all didactic approaches, from more expository activities to those focused on citizenship. This multidimensional condition motivated the creation of a less traditional diagram with greater flexibility, allowing a single activity to be classified in more than one approach.

Thus, with this intentionality and inspired by Eastern cultures, we developed a representation in the form of a mandala (Fig. 8.3), which allows us to visualize the integration and complementarity of multiple approaches in DRRE practices (see more on Matsuo and Silva 2021).

[5] Mention to the popular samba song from 1963 "Chove chuva sem parar" by Jorge Ben.

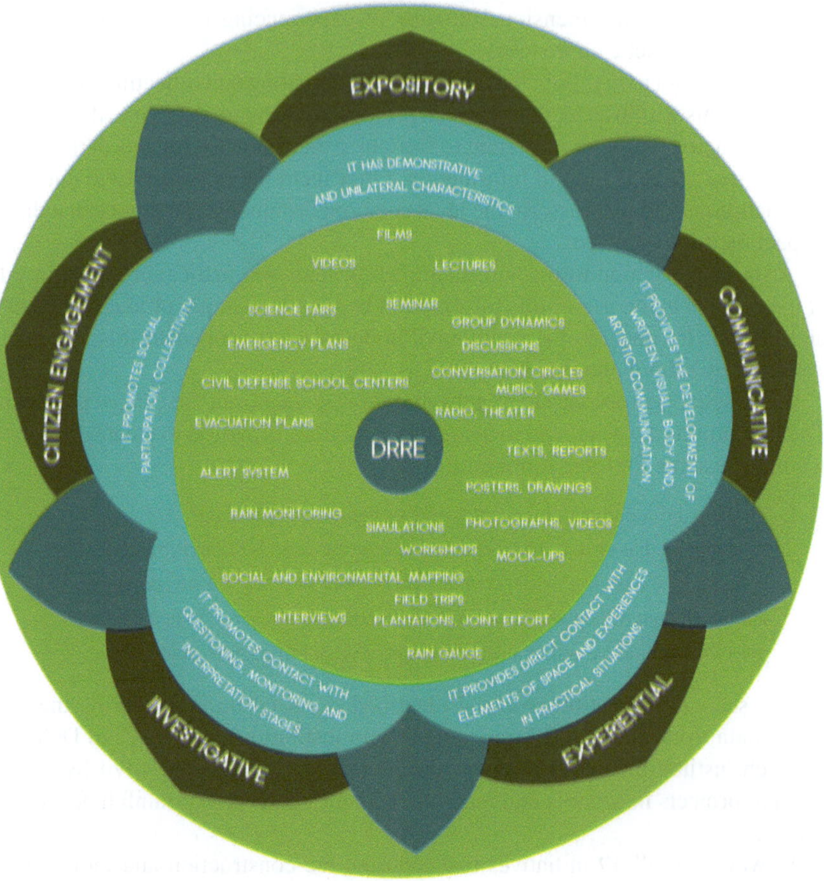

Fig. 8.3 Mandala of didactic approaches in DRRE, description and examples of the teaching modalities. *Source* Matsuo and Silva (2021)

The mandala also plays a significant role as a tool for planning, monitoring, and evaluating DRRE practices. It can be used to analyze the project or actions of a school at a particular moment or over several years, in a long-term analysis. This resource assists in the process of reflection, enabling the identification of progress and setbacks over time.

8.5.3 Collaborative and Contextualized Practices

Disaster is a recent and challenging theme in Brazil, many of these school initiatives have emerged through partnerships with institutions related to risk management, such as Civil Defenses, research institutes, universities, as well as residents affected by disasters. The integration of schools with this vulnerable group in local disasters was identified through practical activities such as field trips, interviews, and mapping of risk areas.

Risk mappings are often well-received by the community, enrich learning processes about DRR and facilitate problem-solving (Petal 2008, Takeuchi et al. 2011). This is because they allow interactions among several local and external social actors, with exchanges of knowledge, experiences, and emotions.

When mappings adopt a participatory approach, it enables the integration of experiences from long-time residents in locating not only areas susceptible to floods, flash floods, and landslides but also potential shelter points, people with disabilities, escape routes, and identification of strengths and weaknesses of the place where they live (Fig. 8.4).

These practices are contextualized with the local reality as they allow the identification and problematization of potential socio-environmental risks and disasters affecting the surroundings of schools. They contribute to a significant, critical, and emancipatory learning process. The adoption of this local perspective is strategic, considering that the greatest impacts of disasters are felt at the neighborhood scale. It is the local populations who have their ways of life in these spaces that suffer the direct consequences of the damages. These populations need to create strategies to face, overcome, and adaptively transform in the face of risk scenarios.

In addition, these practices facilitate the exchange of experience and interdisciplinary knowledge, both from formal and informal spaces. For example, they can involve healthcare workers, community leaders, religious leaders, and women's

Fig. 8.4 Maps produced during the participatory risk mapping activity. *Source* School from São Paulo State

groups living in risk areas. The mapping stages, such as field trips and spatial localization, allow the affected community to have an active voice to express their memories of the events they experienced. This can contribute to the creation of preventive measures that are appropriate to the reality and history of disasters in the region.

The construction of collective memories is strategic in implementing preservation measures and sharing disaster experiences among generations and social groups, strengthening society's perception of disaster risks (Monteil et al. 2020). The participation of these residents in activities such as participatory mapping also brings together multiple views of the same reality, enables collaborative, interdisciplinary, and intergenerational learning, and reinforces the contextuality in these educational practices.

8.6 Potential and Gaps for Strengthening DRRE in Brazil

The monitoring activity using handmade rain gauges is a simple DRRE practice but with significant potential for social transformation, particularly in affected communities and it can save lives. This was exemplified by schools in the municipality of Jaboatão dos Guararapes in the state of Pernambuco, located in the Northeast region of Brazil. These schools actively participated in the Campaign #AprenderParaPrevenir in collaboration with Civil Defense for consecutive years. Students engaged in practical and investigative activities, learning about local risks and vulnerabilities associated with rainfall. During heavy rains in 2022, they were able to identify the risk of landslides by using their rain gauges. They took proactive measures to convince their neighbors to leave their homes, some of which were destroyed hours after the evacuation (Publica 2024). In essence, the students of these schools played a pivotal role in saving lives within their communities through their involvement in DRRE activities at their schools.

This serves as an important case to show the transformative impact that hands-on, experiential learning can have in empowering individuals, especially students, to effectively respond to and mitigate the impacts of disasters in their communities. It also demonstrates the ability to address complex and negative topics in a teaching that is not only contextualized but also critical and empowering. These practices would allow the identification and problematization of risks and disasters affecting their school communities and provide an opportunity to develop practices based on the perspective of critical environmental education in an adaptive and contextualized way.

This chapter presents a brief systematization of the accumulated set of DRRE practices developed in formal education. The results of this first national analysis can serve as base information to expand the integration of this theme into new dialogue spaces, new policies, programs and projects, with new interfaces.

In order to educate in societies of risk and disasters, we point out some reflections for strengthening the field of DRRE articulated with the field of environmental education:

- Denaturalization of disasters: understanding that disasters in Brazil are not natural, but socio-environmental, resulting from a complex network of systemic factors that contribute to the maintenance of unsustainable scenarios of social inequalities.
- Critical formation: going beyond the concepts and behaviors to be adopted in the face of disasters. We need to develop initiatives that contribute to the formation of a critical and resilient society, capable of living with life's uncertainties and reducing social asymmetries. Only by asking ourselves: "Why is my school community affected every year by floods?"—it will be possible to break the unjust cycle of social inequalities and the occurrence of these disasters.
- Formalization of projects and partnerships: establishing ways of formalization of these actions in the education system (curriculum, teacher training, secure spaces, supporting materials) to ensure the long-term continuity and strengthening of initiatives created by teachers in their schools.
- Recognition of schools as transformative social agents: schools have played a crucial role as agents of social transformation, breaking certain barriers and establishing bridges, especially with Civil Defense, in the process of co-constructing knowledge and pioneering school practices for disaster prevention based on their socio-environmental reality. This recognition should not be limited to institutions involved in risk management, but also within the school community itself, which often underestimates its potential role in this context.
- Despite the absence of formalization of disaster risk reduction in the national curriculum,

DRR in Brazil is an emerging field that requires the involvement of a wide network of social actors, such as educators, researchers, public agents, technicians, and politicians, to contribute to new reflections and initiatives based on theoretical frameworks that dialogue with the social approach to risks and disasters, such as decolonial, gender, racism and environmental injustices perspective. The questioning of the economic and political systems that perpetuate social asymmetries and the root causes of disaster risk scenarios becomes especially relevant, as disasters mainly affect vulnerable populations and disaster scenarios are increasingly present in our societies of risks and disasters.

Despite the progress made, there is still a long way to form a new generation of students protected and prepared for these events that intensify with the climate emergency, and also to create a culture of prevention and disaster risk reduction in this tropical and continental country.

Acknowledgements Specially to all schools, teachers and students that participated at the Campaign #AprenderParaPrevenir, Universidade de São Paulo, Cemaden, Conselho Nacional de Desenvolvimento Científico e Tecnológico (CNPq), Fundação Coordenação de Aperfeiçoamento de Pessoal de Nível Superior (Capes) and to grant nº 2023/16333-0, São Paulo Research Foundation (FAPESP).

References

Agência Câmara de Notícias (2023) Defesa Civil aponta 14,5 milhões de afetados e gastos de R$1,4 bilhão com desastres climáticos em 2023. Meio ambiente e energia. https://www.camara.leg.br/noticias/1025455-defesa-civil-aponta-14-5-milhoes-de-afetados-e

Barbosa ML, Haddad IA, Da Silva NAL, Máximo SG, Veiga RM, Hoffmann TB, Rosane de Souza A, Dalagnol R, Streher AS, Souza Pereira FR, Aragão LEOC, Anderson LO (2022) Compound impact of land use and extreme climate on the 2020 fire record of the Brazilian Pantanal. Glob Ecol Biogeogr 31:1960–1975. https://doi.org/10.1111/geb.13563

Barcellos C, Matos V, Lana RM et al (2024) Climate change, thermal anomalies, and the recent progression of dengue in Brazil. Sci Rep 14:5948. https://doi.org/10.1038/s41598-024-56044-y

Beck U (2018) Metamorfose do mundo: novos conceitos para uma nova modernidade. Zahar, Rio de Janeiro

Brasil (2012a) Lei Nº 12.608, de 10 de Abril de 2012. Institui a Política Nacional de Proteção e Defesa Civil—PNPDEC

Brasil (2012b) Resolução Nº 2, de 15 de Junho de 2012. Estabelece as Diretrizes Curriculares Nacionais para a Educação Ambiental

Brasil (1999) Lei nº 9.795, de 27 de abril de 1999. Dispõe sobre a Educação. Ambiental, institui a Política da Educação Ambiental e dá outras providências

Brasil (2018) Base Nacional Comum Curricular. Área de Ciências da Natureza - Ensino Fundamental. Dezembro de 2018

Calim Costa M, Marengo JA, Alves LM et al (2024) Multiscale analysis of drought, heatwaves, and compound events in the Brazilian Pantanal in 2019–2021. Theor Appl Climatol 155:661–677. https://doi.org/10.1007/s00704-023-04655-2

Carniatto L, Fiedler L, Ottaviano S (2018) O papel da educação ambiental e da gestão de risco de desastres no desenvolvimento de cidades sustentáveis e resilientes. Pensamento Educacional, Curitiba, Número Especial, Cadernos de Pesquisa, pp 278–300

Centro Nacional de Monitoramento e Alertas de Desastres Naturais (Cemaden) (2023) Educação em clima de riscos de desastres. Cemaden, São José dos Campos, SP. https://educacao.cemaden.gov.br/midiateca/educacao-em-clima-de-riscos-de-desastres-2/

Centro Nacional de Monitoramento e Alertas de Desastres Naturais (Cemaden) (2024) 81% dos eventos registrados pelo Cemaden em 2023 ocorreram nas regiões Sul e Sudeste do país. 01 fev. https://www.gov.br/cemaden/pt-br/assuntos/noticias-cemaden/81-dos-eventos-registrados-pelo-cemaden-em-2023-ocorreram-nas-regioes-sul-e-sudeste-do-pais

Clarke B, Barnes C, Rodrigues R et al (2024) Climate change, not El Niño, main driver of extreme drought in highly vulnerable Amazon River Basin. Scientific Report. https://doi.org/10.25561/108761

Confederação Nacional de Municípios (CNM) (2023) Desastres forçaram mais de 4,2 milhões a deixarem suas casas nos últimos 10 anos. Boletim CNM, Agosto, pp 8–9. https://cnm.org.br/storage/biblioteca/2023/Boletins/202308_BOL_GMUN_Desastres_Naturais.pdf

Coordenadoria Estadual de Proteção e Defesa Civil (CEPDEC) (2023) Defesa Civil do Estado reúne 400 pessoas em Simulado de Evacuação Emergencial, em São Sebastião. Agosto. https://www.defesacivil.sp.gov.br/

Da-Silva-Rosa T, Mendonça MB, Monteiro TG, Souza RM, Lucena R (2015) A educação ambiental como estratégia para a redução de riscos socioambientais. Ambient & Soc, São Paulo 18(3):211–230

Freire P (1987) Pedagogia do oprimido, 17ª. Paz e Terra, Rio de Janeiro

Freire P (2000) Pedagogia da indignação: cartas pedagógicas e outros escritos. Editora UNESP, São Paulo, p 134

González-Gaudiano E, Maldonado-González AL (2017) Amenazas y riesgos climáticos en poblaciones vulnerables. El papel de la educación en la resiliencia comunitaria. Teoría Educativa 29(1):273–294

Gomes L (2022) Escravidão: da independência do Brasil à Lei Áurea, vol. 3. Globo Livros, Rio de Janeiro, p 592

Jimenez JC, Marengo JA, Alves LM et al (2019) The role of ENSO flavours and TNA on recent droughts over Amazon forests and the Northeast Brazil region. Int J Climatol:1–20. https://doi.org/10.1002/joc.6453

Libonati R, Geirinhas JL, Silva PS et al (2022) Assessing the role of compound drought and heatwave events on unprecedented 2020 wildfires in the Pantanal. Environ Res Lett 17:015005. https://doi.org/10.1088/1748-9326/ac462e

Marchezini V, Muñoz VA, Trajber R (2018) Vulnerabilidade Escolar frente a Desastres no Brasil. Territorium—Internacional de Riscos, v. 25(II): Riscos e Educação. Editores: RISCOS—Associação Portuguesa de Riscos, Prevenção e Segurança; IUC—Imprensa da Universidade de Coimbra, pp 161–177. https://doi.org/10.14195/1647-7723_25-2_13

Maskrey A (ed) (1993) Los desastres no son naturales. Red de Estudios Sociales en Prevención de Desastres en América Latina, Panamá

Mataveli G, de Oliveira G, Silva-Junior CHL, Stark SC, Carvalho N, Anderson LO, Gatti LV, Aragão LEOC (2022) Record-breaking fires in the Brazilian Amazon associated with uncontrolled deforestation. Nat Ecol & Evol 6(12):1792–1793. https://doi.org/10.1038/s41559-022-01945-2

Matsuo PM, Silva RLF (2021) Desastres no Brasil? Práticas e abordagens em educação em redução de riscos e desastres. Dossiê Educação Ambiental e a Escola Básica-contextos e práticas. Educar em Revista 37(e78161):1–23. https://doi.org/10.1590/0104-4060.78161

Matsuo PM, Panzeri CG, Trajber R, Olivato D, Velloso MFA, Marchezini V (2021) Levantando a lebre da redução de riscos de desastres: aprendizagens da Campanha #AprenderParaPrevenir. Rev HumIdades & Inovação 8(44):322–339. www.revista.unitins.br/index.php/humanidadeseinovacao/article/view/3986

Ministério da Ciência, Tecnologia e Inovação (MCTI) (2024) Em 2023, Cemaden registrou maior número de ocorrências de desastres no Brasil. www.gov.br/mcti/pt-br/acompanhe-o-mcti/noticias/2024/01/em-2023-cemaden-registrou-maior-numero-de-ocorrencias-de-desastres-no-brasil

Monteil C, Barclay J, Hicks A (2020) Remembering, forgetting, and absencing disasters in the postdisaster recovery process. Int J Disaster Risk Sci 11:287–299. https://doi.org/10.1007/s13753-020-00277-8

Oliver-Smith A, Alcántara-Ayala I, Burton I, Lavell A (2017) A construção social do risco de desastres: em busca das causas básicas. In: Marchezini V, Wisner B, Saito SM, Londe LR (ed) Reduction of vulnerability to disasters: from knowledge to action. Rima Editora, São Carlos, pp 94–114

Panzeri CG, Matsuo PM, Trajber R, Olivato D, Velloso MFA, Sato AM, Lucena R, Barbosa MS, Pinheiro AG (2020) Campanha #AprenderParaPrevenir: inspirações para reduzir riscos de desastres. In: Magnoni Júnior L et al. (Orgs.) Redução do risco de desastres e a resiliência no meio rural e urbano, 2 ed. Centro Paula Souza, São Paulo, pp 10–26

Petal M (2008) Disaster prevention for schools: guidance for education sector decision-makers. UNISDR—Thematic Platform for Knowledge and Education, Suíça, p 54

Portugal S, Sorrentino M (2018) Diretrizes Curriculares Nacionais para a Educação Ambiental e suas contribuições à Escola Sustentável. In: Baganha DE, Vieira ER, Mortella RD, Rosa MA (Org.) Educação ambiental rumo à escola sustentável. UTP, Curitiba, SEED (2018)

Prefeitura Municipal de São Sebastião (PMSS) (2023a) São Sebastião segue em Estado de Emergência. 19 ago. www.saosebastiao.sp.gov.br/emergencia/noticia.asp?ID=N1982023121245

Prefeitura Municipal de São Sebastião (PMSS) (2023b) São Sebastião permanece em estado de alerta e Prefeitura realiza limpeza nas áreas afetadas. www.saosebastiao.sp.gov.br/noticia.asp?ID=N14620239442

Publica (2024) Eles salvaram vidas porque aprenderam sobre mudanças climáticas na escola. https://apublica.org/2024/02/eles-salvaram-vidas-porque-aprenderamsobre-mudancas-climaticas-na-escola/

Sato AM, Leal PJV, Silva WP et al (2017) Curso de capacitação de professores para redução de desastres. In: Marchezini V, Wisner B, Saito SM, Londe LR (Orgs.) Redução de vulnerabilidade a desastres: do conhecimento à ação. Rima Editora, São Carlos, pp 551–566

Sauvè L (2010) Educación científica y educación ambiental: un cruce fecundo. Enseñanza Las Ciencias 28(1):05–18

Silva RLF, Campina NN (2011) Concepções de educação ambiental na mídia e em práticas escolares: contribuições de uma tipologia. Pesquisa Em Educação Ambiental 6(1):29–46

Soares CCFZ (2023) Estado socioambiental de direito e racismo ambiental: debate sobre as repercussões dos eventos climáticos extremos no Brasil (2010–2023). Rev Sociologias Plurais, 9(2):77–109. https://doi.org/10.5380/sclplr.v9i2.92001

Sulaiman SN, Trajber R, Matsuo PM, Ferreira K (2021) Um caminho estratégico: educação e engajamento social na construção de comunidades e cidades resilientes. In: Ministério do Desenvolvimento Regional, Secretaria Nacional de Proteção e Defesa Civil e Programa das Nações Unidas para o Desenvolvimento. GIRD+10: Caderno Técnico de Gestão Integrada de Riscos e Desastres para Gestão de Risco e Desastre no contexto brasileiro, pp 143–151

Takeuchi Y, Mulyasari F, Shaw R (2011) Roles of family and community in disaster education. In: Shaw R, Shiwaku K, Takeuchi Y (ed) Disaster education: Community, environment and disaster risk management, vol 7. Emerald Group Publishing Limited, Bingley, pp 77–94

Trajber R, Pimentel CT, Luz AEO, Matsuo PM, Olivato D, Saito SM (2017) Educação na prevenção de desastres, uma temática emergente e emergencial. In: Magnoni Júnior L, Stevens D, Lopes ESS, Cavarsan EA, Vale JMF, Magnoni MGM, Figueiredo WS (Org.) Redução do risco de desastres e a resiliência no meio rural e urbano. 1ed, vol 1. Centro Paulo Souza, São Paulo, pp 45–57

Trajber R (2019) Pensar fora da caixa: transição sustentável e resiliente. In: Raymundo MHA, Branco EAB, Biasoli SA, Sorrentino M (Orgs) Avaliação e monitoramento de políticas públicas de educação ambiental no Brasil: transição para sociedades sustentáveis. MH-Ambiente Natural, Piracicaba, pp 55–68

Trajber R (2018) Cunha: educação e participação na prevenção de desastres. In: Sulaiman SN, Jacobi PR (Org.) Melhor prevenir: olhares e saberes para a redução de risco de desastre. IEE-USP, São Paulo, pp 116–121

United Nations Educational, Scientific and Cultural Organization (UNESCO); United Nations Children Fund (UNICEF) (2012) Disaster risk reduction in school curricula: Case studies from thirty countries. UNESCO, Paris/UNICEF Geneva, p 207

United Nations Office For Disaster Risk Reduction - Regional Office For Asia and Pacific (UNISDR AP) (2010) One million safe schools and hospitals campaign: advocacy guide, p 44

United Nations Office For Disaster Risk Reduction (UNISDR) (2006) 2006–7 Disaster reduction begins in schools

United Nations Office For Disaster Risk Reduction (UNISDR) (2015) Global Assessment Report on Disaster Reports on Disaster Risk Reduction (GAR), Risk reduction making development sustainable: the future of disaster risk management. UNISDR, Geneva, Switzerland, p 352

Part II
Community Education

Chapter 9
Community Education and Preparedness for Disaster Risk Reduction

Ayse Nuray Karanci, Gözde Ikizer, and Canay Doğulu

Abstract Disasters, which are rapidly increasing worldwide, have negative consequences due to the social, economic, and environmental impacts they create. Thus, disasters constitute a significant agenda due to their significant disruption of societal functioning and the inability of affected communities to cope with the adverse consequences by using their own resources. For sustainable and risk-informed development, engagement of all actors effectively in disaster risk reduction processes is of vital importance. Community involvement is central to these processes, thereby, devising methods to increase risk perception and developing positive outcome expectancy, self- and community efficacy, and skills for preparedness in communities are considered essential for strengthening risk reduction efforts. In this context, accompanied by investments in community education activities to foster necessary cognitions and behaviors for preparedness, it is also important to specifically target community engagement in disaster preparedness as part of disaster risk management. This chapter will focus on various theoretical models for disaster preparedness and research outcomes related to community education and engagement as well as provide a discussion of methods for the training of community members.

Keywords Community involvement · Preparedness · Education · Disaster risk reduction

A. N. Karanci (✉) · G. Ikizer
Department of Psychology, TOBB University of Economics and Technology, Sogutozu Cad. No: 43 06560, Sogutozu, Çankaya, Ankara, Türkiye
e-mail: akaranci@etu.edu.tr

G. Ikizer
e-mail: gikizer@etu.edu.tr

C. Doğulu
Department of Psychology, TED University, Ziya Gökalp Cad. No: 48 06420, Kolej, Çankaya, Ankara, Türkiye
e-mail: canay.dogulu@tedu.edu.tr

A. Yildiz and R. Shaw (eds.), *Disaster and Climate Risk Education*, Disaster Risk Reduction, https://doi.org/10.1007/978-981-97-5987-3_9

Disasters are occurring more frequently over time, and societies are more seriously affected by their negative consequences. For effective management of hazards and possible disasters at both micro and macro levels, damage reduction and preparedness is critical. In this context, the importance of disaster policies at different geographical scales and societal levels in solving the "disaster" problem has been widely acknowledged on national and international platforms. The "Hyogo Framework for Action 2005–2015" and the "Sendai Framework for Disaster Risk Reduction 2015-2030" are prominent international disaster policies. Among the main pillars of these policies are damage reduction and preparedness.

The "Hyogo Framework for Action 2005–2015" was adopted by 168 states at the United Nations (UN) World Conference on Disaster Reduction in Japan in 2005 (UNISDR 2007). This framework identified five priority areas to significantly reduce disaster losses by building resilience at the national and community levels. Preparedness is one of these priority action areas. According to the Hyogo Framework, strengthening preparedness for effective intervention at all levels is essential. The "Sendai Framework for Disaster Risk Reduction 2015–2030" (UNISDR 2015) is the most up-to-date international policy platform in the field of disasters. The Sendai Framework, evolving from the Hyogo Framework, indicates a transition from disaster management to the understanding of disaster risk management for both natural and human-induced hazards. Adopted by UN member states for examining and evaluating the practices within the Hyogo Framework, the Sendai Framework emphasizes the need for appropriate steps to be taken at all levels and within and between sectors to enhance preparedness for effective intervention. The framework further highlights the necessity of taking these steps for a more effective recovery process (UNISDR 2015).

9.1 Disaster Preparedness and Community Participation

Raising awareness about hazards and potential disasters is an important component of reducing the vulnerability of individuals and communities, as well as developing and reinforcing preparedness behaviors. While technological advancements contribute to the mitigation of disaster consequences, they are not sufficient on their own. In addition to technological developments and technical knowledge, increasing community involvement in disaster risk management processes through awareness and education campaigns and fostering preparedness behaviors are essential. Widespread dissemination of knowledge, skills development, disaster education, and community-based risk communication would contribute to strengthening community participation in disaster risk management activities.

Considering the critical role preparedness has in disaster risk management, as highlighted by the international policy frameworks of Hyogo and Sendai, various disciplines within the social sciences, such as psychology, has gained significance. Empirical studies that reveal the role of psychosocial factors in influencing disaster awareness in communities and the development and reinforcement of preparedness behaviors are crucial for ensuring the effective implementation of educational activities aimed at creating disaster-ready communities. Hence, efforts to increase the impact of disaster education programs should aim to utilize the findings of scientific studies and models in developing the content and objectives of such programs. Nevertheless, a large proportion of studies on social vulnerability, resilience, and adaptation do not explicitly articulate theoretical underpinnings (Kuhlicke et al. 2023), highlighting the need for theoretical grounding in such research. This would entail researchers to formulate clear-cut messages to stakeholders involved in disaster risk reduction and resilience building. In addition to centralising scientific findings in design and implementation of disaster education programs, it would also be beneficial to adopt a community-centered approach to preparedness as informed by experiences of practitioners and community members. From an organization/agency perspective, Johnston et al. (2022) note that increasing community engagement for natural hazard preparedness requires understanding communities, exploring and fostering of community connections, building capacities and competencies for preparation, supporting and sustaining community-led local hazard actions via community programs, and ongoing monitoring-evaluation-learning processes for evidence-based decision-making. Thus, it is essential to invest in building capacity of communities as well as enable scaffolding of relationships between communities and agencies for increasing community members' personalization of risk and their protective actions (Johnston et al. 2022). In this way, it would be possible to reduce the policy-practice gap for community participation in preparedness (Rawsthorne et al. 2023).

9.2 Disentangling Disaster Preparedness

The contribution of preparedness to reducing and eliminating the negative impacts of disasters, thus transforming them positively to enhance resilience against disasters, is vast. The key elements highlighted in preparedness activities within the scope of disaster risk management can be summarized as follows: planning and implementation based on both national and international levels, focusing on both micro and macro scales (local, national, regional, and global), considering sustainability principles, and strengthening and encouraging community participation disaster risk reduction activities. Including these elements can enhance the effectiveness of disaster preparedness efforts (Paton 2019). Notably, it is essential to follow a scientific framework to enhance the effectiveness of disaster education activities. Therefore, understanding factors that promote or hinder preparedness within a scientific framework has the potential to give insights into how preparedness can be facilitated among individuals and communities. In the following sections, theoretical background of disaster preparedness is outlined first. Then, community education for disaster education is explained with respect to risk perception and risk communication, further followed with the elaboration of the topic in the context of building capacities, social empowerment, and community participation.

9.2.1 Theoretical Background of Disaster Preparedness

Preparedness generally refers to protective or adaptive measures, actions, or behaviors aimed at danger, and it is used to express preparations for various disasters, such as earthquakes, floods, and storms. In the development and implementation of public education programs aiming to enhance preparedness for disasters, especially within the fields of social sciences, it is essential to understand and consider the scientific models explaining the pathways to preparedness behaviors. Various theories and models based on scientific studies shed light on the psychosocial processes and concepts underlying preparedness behaviors.

Solberg and colleagues' (2010) review encompasses social psychological variables such as risk perception, norms and social identity, trust, responsibility and control, beliefs related to competence, and fatalism. These variables form the basis for various theories and models that elucidate the psychosocial dynamics of preparedness behaviors, primarily in the context of nature-based hazards such as earthquakes, floods, tsunamis, and fires. Studies conducted within these frameworks seek to understand the role of individual, societal, and institutional factors in preparedness. Research findings indicate that some factors enhance preparedness, while others may decrease or hinder it. Therefore, emphasizing enhancing features in community education programs and simultaneously focusing on hindering elements that could diminish the impact would enhance the effectiveness of these programs.

Until now, various theories and models, mostly from the psychology literature, have been employed for explaining preparedness, community processes, and risk communication (for a review, see Paton 2019). These include Theory of Planned Behavior (Ajzen 1985; Fishbein and Ajzen 1975; McIvor and Paton 2007), Person Relative to Event Model (Duval and Mulilis 1999; Mulilis and Duval 1995, 1997), Protection Motivation Theory (Mulilis and Lippa 1990; Rogers 1983), Disaster Preparedness Model (Paton 2003; Paton et al. 2005), Community Engagement Theory (Paton 2006; 2008), Protective Action Decision Model (Lindell and Perry 2011; Terpstra and Lindell 2012), Community-centered Engagement Model for Preparedness (Johnston et al. 2022), and Community-led Preparedness Model (Rawsthorne et al. 2023). These theories and models demonstrate the deep and complex infrastructure of preparedness behaviors and shed light on the psychosocial dynamics of disaster preparedness. This psychosocial "picture" provides important insights into how preparedness can be increased. Establishing the scientific foundation of disaster preparedness can encourage critical steps to enhance the quality of educational efforts in this field.

Education programs are the most common practice for creating and improving disaster awareness and readiness for various hazards among communities, empowering communities via educational efforts grounded on scientific research is very important. Disaster education does not only teach community members how to deal with disasters but also connects individuals together in their efforts to bounce back after disasters (Feng et al. 2018). According to Rawsthorne et al. (2023), community-led preparedness requires actions across multiple dimensions, namely information, networks, communication, resources, decision-making, self-organising, and inclusion. These dimensions of community action are "interconnected, non-linear, dynamic, adaptable and emergent" (p. 50), thus community preparedness is considered as a complex system (Rawsthorne et al. 2023).

9.2.2 Risk Perception and Risk Communication

Understanding risks and how to mitigate them play are critical for disaster risk management. *The Global Assessment Report on Disaster Risk Reduction* (GAR) is the main report of the United Nations Office for Disaster Risk Reduction (UNDRR), focusing on the global efforts to reduce disaster risks. Published every two years, GAR aims to better understand globally the progress, new trends, and disruptive elements in risk and risk reduction, with contributions from all relevant stakeholders, including science. The fifth edition of GAR was released in 2019, four years after the adoption of the Sendai Framework for Disaster Risk Reduction 2015–2030 (UNDRR 2019). Referred to as GAR19, this report provides an update on the implementation of the Sendai Framework and the Sustainable Development Goals related to disasters, including the outputs, objectives, goals, and priorities. It also addresses progress in addressing the multidimensional nature of risk across different dimensions, scales, and impacts (UNDRR 2019). The report highlights the transition from the *disaster management* approach (that evolved through hazard-based approaches, social dimensions of disasters, and social construction of risk) to the *risk management* approach characterised by systemic risk and system-based outlook (following the adoption of agreements such as the Sendai Framework 2015–2030, Sustainable Development Goals 2030, and the Paris Climate Agreement in the post–2015 period). Along with this transition, the importance of risk-informed sustainable development is emphasized. In this context, the report highlights the need for collaboration among different disciplines, including social sciences, in scientific studies related to the topic (UNDRR 2019).

The importance of understanding risk and risk reduction, which forms the basis of GAR19, aligns with the shift observed in the risk perception research in the field of psychology—a shift that moves from the traditional "risk perception" studies characterized by predominance of individual cognitive elements to the "representation of risk" characterized by inclusion of emotional and sociocultural elements (Joffe and O'Connor 2013). As the transition to *risk experience (i.e., public engagement)* became prominent, research bearing on individuals' subjective risk perceptions started to additionally focus on elements that reflect emotional and sociocultural aspects (e.g., self-efficacy, collective efficacy, outcome expectation, corruption perceptions, responsibility attributions, fatalistic beliefs, and societal trust) and how these factors affect disaster preparedness. In this respect, for enabling community engagement in disaster risk reduction, it is important that these elements relevant to *risk experience* are incorporated into risk communication activities, particularly community education programs aimed at enhancing disaster preparedness. This would further contribute to shared accountability for risk reduction (both risk personalization and protective actions) among community members as well as organizations (Johnston et al. 2022; UNDRR 2019).

The social-cognitive processes individuals experience in perceiving and communicating about risks (e.g., "*Is there a risk, and if so, will I be adversely affected?*") can be considered as the initial step on the way to preparedness. Low levels of risk perception and ineffective or unhealthy risk communication have the potential to impede preparedness. Conversely, high and realistic risk perception, coupled with the successful completion of social-cognitive processes in risk communication, appears to be a significant first step in disaster preparedness. However, studies (e.g., Lindell and Whitney 2000; Paton 2003; Solberg et al. 2010) indicate that increasing risk perception alone does not necessarily elevate preparedness to the desired levels. Even though efforts to increase risk perception may be successful, the actual preparedness behaviors remain limited and narrowly focused. Therefore, with insights from research on the psychological predictors of preparedness, understanding the intersections where individual and societal psychological factors play a role in the path from risk perception to disaster preparedness becomes essential.

9.2.3 Building Capacities, Social Empowerment, and Community Participation

In addition to being aware of the danger and having a risk perception, enhancing psychological capacities such as self-efficacy, collective efficacy, response efficacy, outcome expectations, and problem-focused coping are influential in strengthening preparedness. Self-efficacy involves individuals feeling capable of implementing preparedness behaviors. Collective efficacy, on the other hand, is the confidence individuals have in their ability to take necessary actions as members of a community. Furthermore, it is important for community members to believe that the actions taken in preparation would be sufficient and effective in reducing the negative effects of a disaster. Community education programs should focus on developing these capacities as a primary objective.

Another crucial intersection on the path to preparedness is social empowerment and community participation (Karanci 2009; 2013). Fostering a sense of community in individuals, building trust-based relationships among community members and relevant governing bodies, sustaining this trust, empowering the community, and encouraging their participation in disaster preparedness activities as a community can enhance motivation for preparedness. In other words, facilitating individuals' involvement in disaster risk management processes through community channels (e.g., neighborhoods, civil society organizations) to promote trust, transparency, empowerment, and participation is a significant step in increasing preparedness.

In addition to factors that promote preparedness in disaster education, it seems important to focus on and reduce factors that hinder preparedness. Addressing beliefs that negatively impact preparedness motivation, such as denying hazards and possible damages, externalizing responsibility for taking precautions (attributing responsibility to others), and fatalism, is essential. Ensuring that individuals take responsibility for disaster preparedness and risk reduction (rather than relying entirely on others—government, administration) as well as raising awareness that beliefs reflecting optimism ("*Nothing will happen to me*") and fatalism ("*If it's in my fate, it will happen*") are ineffective in reducing disaster harms would foster preparedness.

Notably, community education programs based on evidence-based controlled interventions conducted in real life settings provide valuable information for enhancing disaster preparedness in a science-informed way. In fact, in a longitudinal cross-cultural study conducted in USA (Seattle) and Türkiye (İzmir), representing two communities that are different culturally but similar in their seismic risk characteristics, Joffe et al. (2019) found that a face-to-face intervention that consisted of workshops was effective in increasing multihazard household preparedness (for earthquakes and home fires). The workshop design and content were informed by preparedness theories and behavior change research. Particularly, it was tailored to target efficacy beliefs related to mitigation measures via a series of interactive, participatory, and discussion-based activities as well as through formal information delivery from trusted experts. This controlled intervention with a one-year follow-up using observation measures showed that there was a significant increase in overall preparedness levels of the intervention group participants in both Seattle and İzmir, compared to their baseline assessments (i.e., before the intervention). Joffe et al. (2019) further found that having higher levels of outcome expectancy (i.e., believing that one's preparedness behavior will be effective in reducing their risk of being negatively affected by hazards) and being homeowner at baseline assessment predicted this change observed after 12 months in both intervention groups. Moreover, in line with past findings revealing fatalism as hindering earthquake preparedness, at baseline assessment, higher levels of fatalism were associated with lower levels of preparedness for İzmir participants. Taken together, Joffe et al. (2019) point out that fostering a sense of agency of individuals and communities regarding preparedness are promising for behavioral change. Thus, designing education programs to facilitate individual and community efficacy beliefs as well as of decrease fatalistic thinking on disasters can contribute to desired behavioral changes for natural hazard preparedness. In this way, the complex path from risk perception to disaster preparedness can be strengthened, thereby, increasing the likelihood of risk perception translate into adoption of preparedness behaviors (also see Karanci 2013; Paton 2019).

9.3 Community Education Programs for Increasing Preparedness

In the context of disaster preparedness education, it is essential to consider the scientific context in the planning and implementation of activities aimed at informing, raising awareness, and improving or enhancing the preparedness levels of communities to disasters. Particularly, educational efforts conducted to increase disaster preparedness based on scientific findings (e.g., Joffe et al. 2016; Joffe et al., 2019; Karanci et al. 2005) are highly informative due to their theoretical underpinnings.

In general, Twigg (2004) suggests detailed planning before any application or project aimed at reducing disaster risk. It is emphasized that any application (or project) should go through the stages of identification, design (considering different target groups), implementation, monitoring, and evaluation in light of the disaster risk plans and established policies. Particularly, the importance of evaluations conducted after the completion of applications, often performed after the applications are completed, is highlighted as they serve to guide future similar applications and provide information for general policies. In addition, community education should be taken as a collaborative, planned, and long-term process rather than a single training event.

Moreover, when developing community education programs, it is necessary to plan the content and process simultaneously. Program content is related to what topics the training will focus on, what messages will be conveyed to the target audience, and how detailed and in what manner these messages will be delivered. On the other hand, concerning the process of training programs, issues such as which individuals or groups are intended to benefit from the training, the methods to be used during training, whether evaluation and monitoring studies for the program will be conducted, and if so, how they will be carried out, should be thoroughly planned.

9.3.1 *Scope of Training*

Social scientists and communication experts have conducted numerous studies on how messages aimed at changing the attitudes and behaviors of community members should be. It has been shown that the source and the content of the message, and the characteristics of the target audience to whom the message is conveyed are important to consider.

It has been previously discussed that the level of persuasiveness of the message increases when it is thought that the purpose of the message is not to manipulate the masses intentionally (Hogg and Vaughan 2010). Additionally, messages conveyed using 'strong' language and arguments can be beneficial in persuading individuals. Fear can also be a powerful tool in changing both attitudes and behaviors. It has long been thought that messages creating fear can be effective in changing individuals' behavior. Therefore, in educational campaigns aimed at increasing preparedness for earthquakes, showing collapsed buildings and people trapped under debris have been thought to trigger fear in the target audience, thereby increasing preparedness behavior. However, some studies in the field of psychology have shown a reverse relationship between the level of fear arousal and behavior change. Low-fear messages in advertisements have been shown to lead participants to perform the desired behavior more frequently than those receiving high-fear messages (Hogg and Vaughan 2010). The reason for this is that the panic and anxiety created by the message can cause individuals to miss important points in the message, and in some cases, individuals may completely avoid the message. The inhibitory effect of excessive anxiety is also emphasized in Paton's (2003) disaster preparedness model. Moderate levels of fear and anxiety have a motivating effect and are known to be fundamental factors that lead individuals to be prepared, as stated in the Citizen Corps (2005); however, excessive anxiety can have an inhibitory effect when it reaches very high levels. Considering these findings, it is crucial to carefully manage the dose of fear content in messages presented in campaigns conducted through the media so that community members do not miss important information and even avoid the message.

Traditionally, community education programs have focused on various methods to convey information to the target audience and thus increase awareness of danger and risk and preparedness. However, it is observed that training programs focus less on elements that may be important for preparedness, such as community participation, empowerment, giving responsibility, trust, and coping. These elements are generally provided outside of training programs through general policies and strategies. However, it is also important for training programs to include skill development and applications that enable the necessary practices. For example, skills such as securing items, identifying safe places, and preparing a disaster emergency plan should be included in education programs for disaster preparedness. Additionally, skills related to first-aid and search and rescue activities that may be needed after a disaster, as well as skills related to health and nutrition in possible temporary shelter situations, should also be included in education programs. In addition to these, practical applications that enable the implementation of necessary practices should also be included in education programs.

Finally, considering the interrelations among all stages of disaster risk management, it is important to cover not only pre-disaster but also training for during and post-disaster phases for the public and other target audiences. Including topics such as what disasters are, the difference between hazard and risk concepts, how disaster damages can be reduced, how to be prepared for disasters, and what to do/not to do during and after a disaster in education programs is highly important.

9.3.2 Target Audience

The identification of target audiences is one of the crucial aspects in planning disaster education programs, especially in determining the characteristics of the target audience to whom the training programs would be applied, and messages about disaster preparedness and risk communication would be conveyed. Understanding how the target audience perceives disaster risk is essential to increase preparedness levels and ensure appropriate risk communication. Understanding the needs of target audiences and gaining general information about these audiences are important steps in implementing practices. Various factors are considered to influence risk perception in individuals. In addition to personal factors, the features of messages conveyed to form risk perception should also be considered.

Primarily, when designing any disaster education program, evaluating the characteristics of the target audience in a way that aligns with the purpose of the training is of great importance. In training programs, considering the target audience's awareness of disaster risk, literacy status, and formal education level, and delivering messages about disaster preparedness at a level suitable for the audience's level are important conditions for the success of training programs. In addition, Bogdan et al. (2021) asserted that designing educational programs with local community members and ensuring that the programs are locally relevant are essential for the success of any program. They also showed that when connections between community members are already established, community disaster planning is more effective.

In addition, it should be remembered that various psychological factors are at play in convincing individuals during disaster preparedness training. Among these factors, attitudes and persuasion stand out. Attitudes are general and relatively stable evaluations that individuals have about others, objects, or ideas (Petty et al. 2003). Attitudes can be very broad and abstract (e.g., preparedness) or very specific and concrete (e.g., materials to be stocked at home). How attitudes can be changed and whether individuals' behaviors are consistent with their attitudes are generally topics that psychologists focus intensely on. While it is assumed that individuals with positive attitudes about disaster preparedness can change their behavior through training, it is known that having a more positive attitude does not always lead to behavior change. Therefore, those who express a positive attitude towards disaster preparedness may not take any preparedness measures later. Various factors increase the likelihood of attitudes being consistent with behaviors (Kenrick et al. 2010). These factors include knowledge, personal interest, accessibility of attitudes, and behavioral intentions. More explicitly, as individuals' knowledge on a subject increase, if the subject is relevant to them, if attitudes about the subject come to their minds quickly when asked, and if individuals express an intention to show a specific behavior when asked about attitudes, the likelihood of attitudes being consistent with behavior increases.

Another psychological factor that motivates individuals to act in a certain direction is cognitive dissonance. Cognitive dissonance is the tension or discomfort that arises when an individual becomes aware of two cognitions that are inconsistent with each other (Festinger 1957; Myers 2015). The theories explaining cognitive dissonance, where individuals try to adapt their thoughts to the situation to relieve the discomfort caused by the contradiction, have explanatory power in explaining why the desired level of success is not achieved in various trainings targeting certain individuals or groups. For example, if community members within the target audience believe that preparing for a predicted earthquake is difficult, time-consuming, and costly, even if they learn in training that taking precautions is important and believe in the accuracy of this information, they may tend to not take any precautions because they think others are not prepared or there is a low probability of such an earthquake happening to them. If a person believes that preparing for a disaster is "easy" while not taking any steps towards preparedness, it is known that this will create cognitive dissonance. In this case, if the person believes that preparing is actually very difficult, it would resolve the contradiction and relieve the tension caused by the dissonance. Therefore, in disaster preparedness training programs, gradually encouraging individuals to prepare for disasters, starting with simple measures (e.g., having a flashlight and radio with batteries at home), can reduce the likelihood of cognitive dissonance occurring.

In addition to psychological factors, the personal characteristics of individuals in the target audience can also facilitate the acceptance of messages conveyed in training about preparedness. Several studies have shown that younger individuals can be more easily convinced compared to older individuals (Hogg and Vaughan 2010). The reason for this is thought to be that attitudes adopted in youth are not easily changed later. However, as attitudes also tend to weaken with age, addressing the negative attitudes of older individuals in training about disaster preparedness may also be one approach. Another personal characteristic is self-esteem. Individuals with low self-esteem may have difficulty understanding discussions, and individuals with high self-esteem may be reluctant to change their attitudes and behaviors. Therefore, it is suggested that individuals with moderate self-esteem are easier to convince (Myers 2015). The thoughts of the target audience who received the training about the conveyed messages and arguments also play a role in the success of the training program. It is possible to make people think about the messages by asking questions, repeating the messages, ensuring that people do not get distracted, making people feel responsible for evaluating and spreading the message, and allowing multiple speakers to present discussions and arguments. This helps trigger thinking, making strong messages more convincing and weak messages less convincing (due to counterarguments) (Myers 2015).

It should be considered that the preparedness levels of the target audience before training may be different. Prochaska and DiClemente (1982; 2002) stated in the transtheoretical model that there are six stages of behavior change: precontemplation, contemplation, preparation, action, maintenance, and termination. According to this model, in the precontemplation stage, individuals are not ready for change. They do not think that their behavior needs to change. In the contemplation or intention stage, individuals become aware that their behavior needs to change, and they evaluate their situation by seeking information. Individuals in the preparation stage start to take small steps toward change and are ready to start action in the action stage. In the maintenance stage, while the changed behavior is maintained, in the termination stage, individuals' efforts for change come to an end, or their behavior or thoughts have permanently changed. In this model, it is observed that change does not progress in a straight line for everyone, and some may revert to the precontemplation stage from the intention stage. That is, the intention to change does not always end with change. Thomas and colleagues (2018) showed that individuals in the precontemplation, intention, and preparation groups who received disaster preparedness training progressed at least one stage three months after the training, and they suggested that training programs should be adapted to the readiness level for change of the target audience. When designing education programs, it should be taken into account that individuals in the intention and preparation stages may find it easier to take action for disaster preparedness compared to those in other stages.

9.3.3 Methods of Education

The International Federation of Red Cross and Red Crescent Societies (IFRC 2011), in its guide on public awareness and education for disaster risk reduction published in 2011, discusses four different methods for increasing community awareness of disaster risk and providing education to community members. These methods are as follows:

- Campaigns
- Participatory learning
- Informal education
- Formal school-based interventions.

Campaigns aim to deliver standardized messages to create a large-scale impact in the community. Successful campaigns often involve sustained, repeated, and consistent messaging on a specific theme, typically delivered under an inclusive slogan. These campaigns can be conducted through events such as memorial days, festivals, fairs, and demonstrations. Planning, executing, and sustaining campaigns can be challenging due to the high media attention, the size of the target audience reached, and the strength of the partners involved. It is recommended not to terminate campaigns prematurely, especially before achieving their intended goals. Another education method is participatory learning. Engaging individuals in the disaster risk management process is considered positive as motivation increases when people believe they contribute to a solution and that their actions reflect their own ideas. Participatory learning can be achieved through vulnerability and capacity assessments, disaster management plans, follow-up, and improvement of plans through simulations or exercises. Informal education, the most flexible method offered in communities, involves a wide range of tools such as written materials, websites, and games. These tools are selected based on the characteristics and needs of the target audience. Peer-to-peer applications make it possible to reach individuals of all ages. Lastly, formal education in schools, with lessons added to the curriculum related to disaster risk management, is deemed significant. The mentioned programs are not mutually exclusive, and the key is to align the chosen method with goals, characteristics of the target audience, and existing strengths and resources (IFRC 2011).

The guide also presents communication forms that can be utilized for the mentioned education methods. When creating educational programs, it is believed that using various methods together and repeating messages for maximum effectiveness is crucial (IFRC 2011). These forms are further categorized by Lundgren and McMakin (2013) into seven headings:

- Visualizing risk: Using graphic elements and minimal text to communicate simple risk messages, including photographs, posters, advertisements, videos, and television.
- Face-to-face communication: Direct interaction with the target audience, including presentations, school activities, other training sessions, tours, and demonstrations.
- Stakeholder involvement: Active participation of the target audience in discussing, analysing, and managing risks, often through focus groups, advisory boards, and workshops.
- Technology-supported communication: Frequently using computer-based technology to disseminate risk-related information or allow members of the target audience to obtain information or inquire about risks.
- Information materials: Designed for the target audience to read, including brochures, booklets, posters, articles, technical reports, newsletters, and various games.
- Social media: Utilizing the Internet through text, graphics, or videos to share ideas and other information related to the target audience.
- Press (mass communication): Using sources such as newspapers, magazines, television, radio, and the Internet to reach wide target audiences as powerful and trustworthy sources.

In recent years, the method of presenting information materials through websites has become increasingly prevalent in education. Individuals often access emergency-related information online, and many government and non-government organizations have websites dedicated to disaster preparedness. However, these websites are often classified as 'difficult' to read, emphasizing the importance of readability and visual appropriateness (Friedman et al. 2008).

Regarding the characteristics of those delivering the education, research suggests that messages delivered by experts, popular, and appealing individuals, or those who speak fluently and quickly tend to be more persuasive (Hogg and Vaughan 2010). It is essential to consider that not all participants in the target audience may focus on the content of the messages, as their attention may be directed towards other environmental features. Therefore, involving trusted individuals from the community (e.g., disaster experts, scientists, artists, religious leaders) in education and conducting programs in locations where the target audience regularly visits (e.g., schools, mosques) would be beneficial practices.

Empirical studies suggest that not all disaster risk reduction activities are effective for all communities. In a Republic of Korean aid-supported project in the Myanmar region, Gouda and Yang (2023) was shown that while disaster awareness meetings and trainings, first-aid trainings, distribution of informative flyers, and personal visits to households were highly related to subjective disaster preparedness levels, drills or community or volunteer activities were not. Hence, it seems important to tailor the education methods to characteristics and needs of specific communities.

9.3.4 Evaluation and Monitoring Methods

Due to the limited number of assessment studies related to global applications, there is insufficient knowledge about the methods to be used in evaluation and monitoring studies after disaster preparedness training programs. In such studies, research methods such as surveying, interviewing, or counting the number of people reached are often used when assessing individual and societal preparedness levels. However, the actual behavior change needs to be monitored and evaluated.

The base organization strategy developed by the Federal Emergency Management Agency (FEMA) in the United States to enhance collaboration among government and community leaders in all sectors is called Citizen Corps. Citizen Corps aims to contribute to the preparedness, planning, mitigation, response, and recovery phases of the entire community. In the United States, regular research is conducted to increase community resilience, and these are periodically published in a compilation text called the Citizen Preparedness Review (CPR). CPR reviews numerous studies conducted in the field and highlights specific methodological issues that need to be carefully considered in preparedness research (Citizen Corps 2005; 2007). These issues are relevant to any evaluation or monitoring study planned to be conducted.

Firstly, when evaluating the results of preparedness studies, it is essential to consider certain questions (Citizen Corps 2005). These are:

- Are questions asked to everyone? (e.g., certain questions are asked only when positive answers are given to other questions)
- What is being reported? (e.g., results that are significantly different from national averages or results that combine answers to two questions can be reported)
- How specific/clear is the question? (e.g., providing examples when asking questions may increase social desirability)
- Are there meaningful differences between samples? (e.g., the characteristics of the sample group are not carefully considered before the study).

Due to the factors mentioned above, the results of preparedness studies can appear quite different from each other. Additionally, the social desirability effect is prominent in disaster preparedness studies (Citizen Corps 2005). Especially in face-to-face research and telephone surveys, this effect is even stronger. Participants may think that if they report not taking precautions for preparedness, it could lead to a negative judgment about their character, and they may reflect themselves as more prepared than they are.

Hence, it is evident that the method and questions to be used in evaluation and monitoring studies need to be carefully selected. Additionally, in studies, it should not be forgotten that being prepared includes elements beyond knowledge about disasters, having a family emergency plan, and having an emergency kit (Citizen

Corps 2007). These elements include individuals preparing small bags to take with them during evacuation, participating in volunteer work in organizations for disaster preparedness, learning first aid skills, planning how to evacuate pets if any, determining someone to contact in areas outside where they live, specifying certain gathering places after disasters, and knowing how to access emergency messages on the radio or television.

To sum, when developing educational programs, it is important to plan for monitoring and evaluating their effectiveness. During the monitoring and evaluation process, reliable tools are needed to assess attitudes, emotions, and behavior changes while keeping the basic goals and outcomes of the educational program in mind. When assessing the effects of education, it is important for a group not exposed to education, a control group, to participate in the evaluation process for comparison with the educated group. For example, if the goal of the education is to increase awareness of earthquake hazards and risks, the evaluation should focus on how the educated and uneducated groups perceive earthquake hazards, how they assess the damage that earthquake hazards can cause, and whether the two groups differ in these aspects. It is also essential to monitor the educated group and evaluate the effects of education and how changes occur over time. Therefore, it is necessary to develop scales that can assess all the goals of education and apply them to both the educated and uneducated at various time intervals after the end of education. In addition to scales, making unbiased observations, especially focusing on behavior changes, may be recommended during evaluation. In-depth interviews and focus group discussions with target audiences can enrich survey results using qualitative methods.

9.4 Conclusion

It is known that being prepared would mitigate the potential effects of disasters, thus reducing both human and material losses in communities. Reducing losses would also decrease the psychological traumatic effects that disasters can cause, leading to decreased distress in community members. Therefore, developing education programs to build disaster-ready communities based on theoretical models and research findings is of great importance. Fostering awareness about disaster risks in all segments of society is necessary to build a resilient/ready community. This involves not only acquiring information about all types of hazards but also understanding the damages these hazards can cause. Increasing risk awareness is crucial as the first step, and effective risk communication principles should be considered in achieving this awareness.

Effective risk communication includes attracting the attention of different audiences, ensuring understanding, and making messages credible. In addition to risk awareness, a change in attitude towards damage reduction and preparedness is necessary. Developing and reinforcing self-efficacy, collective competence, and personal responsibility are important in this regard. It is essential not only to learn the actions that can be taken for damage reduction/preparedness but also to create the skills and resources needed to perform these actions. This is necessary to motivate and sustain the behaviors of preparedness for reducing disaster damages. Additionally, strengthening community empowerment, fostering community participation, and building social capital and trust are important. Finally, it is necessary to target and reduce factors that hinder preparedness, such as fatalism, optimism bias, and denial, in education programs.

It is evident that there are various factors that can influence the preparedness of communities against disasters. Attention should be given not only to effective and sustainable educational programs that address the needs and learning motivations of different target audiences but also to other important aspects. Developing a sense of responsibility in the community, strengthening trust in public institutions, building belief in the effectiveness and feasibility of measures, and creating resources for all measures are some of these aspects.

Creating frameworks tailored to the characteristics of local stakeholders and local communities for the implementation of education to prevent disaster damages and starting disaster education as early as possible are important issues to be addressed in the future (BRI-GRIPS 2007). Moreover, developing educational templates for individuals and groups unwilling to participate in communities and disaster education would be beneficial. Disaster preparedness education should consist of well-thought-out and carefully prepared strategies in a template. Especially, gaps in policies related to disaster risk management can hinder this. Considering that knowledge accumulation occurs gradually, it should be questioned how lasting knowledge acquisition can be when disaster preparedness education is conducted as a one-time or single training session. Additionally, presenting different pieces of information as part of a whole, i.e., relating them to each other, is crucial.

There are rich examples of good practices that have been widely implemented using appropriate methods and technology for different target audiences worldwide within disaster education programs (for examples of good practices, see BRI-GRIPS 2007; European Commission 2017). However, the main weakness in many of these educational programs is the inadequacy of monitoring and evaluation studies. Despite the presence of numerous good practices to increase disaster preparedness, the failure to conduct monitoring and evaluation studies for many of them is a significant deficiency.

The generic approach proposed by Twigg (2004) as an important aspect in initiatives and efforts to reduce disaster risks is not widely adopted is not widely adopted in disaster preparedness practices in many countries probably due to the high risk of some hazards (e.g., earthquakes in Turkey), the prominence of damages by a particular kind of hazard, and the abundance of work by state institutions, universities, or other institutions on specific hazards. In practice, focusing on multiple hazards can lead to conflicts and unnecessary repetitions among responsible institutions in disaster management (Twigg 2004). However, focusing on a single hazard may ignore the risks of other hazards and the risks arising from them. Hence, practical approaches focusing on multiple hazards and addressing general humanitarian and institutional principles of disaster preparedness would be appropriate.

Another important issue in designing disaster preparedness education programs is the funding for these practices. Humanitarian organizations often make efforts to generate financial resources for the immediate meeting of urgent needs. However, this hinders their efforts to generate sufficient initiatives for disaster risk reduction. Twigg (2004) attributes the limited funding for disaster risk reduction to two reasons. Firstly, governments and funding agencies do not pay enough attention to this issue. Secondly, there are distinctions between budgets allocated for emergency intervention efforts and projects planned for development. Disaster risk reduction and preparedness fall between the two and cannot benefit sufficiently from either budget.

Moreover, pilot studies are essential in planning educational programs to minimize wasting resources and time. Conducting trials with relatively small groups specifically tailored to certain target audiences allows for easier monitoring and evaluation of program adherence and effectiveness. Successful programs from these small pilot groups can then be implemented on a larger scale.

Finally, when creating disaster preparedness training and community engagement programs, collaboration between academic institutions, public health and emergency management agencies, and community leaders is crucial. Technical expertise support for conducting training among stakeholders, providing workforce, allocating training venues, etc., can be facilitated through collaboration between academic institutions and other stakeholders, especially universities. Dunlop et al. (2016) have shown in focus groups with representatives of stakeholder institutions that effective collaboration involves various facilitating factors and barriers. Therefore, when planning training programs, it is important for stakeholders with a role in disaster risk management to come together. Effective and sustainable collaboration should be aimed by considering facilitating factors and attempting to overcome barriers.

In conclusion, it seems important to pursue collaborative practices to increase public awareness and participation in disasters through involvement of different sectors and stakeholders. Additionally, efforts should be made to institutionalize these practices to make them sustainable. Furthermore, developing standards for elements of educational programs, such as scope and method (e.g., educational materials), providing widespread access to target audiences, repetition of training and related activities at certain intervals, and encouraging active participation of the trained audiences are important. Overall, we point out that within the scope of disaster education programs, the following aspects are worthy of consideration for their effectiveness:

(1) Developing and fostering disaster awareness in the entire community, taking into account all target audiences in the community,
(2) Increasing the capacity of the community to cope with disasters by providing the necessary education for disaster preparedness and developing the necessary skills,
(3) Creating resources to take action to reduce disaster damages and be prepared,
(4) Considering factors that hinder preparedness when developing educational programs,

(5) Encouraging the organization of civil society to collectively cope more effectively with disasters,
(6) Deciding and implementing methods and processes for monitoring and evaluation in advance for the work carried out, and
(7) Aiming to revise and improve existing programs with the information obtained from the monitoring and evaluation processes should be considered.

References

Ajzen I (1985) From intentions to actions: A theory of planned behavior. In: Kuhl J, Beckman J (eds) Action-control: From Cognition to Behavior, 1st edn. Springer, Heidelberg, pp 11–39

Bogdan EEA, Roszko AM, Beckie MA et al (2021) We're ready! Effectiveness of community disaster preparedness workshops across different community groups in Alberta. Canada. Int. J. Disaster Risk Reduct. 55:102060

Building Research Institute (BRI), National Graduate Institute for Policy Studies (GRIPS) (2007) Disaster education. BRI/GRIPS, Japan

Citizen Corps (2005) Methodological considerations and key findings in preparedness research. Citiz Prep Rev 1:1–17. https://s3-us-gov-west-1.amazonaws.com/dam-production/uploads/20130726-1854-25045-5078/citizen_prep_review_issue_1.pdf

Citizen Corps (2007) Update on citizen preparedness research. Citiz Prep Rev 5:1–16. https://www.fema.gov/media-library-data/20130726-1854-25045-5729/citizen_prep_review_issue_5.pdf

Dunlop A, Logue KM, Vaidyanatha L et al (2016) Facilitators and barriers for effective academic-community collaboration for disaster preparedness and response. J Public Health Manag Pract 22(3):20–28

Duval TS, Mulilis J-P (1999) A Person-Relative-to-Event (PrE) approach to negative threat appeals and earthquake preparedness: A field study. J Appl Soc Psychol 29(3):495–516

European Commission (2017) Are you prepared for a disaster? Help is at hand. In: Success Stories. TACTIC Project Consortium. https://ec.europa.eu/research-and-innovation/en/projects/success-stories/all/are-you-prepared-disaster-help-hand

Feng S, Hossain L, Paton D (2018) Harnessing informal education for community resilience. Int. J. Disaster Risk Reduct. 27(1):43–59

Festinger L (1957) A theory of cognitive dissonance. Stanford University Press, California

Fishbein M, Ajzen I (1975) Belief, attitude, intention, and behavior: An introduction to theory and research. Addison-Wesley, Boston, Reading, MA

Friedman DB, Tanwar M, Richter JVE (2008) Evaluation of online disaster and emergency preparedness resources. Prehosp Disaster Med 23(5):438–446. https://doi.org/10.1017/s1049023x00006178

Gouda M, Yang Y (2023) Investigating the impact of a community disaster awareness training on subjective disaster preparedness: The case of Myanmar's Ayeyarwaddy region. Disasters 47(4):1047–1068. https://doi.org/10.1111/disa.12575

Hogg MA, Vaughan GM (2010) Essentials of social psychology. Pearson Education, Edinburg

https://www.humanitarianlibrary.org/sites/default/files/2014/06/Disaster%20risk%20reduction%20Mitigation%20and%20preparedness%20in%20development%20and%20emergency%20planning.pdf

International Federation of Red Cross and Red Crescent Societies (2011) Public awareness and public education for disaster risk reduction: A guide. Switzerland, Geneva

Joffe H, O'Connor C (2013) Risk society and representations of risks: Earthquakes and beyond. In: Joffe H, Rosetto T, Adams J (eds) Cities at risk: Living with perils in the 21st century. Springer, New Yok, pp 9–23

Joffe H, Perez-Fuentes G, Potts HW, Rossetto T (2016) How to increase earthquake and home fire preparedness: The fix-it intervention. Nat Hazards 84(3):1943–1965. https://doi.org/10.1007/s11069-016-2528-1

Joffe H, Potts HW, Rossetto T, Doğulu C, Gul E, Perez-Fuentes G (2019) The Fix-it face-to-face intervention increases multihazard household preparedness cross-culturally. Nat Hum Behav 3(453):453–461. https://doi.org/10.1038/s41562-019-0563-0

Johnston KA, Taylor M, Ryan B (2022) Engaging communities to prepare for natural hazards: a conceptual model. Nat Hazards 112:2831–3285. https://doi.org/10.1007/s11069-022-05290-2

Karanci AN (2009) Psychosocial aspects of disaster risk management. In: Gülkan P, Başbuğ Erkan B (eds) Perspectives in Disaster Management. METU Press, Ankara, pp 39–66

Karanci AN (2013) Facilitating community participation in disaster risk management: Risk perception and preparedness behaviors in Turkey. In: Joffe H, Rosetto T, Adams J (eds) Cities at risk: Living with perils in the 21st century. Springer, New Yok, pp 93–108

Karanci AN, Akşit B, Dirik G (2005) Impact of a community disaster awareness training program in Turkey: Does it influence hazard related cognitions and preparedness behaviours? Soc Behav Pers 33(3):243–258. https://doi.org/10.2224/sbp.2005.33.3.243

Kenrick DT, Neuberg SL, Cialdini RB (2010) Social psychology: Goals in interaction. Pearson Education, Boston

Kuhlicke C, de Brito M, Bartkowski B, Botzen W, Doğulu C, Han S, Hudson P, Karanci AN, Klassert, CJ, Otto D, Scolobig A, Soares TM, Rufat S (2023) Spinning in circles? A systematic review on the role of theory in social vulnerability, resilience and adaptation research. Glob Environ Change 80. https://doi.org/10.1016/j.gloenvcha.2023.102672

Lindell MK, Perry RW (2011) The protective action decision model: Theoretical modifications and additional evidence. Risk Anal 32(4):616–632. https://doi.org/10.1111/j.1539-6924.2011.01647.x

Lindell MK, Whitney DJ (2000) Correlates of household seismic hazard adjustment adoption. Risk Anal 20(1):13–26. https://doi.org/10.1111/0272-4332.00002

Lundren RE, McMakin AH (eds) (2013) Risk communication: A handbook for communicating environmental, safety, and health risks. Wiley-IEEE, New Jersey

McIvor D, Paton D Preparing for natural hazards: Normative and attitudinal influences. Disaster Prev Manag 16(1):79–88. https://doi.org/10.1108/09653560710729839

Mulilis JP, Duval TS (1995) Negative threat appeals and earthquake preparedness: A Person-Relative-to-Event (PrE) model of coping with threat. J Appl Soc Psychol 25(15):1319–1339

Mulilis JP, Duval TS (1997) The PrE model of coping and tornado preparedness: Moderating effects of responsibility. J Appl Soc Psychol 27(19):1750–1766

Mulilis JP, Lippa L (1990) Behavioral change in earthquake preparedness due to negative threat appraisals: A test of protection motivation theory. J Appl Soc Psychol 20(8):619–638

Myers DG (ed) (2015) Exploring social psychology. McGraw-Hill, New York

Paton D (2003) Disaster preparedness: A social-cognitive perspective. Disaster Prev Manag 12(3):210–216

Paton D (2008) Risk communication and natural hazard mitigation: How trust influences its effectiveness. Int J Glob Environ Issues 8(1/2):2–16

Paton D (2019) Disaster risk reduction: Psychological perspectives on preparedness. Aust J Psychol 71(4):327–341. https://doi.org/10.1111/ajpy.12237

Paton D, Smith L, Johnston D (2005) When good intentions turn bad: Promoting natural hazard preparedness. Aust J Emerg Manag 20(1):25–30

Paton D (2006) Disaster resilience: Building capacity to co-exist with natural hazards and their consequences. In: Paton D, Johnson D (eds) Disaster resilience: An integrated approach. Charles C Thomas, Springfield, pp 3–10

Petty RE, Wheeler SC, Tormala ZL (2003) Persuasion and attitude change. In: Lerner MJ, Weiner IB (eds) Personality and social psychology. Handbook of psychology, vol 5. Wiley, New Jersey, pp 369–382

Prochaska JO, DiClemente CC (2002) Transtheoretical therapy. In: Kaslow F, Magnavita J, Lebow JL (eds) Comprehensive handbook of psychotherapy, integrative/eclectic, vol 4. Wiley, New Jersey

Prochaska JO, DiClemente CC (1982) Transtheoretical therapy: Toward a more integrated model of change. Psychol Psychother 19(3):276–288. https://doi.org/10.1037/h0088437

Rawsthorne M, Howard A, Joseph P, Sampson D, Harris MK (2023) Understanding community-led disaster preparedness. Aust J Emerg Manag 38(2):49–54. https://doi.org/10.47389/38.2.49

Rogers RW (1983) Cognitive and physiological processes in fear appraisals and attitude change: A revised theory of protection motivation. In: Cacioppo JT, Petty RE (eds) Social psychophysiology: A sourcebook. The Guilford, New York, pp 153–156

Solberg C, Rosetto T, Joffe H (2010) The social psychology of seismic hazard adjustment: Re-evaluating the international literature. Nat Hazards Earth Syst Sci 10:1663–1677. https://doi.org/10.5194/nhess-10-1663-2010

Terpstra T, Lindell ML (2012) Citizens' perceptions of flood hazard adjustments: An application of the protective action decision model. Environ Behav 45(8):993–1018. https://doi.org/10.1177/0013916512452427

The United Nations International Strategy for Disaster Risk Reduction (UNISDR) (2007) Hyogo framework for 2005–2015: Building the resilience of nations and communities to disasters. https://www.unisdr.org/files/1037_hyogoframeworkforactionenglish.pdf

The United Nations International Strategy for Disaster Risk Reduction (UNISDR) (2009) UNISDR terminology on disaster risk reduction. http://www.unisdr.org/files/7817_UNISDRTerminologyEnglish.pdf

The United Nations International Strategy for Disaster Risk Reduction (UNISDR) (2015) The sendai framework for disaster risk reduction 2015–2030. http://www.unisdr.org/files/43291_sendaiframeworkfordrren.pdf

Thomas TN, Sobelson RK, Wigington CJ, Davis AL, Harp VH, Leander-Griffith M, Cioffi JP (2018) Applying instructional design strategies and behavior theory to household disaster preparedness training. J Public Health Manag Pract 24(1):e16–e25. https://doi.org/10.1097/PHH.0000000000000511

Twigg J (2004) Disaster risk reduction: Mitigation and preparedness in development and emergency programming. Good practice review by Humanitarian Practice Network (Number 9). Overseas Development Institute, London, UK

United Nations International Strategy for Disaster Risk Reduction (UNDRR) (2019) Global assessment report on disaster risk reduction. Geneva, Switzerland. United Nations Office for Disaster Risk Reduction (UNDRR). https://gar.undrr.org/report-2019.html

Chapter 10
School Community Engagement in Japan for Proactive Risk Reduction

Yafang Wen, Yinan Gao, Tomo Kawane, and Rajib Shaw

Abstract In Japan, school plays a vital role in the community. A city is usually divided into school district, where school is the core of community. Regular joint disaster drills happen in school and school also has the community storage of disaster prevention goods (bousai shoko in Japanese). After the Kobe earthquake of 1995, the city of Kobe started Boukomi (bousai fukushi community, which is disaster prevention and social welfare community) around the schools in Kobe. After that several cities and rural areas also started school community joint programs on disaster risk reduction. Disaster memories fade over time. However, it remains in the people who experienced the disaster. It is very important for the lessons and memories to be transferred over years and generation. Thus, the legacy tellers play important roles in transmitting the experiences to the next generation, and this chapter will highlight some of these experiences from Japan. This chapter will summarize some of the key lessons emerging from school and community engagement in disaster and climate change.

Keywords Community engagement · Disaster risk reduction · Japan · School community linkage

10.1 Introduction

Schools, as an important part of the Japanese community, occupy a pivotal position in the composition of the Japanese community. In the late 1990s, Japan implemented the highly distinctive education committee system in education and the school committee system of shared governance between schools and the community, and the implementation of these two systems has to a great extent promoted community participation in school governance in Japan and enhanced the status of schools in the Japanese community. Particularly after the educational load reduction at the beginning of the twenty-first century, Japan's Ministry of Education, Culture, Sports, Science and

Y. Wen · Y. Gao · T. Kawane · R. Shaw (✉)
Graduate School of Media and Governance, Keio University, Fujisawa, Japan
e-mail: shaw@sfc.keio.ac.jp

A. Yildiz and R. Shaw (eds.), *Disaster and Climate Risk Education*, Disaster Risk Reduction, https://doi.org/10.1007/978-981-97-5987-3_10

Technology (MEXT) put forward the famous community school concept, which calls for schools to play a more diversified and compounded role in the life of the community.

In Japan, school community involvement in active risk reduction is a key activity, which includes Community-Based Disaster Risk Reduction Education (CBDRR), which emphasizes the reduction of risk to vulnerable groups and communities, increasing their capacity to cope with, prevent, or minimize loss of life, property, and the environment. Since the 1990s, from the Yokohama Action Plan to the Sendai Framework for 2015–2030, Japan has undergone dramatic changes in concepts, approaches, and methodologies for reducing the impact of disasters. Schools, teachers, parents, and community members are involved in learning about disaster risk and practicing early warning simulation drills and evacuations in response to anticipated and recurring disasters. These activities are designed to ensure that school sites are selected through risk assessments so that schools are disaster-resistant and multi-hazard resilient. In Japan, the involvement of the school community in active risk reduction practices is seen as a central part of reducing the impact of disasters. This engagement typically includes the following:

(1) Community-based Disaster Risk Reduction Education (CBDRR): this approach aims to reduce risk to vulnerable groups and communities and increase their ability to cope with, prevent, or minimize loss of life, property, and the environment. CBDRR also emphasizes minimizing human suffering and accelerating recovery; (2) Self-help, Mutual Aid, and Public Help: Japan has three types of Disaster Risk Reduction (DRR). After the 2011 Great East Japan Earthquake and Tsunami Disaster, the capacity of municipal governments was severely impacted, and their ability to help people was paralyzed. In this situation, community residents were responsible for supporting each other's survival, for example, by evacuating from tsunami-affected areas, operating evacuation shelters, and ensuring adequate supplies of food and water. As a result, communities become first responders. The importance of self-help and intra-community cooperation was re-emphasized as the limitations of public help were re-recognized in the aftermath of this disaster; (3) Risk assessment and risk information exchange: Accurate risk assessment and interactive risk communication between government, experts and community residents were re-emphasized to better understand local disaster risks and to take appropriate action in emergencies; (4) Disaster-resilient communities at the center of schools: Safe school sites are selected through risk assessments to ensure that schools are disaster-resilient and multi-hazard resilient. Students, teachers, parents, and community members are learning about disaster risks and practicing early warning simulation drills and evacuations in response to anticipated and recurring disasters. These practices raise the community's awareness of disaster risk and enhance the community's ability to cope in the event of a disaster. They also demonstrate Japanese schools' increasingly important role in community disaster preparedness, a Japanese model of mutual influence and support between the community and the school that is characteristic of Japan.

10.2 Literature Review

Uniquely Japanese Community Participation in Community Disaster Avoidance Centered on Community Schools. The community school, also known as the school management committee system, is a uniquely Japanese community system in which the community participates in the management of the school, and the school then feeds back to the community. Japan is a country that has long suffered from disasters, and mega-disasters such as the East Japan Earthquake and Tsunami (EJET) have had devastating effects on schools and communities. Not only have they disrupted educational activities, but they have also weakened community ties through the displacement of affected populations, exacerbating the chronically low birth rate aging problem faced by many rural communities in Japan (Shohei and Rajib 2013). Japan's neighborhood councils, although long established and rooted in the community, have seen their capacity eroded and increasingly face the prospect of disappearing due to urbanization, diversification of lifestyles, and an aging population (Shohei and Rajib 2013). The Ministry of Education, Culture, Sports, Science and Technology (MEXT), taking into account the critical role that schools play in the life of Japanese nationals, believes that schools can be viewed as multifunctional facilities, including, but not limited to, the enhancement of community education, represented by the training of residents in the neighborhoods, community services, defined by emergency relief, and community participation, represented by the exchange of residents, between the school and the community connection (Matsuura and Shaw 2015).

The tradition of centralized control of education and specialization of educators is deeply rooted in Japanese education. In this culture, there is no real need for parents and communities to be involved in education. However, schools today are experiencing increasing difficulties in fulfilling their responsibilities. There is a growing call for a new partnership between schools and communities. (National Institute for Educational Policy Research 2016). Community schools not only focus on students' academics but also work to meet other community needs, such as healthcare and emergency relief. This comprehensive school model helps to strengthen community connections and increase community resilience and coping capacity (Taylor 2020). Regarding specific community school practices, Hachimagawa Elementary School in Koto-ku, Tokyo, has been implementing school-wide disaster adaptive education since 2010. Since then, the school has been noted for implementing several pioneering initiatives (Pang et al. 2003).

10.3 Importance of School Community Linkages

Community-based disaster risk prevention has recently become very popular in Japan and is vital in disaster prevention. Kameda and Okada (2001) pointed out that since it is difficult for governments to take all disaster risk prevention measures, a community-based disaster risk reduction approach is necessary. In addition, the

Japanese government attaches great importance to disaster education due to the lessons learned from the 1995 Great Hanshin-Awaji Earthquake. Since the major earthquake in 1995, Japanese universities, middle schools, and primary schools have begun introducing disaster and environmental courses to cultivate residents who can contribute to disaster management (Shaw 2015). At the same time, since adopting the Hyogo Framework for Action in 2005, the importance of schools for disaster risk reduction, including disaster education, has been widely agreed upon. Whether it is a community-based disaster management law or a method that emphasizes disaster education in schools, the Japanese government is making efforts to control disaster risks. Japan's community participation in disasters and school disaster education are separate but have essential connections. Cooperation between the two is complementary in disaster risk reduction (Shiwaku and Shaw 2016). This section will illustrate the importance of school-community linkage through examples of school-community engagement in Japan for proactive risk reduction.

10.3.1 Case Study1: Unnan City, Shimane Prefecture

The critical role played by disaster education disseminated in schools is widely recognized by society, and disaster prevention knowledge in schools can be conveyed to parents through students, which is a very effective and direct way of dissemination. (Shaw et al. 2004). From a community perspective, although community-based disaster prevention methods can cover more residents, the drawback is the need for more professionalism in disaster prevention knowledge (Shiwaku and Shaw 2016). An example of disaster knowledge promotion in two elementary schools in Unnan City, Shimane Prefecture, will demonstrate this point.

Unnan City is located in the eastern mountainous area of Shimane Prefecture, with a population of 34,488 (as of 2023). The geology of the area is mainly composed of granite, and a river called "Hii River" in the center of the city flows into a lake located in the north of the city. A plain area spreads out at the confluence of the Akagawa, Kunogawa, and Mitoyagawa rivers. Historically, the area has been prone to disasters such as mudslides and landslides (The page of Unnan City government; Ichiro M., 2023). This disaster prevention education uses disaster prevention models to popularize knowledge about debris flow disasters for students at Kitsugi Elementary School and Wabii Elementary School in Unnan City. Before the disaster simulation experiment started, a questionnaire survey was conducted on the awareness of disaster prevention among sixth-grade students at Kitsugi Elementary School and fifth-grade students at Wabii Elementary School. About the question: When asked whether they had done anything related to disaster prevention, 92% of the students at Kitsugi Elementary School (sixth grade) said they had done so, and 56% of the students at Wabii Elementary School (fifth grade) said they had not done so. The disaster prevention course first introduces the weather conditions when disasters such as landslides occur, the overall situation of landslides, and subsequent rescue work. Then, students began to use disaster prevention models to simulate slope damage

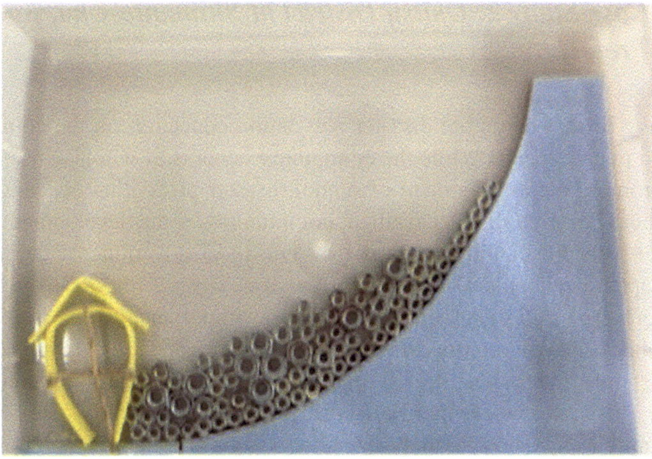

Fig. 10.1 An example of a disaster prevention model (landslide) used in a disaster prevention class at an elementary school in Unnan City, Shimane Prefecture

disasters and debris flow disasters caused by rainfall. After the experiment, a post-test was conducted on the students using a questionnaire. In the question "What do you think of today's disaster prevention course?" most students in the two primary schools said that the experiment was fascinating and meaningful, and they would tell their families and use it in their lives. There are also many sixth-grade students who expressed interest not only in the experiments of their own group but also in the experiments conducted by other groups (Ichiro 2023) (Fig. 10.1).

Judging from the results of the questionnaire, this disaster education course was very successful. It stimulated students' interest in studying debris flow and heavy rain disasters and improved their awareness of disaster prevention. The most valuable thing is that they learned professional disaster knowledge. At the same time, students will convey the knowledge and experiences they have learned in class to their parents. In this process, disaster knowledge is not limited to the school but also spreads to the community, bringing professional disaster awareness and disaster risk prevention knowledge to community residents. In addition, the promotion of disaster knowledge is not limited to schools. Local efforts include the establishment of a voluntary disaster prevention organization in Kamo Town, Unnan City, disaster prevention and mitigation education for local residents, and river flooding. The local government has also realized that it is indispensable to establish disaster prevention links between schools and communities, so the local government is gradually exploring cooperation between schools, communities, and government departments to contribute to disaster management. School-community linkages can relieve the pressure on government disaster reduction activities and resources while improving social disaster response capabilities. The professional knowledge students learn from school is spread to the community through family, friends, etc. The community organizes escape drills to apply professional knowledge in real life.

10.3.2 Case Study 2: Okitsu District of Shimanto-Cho, Kochi Prefecture

The positive impact of school disaster prevention education on the community is reflected in many aspects. While the community benefits, it will also use its advantages to promote the development of school disaster prevention education. Establishing school-centered disaster-resilient communities is the development direction for disaster risk prevention proposed by some local governments in Japan after the East Japan Earthquake (Matsuura 2015). Toni District of Kamaishi and Saijo, Ehime Prefecture, and other places attach great importance to the joint development of school and community disaster risk prevention and are committed to establishing a school-centered disaster prevention system to enhance community disaster resilience. This section takes the Okitsu district of Shimanto-cho, Kochi Prefecture, as an example to focus on the importance of school disaster education to community disaster risk prevention and the advantages of joint school-community disaster prevention (Shiori and Katsuya 2020).

The Okitsu district of Shimanto-cho, Kochi Prefecture, is located in the southwestern part of Kochi Prefecture, facing the Pacific Ocean, and connected to the nearest central area only by a steep mountain road. It has a population of approximately 800 people and an aging rate of roughly 55% (as of April 2019). The main activities are agriculture (facilities and horticulture) and fishing. According to assumptions regarding a Nankai Trough earthquake, in the worst-case scenario, the area would be hit by vigorous shaking with a seismic intensity of 6+ , a tsunami would hit the coast within 15 to 20 min after the earthquake, and the maximum tsunami height would be 25. It is possible that almost all residential areas could be submerged under water (Shiori and Katsuya 2020; Kei 2019). Due to the high probability of disasters in this area and the vast dangers that disasters may cause residents in this area have a strong awareness of disaster prevention. Okitsu Elementary School in the district has been actively promoting disaster risk education and has achieved excellent results nationwide. In addition, Yoshino Elementary School, Motoyama Elementary School, Gomen Noda Elementary School, etc. In recent years, disaster prevention education has been implemented at Kochi Prefecture based on children's production of disaster prevention maps.

First, the long-term development of disaster prevention education in the area has promoted the improvement of disaster prevention hardware facilities in the community. Many primary schools in the area mentioned above draw hazard maps every year. In national competitions, these maps have received high evaluations. One of the essential reasons is that these maps are not limited to paper drawings but actual drawings. They are used to improve community disaster prevention hardware facilities. One of the representative buildings is the nursery, and the elderly welfare center has moved to a high platform, as shown in Fig. 10.2.

The relocation of this facility was attributed to the disaster prevention map produced by children during the disaster prevention study in 2007, which pointed out that the site (original address) had disaster risks. The local government took this

Fig. 10.2 Nursery school and day service complex relocated to higher ground (Okitsu district, Shimanto Town)

matter very seriously and planned to relocate the facility to higher ground. Disaster prevention education at Okitsu Elementary School plays an important role in optimizing disaster prevention hardware facilities in the entire region. At the same time, the local government attaches great importance to the results of disaster education in primary schools and adopts them to recognize and promote disaster education in primary schools.

Secondly, the long-term development of disaster education in schools has also promoted the participation rate of residents in the community's evacuation drills. The researchers analyzed the relationship between tsunami disaster prevention drills implemented once or twice a year in the Okitsu area from 2012 to 2016 (data was lost in 2013) and disaster prevention education in the area. The results showed that among the residents who participated in evacuation drills every year, those who had primary school children at home were more likely to participate in the drill than those who did not have primary school children at home. The study shows that educating children about disasters also affects adults. Parents will actively participate in evacuation drills at the initiative of their children to improve their ability to cope with disasters (Shiori and Katsuya 2020).

Schools and communities share responsibility for disaster response in the face of disasters. School buildings are more tolerant of disasters than ordinary residential buildings in the community, and there are relatively open areas to accommodate the stricken residents. In addition, schoolteachers are more experienced in evacuating students during disasters. In the Great East Japan Earthquake in 2011, students were in class when the earthquake struck. More than 600 students and teachers died in the earthquake after the disaster. However, looking at the overall proportion, the proportion of student casualties was still far lower than that of society (Fujioka 2016).

Since most schools are located in the center of communities and have relatively large spaces and solid buildings, schools have become the main refuge places for Japanese communities after experiencing several major disasters (Shiwaku and Shaw 2016). The school's mature and complete disaster education is conveyed to families through students and even affects disaster management in the community. Strengthening the joint school-community disaster management model is one of the crucial paths to future disaster risk prevention.

10.4 Role of Legacy Teller in Disaster Museums

10.4.1 Disaster Prevention and Social Welfare Interwoven with Community Ethos

The legacy tellers play essential roles in transmitting the experiences to the next generation and help sustain and develop community countermeasures. They could also serve as reminders to the survivors of the region, which would help the local populace prepare themselves for future disasters. Gerster and Maly (2022) consider disaster memorial museums to be "reliable, objective institutions of memory." Their role is to carry forward the lessons learned in disasters. The messages are not only for the people who are affected in the region. Pierre Nora (1989) maintains that lieux de mémoire has occurred at a particular historical moment with the consciousness that the memory has been torn where a sense of historical continuity persists even without real environments of memory—mileux de mémoire. Ono et al. (2021) point out that lieux de mémoire is created by the interaction between memory and history. To make a site function, it is important to establish communication among people who hold memories of something and the people who do not. Joint community programs to transfer lessons learned through past disasters and memories give opportunities to the community members to learn how to cope with disaster situations by themselves and amongst themselves.

10.4.2 Understanding Japanese Society: Schools as the Core of DRR

The history behind the disaster, including data, records, and peoples' stories; scientific explanations of natural hazard mechanisms and predictions; personal actions that can be taken to prepare for disasters; and the option to talk to survivors at the facility (Gerster and Maly 2022) could consist of the backbone of legacy tellers.

In Japan, schools play a vital role in the community. Schools, as vital educational infrastructure in a community in the context of disaster risk reduction (DRR), are expected to function as a pacifying catalyst to ease the difficulties of the concerned community in coping with disaster situations. The involvement of schools in disaster risk reduction (DRR) through legacy tellers could be divided into the following three phases, namely, before, during, and after the disaster.

10.4.2.1 Before the Disaster—Community-Based Safe Measures with Schools Involved

Disaster drills at schools are part of the safety measures of the community in the locality where a school exists. Residents are informed of the passage to a safe evacuation much before any disaster impacts their locality. It fundamentally focuses on saving the lives of people. Disaster prevention goods stored at schools could be lifesavers for the students and teachers of the school as well as the local people. Therefore, it is the responsibility of the local government in Japan to inform the general public about such arrangements for disaster preparedness involving school and community people. Thus, schools are considered as the facilitators for the people in the school as well as the local community to assemble to support each other for safe evacuation.

10.4.2.2 During the disaster—for the First 72 Hours for Self-Survival

It is said that the first 72 h are most important to ensure the survival of a person whosoever faces disaster situations. Social welfare in the community is embodied through such practices as disseminating information and physical measures that could be useful for critical moments left for survival. The 3.11 calamity reemphasized "the importance of self-help and cooperation within each community" as the prime focus of resilience post-disasters. "Owing to the massive scale of the 2011 Great East Japan Earthquake (GEJE) and tsunami disaster, municipal governments were severely affected, and their ability to help people was paralyzed. In this situation, community residents were responsible for supporting each other's survival, such as evacuating from tsunami-affected areas, operating evacuation shelters, and securing adequate supplies of food and water. In essence, communities became first responders" (Sakurai and Ito 2022).

10.4.2.3 After the Disaster—Passing on the Lessons Learnt

Disaster heritage and narratives are not necessarily kept intact or transmitted only because that could not be the primary focus of the survivors to get themselves ready for a transition to come out of the post-disaster trauma. So often, the much-needed transition to a life post-disaster lacks self-reflection on the past mistakes or missed moments that could have been fatal to reduce or erase the chance of the survival rate of the affected people.

10.4.3 Some Examples of Legacy Tellers in Japan

The legacy tellers play important roles in transmitting the experiences to the next generation. Some of the experiences could be shared in the following section.

10.4.3.1 Kobe

An exceptional impact of the Hanshin-Awaji Earthquake Memorial, Disaster Reduction and Human Renovatino Institution (DRI) in Kobe (DRI 2020) was entrusted with two goals as:

(1) Collect, preserve, and educate using artifacts from the Hanshin Awaji earthquake
(2) Research and communicate knowledge about disaster-related studies.

It was started as the Hanshin-Awaji Earthquake Memorial Project in the year 1995 itself, approved by the National Diet, and the institution was inaugurated in April 2022 funded by the Government of Japan and the Hyogo Prefectural Government. The name was chosen with 7612 votes as it signifies the year of volunteerism in Japan (Sakamoto, 2020). The DRI's several established features became a model for subsequent disaster memorial museums. Sakamoto (2020) reports that 67% of visitors are school students born post the Great Hanshin-Awaji Earthquake and out of them, 22% of them wished to know more about the Great Hanshin-Awaji Earthquake, followed by 10% of them their visit was because it was a part of a field trip. Maly and Yamazaki (2021) find that, out of seven pre-3.11 disaster museums, DRI signifies disaster education as a major function of the museum where many programs were there for school children.

10.4.3.2 Tohoku: Official Narratives as Legacy Tellers Post 3.11

The 3.11 Great East Japan Earthquake, tsunami, and nuclear disaster has become a turning point for remembering disasters in Japan. Creating official narratives through 17 government-established 3.11 museums, public disaster museums with the question of separating man-made and artificial disasters not able to offer comprehensive pictures of the compound disaster, understanding and effective disseminations of the lessons learnt (Gerster and Maly 2022). It is important to note that aging society and depopulation issues had already plagued Tohoku region: the region was commercially promoted through a certain degree of musification by depicting it as the authentic past of Japan with folklore, tradition through anti-modernist yet commercial moves to actively promote the countryside.

10.4.3.3 Selecting Legacy Tellers Fit for the Public

"3.11 *Densho* Road" authenticates selected narratives as common heritage. The concentration of the disaster museums in the coastal side of Tohoku region reveals that they have been established as the platforms for public narratives. While it could be a point of reference, it is necessary to understand how '3.11 *Densho* Road' would continue disseminating the learning and preparation for disaster preparedness within Japan and abroad, not to forget the past disasters by networking these facilities. These facilities in the affected areas by the Great East Earthquake and Tsunami have the role to inform the realities of the damages and lessons from the disasters. They come under the Committee of the Association of *Shinsai Densho Nettowaaku* (The Secretariat of the Association of Legacy Teller (*Densho*) of Disasters (*Shinsai*), which is housed in the office building of the City of Sendai (仙台合同庁) (Kimura 2015). (Shinsai Densho Nettowaku Kyogikai—Department of Tohoku Region Development, Ministry of Land, Infrastructure and Transport, Government of Japan n.d.). How the authority determines what could be kept as the public memories by national and local authorities could be explained with Fig. 10.3:

In the context of "Dark Tourism" where people visit disaster heritages, feeling the awe towards nature helps people to think seriously about disaster preparedness: the reconstruction process from the long-term perspectives requires to awaken local populace about the hidden values in the locality impacted by the disaster (Ide 2015). In fact, the disaster accelerated "further mumification.

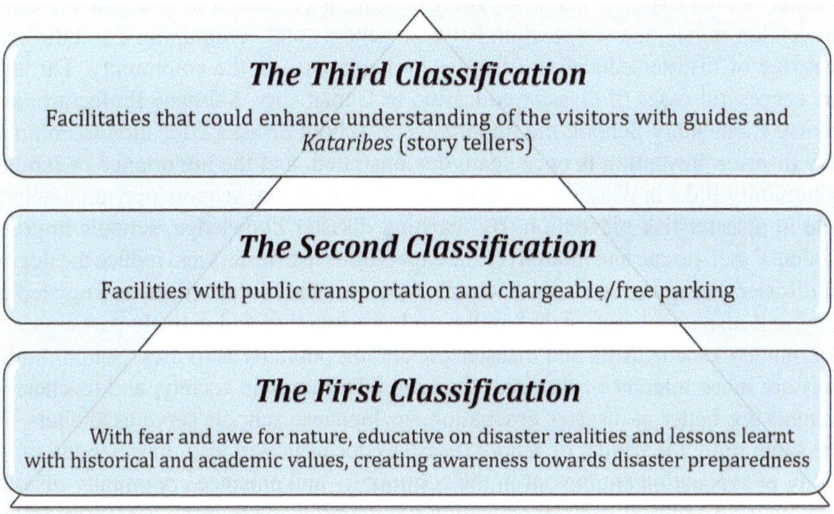

Fig. 10.3 Classifying Legacy Tellers (*Source* Authors). *Note* Reconstructed from the three-step classification of legacy tellers as indicated by Ministry of Land, Transport and Roads (2024)

10.4.3.4 Education for Sustainable Development Considering Climate Change

The global impact of climate change makes it necessary for schools to help learners to have informed decisions could make it possible for the learners to take next steps, namely, to make actions, learn values and skills fit for the transition to more sustainable climate resilient societies (UNESCO 2016). In this context, the learners are not limited to students, embracing the entire community in the locality and envisaging community resilience. While examining climate change require to have (1) local and global perspectives, (2) gender perspectives, (3) past, present and future perspectives, (4) emotional, value-based, and fact-based perspectives, (5) human and non-human perspectives, and (6) perspectives of powerful and marginalized individuals, groups, and countries (UNESCO 2016), it is necessary to make a transition from community resilience to societal resilience. Disaster heritage through legacy tellers awaken the general public to create a resilient ecosystem for the local community where schools embrace people as co-learners as well as co-educators in society.

10.5 Discussion

Disasters in Japan are widespread, and the types of disasters are complex, so the Japanese government attaches great importance to disaster risk prevention. This chapter mainly explains and analyzes school disaster education in Japan's disaster prevention policy, the cooperation between schools and communities, and the critical role of disaster education (disaster heritage, etc.) in the community. Through the successful cases of disaster education in Unnan City, Shimane Prefecture, and Okitsu Elementary School, the importance of school disaster education to community disaster prevention is once again demonstrated, and the importance of school-community links in disaster risk reduction. Among them, schools play an essential role in disaster risk prevention. By teaching disaster knowledge, schools improve students' self-rescue and mutual rescue capabilities in disasters and reduce the degree of disaster damage to students. Secondly, after students receive disaster knowledge, they will also encourage their families to learn together and actively participate in community escape drills and disaster prevention publicity activities. School buildings are more tolerant of disasters than most buildings in society, and teachers in schools are better at disaster evacuation, so Japanese schools serve as shelters. At the same time, the results of school disaster education will also optimize the rationality of evacuation equipment in the community and enhance community disaster resilience. In addition, disaster education is not limited to schools. Disaster museums and other disaster legacy tellers in Japan also play a role in disseminating disaster knowledge. This chapter demonstrates through examples such as the Hanshin-Awaji Earthquake Memorial and the 3.11 Earthquake Museum that disaster heritage plays an important role in improving community residents' disaster awareness and learning and summarizing experiences from past disasters.

10.6 Conclusion: The Challenge and the Way Forward

The importance of disaster education is self-evident. Schools and communities are equally important in reducing disaster risks and improving urban disaster resilience. Japan's school disaster education system has developed to a high level, and society pays enough attention to school disaster education. However, to face more significant disaster risks in the future, the Japanese government still needs to strengthen disaster education for students of all grades and fully Leverage disaster legacies to strengthen collaboration between schools and community engagement. Since schools play the role of refugees, the government must also increase the construction of school refuge facilities and improve teachers' ability to direct emergency evacuation.

In addition, from the perspective of disaster heritage. Since they are expected to form a part of public memory, the focus of the governance comes into the picture. Culture is constructed as heritage, as the sign of the region's 'museumification' (Littlejohn 2020). Thus, ownership of the memory of disaster heritage only partially rests on the shoulders of the public. There is a process of authentication of public memory and common heritage in the tangible ways of expressions to be learned by the contemporary as well as the young generation. Institutionalizing public memories of disaster heritage raises the challenges of objectivity and neutrality of such approved memories and learnings. It is important to note that there is a possible bias and predetermined conceptualization or premeditated standards of evaluation because the narratives of the people are not taken as they are. It is a process of "[r]e-reading the disasters to be woven into disaster heritages (Meskell 2002)." Ide (2015) emphasizes that looking into the fundamental source of grief in Fukushima lacks reflections in multiple ways on the fundamental question of why it was Fukushima that had to accept the nuclear power plant when we think about legacy tellers of the Great East Japan Earthquake. A disaster itself is a tragedy, and it could not be enough to explain the essence of what one may call regional grief. We have always remembered that the schools are given the task of balancing the sentiments of local people with the practical needs of self-help and mutual help under the gaze of public help. What is most important is an open dialogue among the stakeholders to co-create accurate legacy tellers that would also be able to address future crises due to climate change.

References

Central Council for Education (1998) Report on improvement of local educational administration (Japanese) Available at https://www.nier.go.jp/English/educationjapan/pdf/201203EAJ.pdf

Central Council for Education (2003) Interim report on discussions on ways to improve school management (Japanese). https://www.nier.go.jp/English/educationjapan/pdf/2016PCISM.pdf

Fujioka T (2016) Disaster prevention education in the Japanese school curricula in recent years: Current status and future challenges. Disaster Resil Educ Syst: Exp Jpn:39–52

Gerster J, Maly E (2022) Japan's disaster memorial museums and framing 3.11: Othering the Fukushima Daiichi nuclear disaster in cultural memory. Contemp Jpn, 34(2):187–209. https://doi.org/10.1080/18692729.2022.2112479. Accessed on 20 Feb 2024

Taylor H (2020) School-community collaboration: An approach to integrating and democratizing knowledge. Penn GSE Perspect Urban Educ, 17 (Spring), 5(12)

Ichiro M (2023) Case studies and perspectives of school and community initiatives aimed at raising awareness of disaster prevention and mitigation. In: Proceedings of the Japanese society for science education 47. Japan Society for Science Education, pp 351–352. https://doi.org/10.14935/jssep.47.0_351

Ide A (2015) The true value of dark tourism and the process of reconstruction: What lies further ahead of "reconstruction". Fukko [Reconstruction], 7(1): 49–56. Available at https://f-gakkai.net/wp-content/uploads/2020/09/13-1-7IDE-1.pdf

Kameda H, Okada N (2001) Sogo bosai teki shiten (Perspective of integrated disaster management). In: Bosaigaku handbook (Disaster management handbook). Disaster Prevention Research Institute, Kyoto University, Japan, Asakura Publishing Co., Ltd. pp 31–39

Kei S (2019) Regional disaster prevention efforts in the Okitsu district of Shimanto Town—Nankai Trough Earthquake Countermeasures Okitsu District Rural Disaster Countermeasure Development Project. J Rural Plan Soc, 37(4):362–363. https://doi.org/10.2750/arp.37.362

Kimura T (2015) Current situation of disaster heritage of the Great East Japan Earthquake—Focusing on the movement within miyagi prefecture [Higashi Nihon Daishinsai ni okeru Shinsai Iko no Genjyo—Miyagi Ken nai no Doko wo Chushin ni (original in Japanese)]. Fukko (No.13) [Reconstruction], 7(1):11–19. Available at https://f-gakkai.net/wp-content/uploads/2020/09/13-1-2KIMURA.pdf. Accessed on 14 Mar 2024

Littlejohn A (2020) Museums of themselves: disaster, heritage, and disaster heritage in Tohoku. Japan Forum. 33(4):476–496. https://doi.org/10.1080/09555803.2020.1758751

Maly E, Yamazaki M (2021) Disaster museums in Japan: Telling the stories of disasters before and after 3.11. J Disaster Res (Spec Issue Disaster Storytell, Commem 2020 TeLL-Net Forum, Kobe, Jpn) 16(2):146–156. https://doi.org/10.20965/jdr.2021.p0146

Matsuura S (2015) Developing school-centered disaster resilient communities in the aftermath of the East Japan Earthquake and Tsunami. https://doi.org/10.14989/doctor.k19158. Matsuura S, Shaw R (2014) Concepts and approaches of school centered disaster resilient communities. Community Pract Disaster Risk Reduct Jpn, 63–89. https://doi.org/10.1007/978-4-431-54246-9_5

Matsuura S, Shaw R (2015) Exploring the possibilities of school based recovery and community building in Toni district, Kamaishi. Nat Hazards 75(1):613–633

Meskell L (2002) Negative heritage and past mastering in archaeology. Anthropol Q 75(3):557–574. https://doi.org/10.1353/anq.2002.0050

Naoto I, Shunichi F, Masaki S, Nagai T, Ichiro M, Kenya H (2023) Disaster prevention class for elementary school students using a disaster prevention model. Collection of summaries of the 2020 Sabo Society research presentations R3–15. Available at https://www.jsece.or.jp/event/conf/abstract/2023/pdf/80.pdf

Nora P (1989) Between memory and history: Les Lieux de Mémoire. Representations, 26 (Special Issue: Memory and Counter-Memory (Spring 1989), 7–24, published by University of California Press (the Regents of the University of California). https://doi.org/10.2307/2928520. Available at https://www.jstor.org/stable/2928520

Ono Y, Murray M, Sakamoto M, Sato H, Thumwimol P, Thumwimol V, Thongthip R (2021) The role of museums in telling live lessons. J Disaster Res 16(2):135–140. https://doi.org/10.20965/jdr.2021.p0135. (Accessed on 20 Feb 2024)

Pang IW, Isawa E, Kim A, Knipprath H, Mel MA, Palmer T (2003) Family and community participation in education. In: Keeves JP et al. International handbook of educational research in the Asia-Pacific Region. Springer International Handbooks of Education, vol 11. Springer, Dordrecht. https://doi.org/10.1007/978-94-017-3368-7_73

The Great Hanshin Awaji Earthquake Memorial Disaster Reduction and Human Renovation Institution (DRI) (2020). https://www.dri.ne.jp/en/. (Accessed 10 Mar 2024)

Saito Y (2016) Parental and community involvement in school management in Japan (National Institute for Educational Policy Research,) https://www.nier.go.jp/English/educationjapan/pdf/2016PCISM.pdf

Sakamoto M (2020) Disaster memories in museums and disaster recovery—Disaster Reduction Institute and 1995 Hanshin-Awaji Earthquake Recovery. In: 17th world conference on earthquake engineering, 17WCEE Sendai, Japan—September 13th to 18th 2020 Paper N° C000492 Registration Code: A01185. Available at https://wcee.nicee.org/wcee/article/17WCEE/10a-0009.pdf. (Accessed on 20 Feb 2024)

Sakurai A, Ito T (2022) Community-based disaster risk reduction education in Japan. In: Ito T, Tamura M, Kotera A, Ishikawa-Ishiwata Y (eds) Interlocal adaptations to climate change in East and Southeast Asia. SpringerBriefs in Climate Studies. Springer, Cham. https://doi.org/10.1007/978-3-030-81207-2_8. (Accessed on 10 Mar 2024)

Shaw R, Shiwaku Hirohide Kobayashi K, Kobayashi M (2004) Linking experience, education, perception and earthquake preparedness. Disaster Prev Manag: Int J, 13(1):39–49. https://doi.org/10.1108/09653560410521689

Shaw R (2015) Tohoku recovery: reflections on some key lessons. Tohoku Recover: ChallS, Potentials Futur:3–12. https://doi.org/10.1007/978-4-431-55136-2_1

Shinsai Densho Nettowaku Kyogikai—Department of Tohoku Region Development, Ministry of Land, Infrastructure and Transport, Government of Japan (震災伝承ネットワーク協議会—国土交通省東北地方整備局) (n.d.) www.thr.mlit.go.jp. (Accessed on 10 Mar 2024)

Shiori C, Katsuya Y (2020) Essay on implementation and verification of school disaster prevention education from a long-term perspective. Disaster Inf, 18(1):25–33. https://doi.org/10.24709/jasdis.18.1_25

Shiwaku K, Shaw R (2016) Disaster resilience of education systems. Disaster Risk Reduction. https://doi.org/10.1007/978-4-431-55982-5

The page of Unnan City government. Available at https://www.city.unnan.shimane.jp/unnan/

Ministry of Land, Transport and Roads (2024) Shinsai Densho Nettowaku Kyogikai—Kokudo Kotsusho Tohoku Chihou Seibikyoku (震災伝承ネットワーク協議会—国土交通省東北地方整備局) Available at www.thr.mlit.go.jp. (Accessed on 10 Mar 2024)

United Nations Educational, Scientific and Cultural Organization (UNESCO) (2016) Getting climate-ready: A guide for schools on climate action. https://doi.org/10.54675/MOYX1103

Chapter 11
Climate Change and Human Behavior: A Behavioral Economics Perspective

Sanjit Dhami and Mehar Chauhan

Abstract In the context of the pressing need to address climate change, this chapter highlights the transforming power of behavioral economics through education. There is a large literature on an economic approach to climate change using traditional economic policy instruments such as taxes, subsidies, and regulation in a world of risk and unbounded rationality. By contrast, our aim in this chapter is to consider the promise and challenges that arise by using behavioral economics in a 'non-standard setting.' Our non-standard elements are bounded rationality, behavioral risk preferences, behavioral time preferences, true uncertainty, and complex adaptive dynamics. We distinguish between a 'micro approach' to the problem that leverages insights from behavioral economics to influence pro-environmental actions, and a 'macro approach' that considers the system-wide properties of the dynamic system. We discuss the role that can potentially be played by education/persuasion of the lay public, politicians, and policymakers in inducing more environmentally friendly outcomes. We address an interdisciplinary audience, make no attempts at comprehensiveness or generality, and our approach is necessarily selective.

Keywords Climate change · Education/persuasion · behavioral public policy · Loss aversion · Present-focused preferences · Complex adaptive systems · Emergent phenomena · Lock-in effects · Punctuated equilibria

11.1 Introduction

One of the most important challenges for the social, behavioral, and natural sciences is climate change, which requires a deep understanding of the human response to climate change (Stern 2007, 2008, 2022; Nordhaus 2019; IPCC 2014). Stern (2022,

S. Dhami (✉)
Division of Economics, SchoolofBusiness, University of Leicester, London Road, Leicester LE2 1RQ, UK
e-mail: sd106@leicester.ac.uk

M. Chauhan
Department of Psychology and Neuroscience, University of Leicester, Leicester LE1 7RH, UK

A. Yildiz and R. Shaw (eds.), *Disaster and Climate Risk Education*, Disaster Risk Reduction, https://doi.org/10.1007/978-981-97-5987-3_11

p. 1261) lists some of the problems and issues at stake: "…fires associated with heat and drought; severe flooding; hurricanes and typhoons; storm surges; sea level rises; local temperatures at levels dangerous to human life….. Many parts of the world could become uninhabitable…Hundreds of millions, possibly billions, would have to move, likely resulting in severe and extended conflict."

The problem of climate change, and the policy response to it, is too complex to be solved by any single discipline. It requires a multi-pronged approach in which the social, behavioral, and the natural sciences collaborate to make a concerted and complementary effort. We outline a framework to think about issues of climate change from the lens of *behavioral economics.*

Behavioral economics is an interdisciplinary approach to economics that enriches economic models by combining insights from psychology, sociology, evolutionary biology, and neuro- science. As such, it leads to a significant increase in explanatory power relative to the dominant paradigm within economics, neoclassical economics.[1] Yet, the contributions of behavioral economics to issues of climate change have perhaps not been as wide-ranging, and impactful, as one might have expected.

The purpose of this chapter is to briefly outline some of the possibilities offered by a behavioral economics approach to the problem.[2] We also touch upon some issues around the dissemination of the relevant information to the public through formal or informal means. The chapter is not meant to be comprehensive or definitive account of the topic. The paper is written for an interdisciplinary audience that has possibly little training in economic theory or in behavioral economics. As such it also creates familiarity with the language and discourse used in economics we distinguish between two different behavioral perspectives, the micro and macro perspective, to the problems of climate change. Each of them is likely to have different implications for how public education programs may be used in order to inform the public better about the likely risks of climate change.

1. A micro perspective: By a micro perspective, we mean changes in actions or behavior at the level of individuals that can be leveraged by public policy to achieve desirable green outcomes. For instance, how can we induce households to reduce their energy consumption, or recycle better, or contribute towards a better consensus on bearing a higher tax burden arising from increased public expenditures to countering environmental problems. This is where behavioral economics is traditionally likely to feel most comfortable in making valuable policy recommendations.

[1] For a comprehensive introduction to behavioral economics, see Dhami (2016) and Dhami and Sunstein (2022).

[2] We have deliberately omitted the large literature on standard economic instruments such as taxes, subsidies, and regulation, such as the work on carbon taxes and tradeable permits under risk and uncertainty; see Stern and Stiglitz (2022) for some of the references. Our paper is also not related to the large and important literature on integrated asset models (IAMs) that do not use the non-standard macro features used in our paper (Nordhaus 1991, 2019). Throughout, we keep references to a minimum as our aim is to selectively highlight a few key non-standard features and approaches to the problem.

2. A macro perspective: By a macro perspective we mean conditional on people following boundedly rational rules, a study of the underlying dynamic system of the economy. This requires a study of how an economic system behaves over time and how public policy can influence such behavior.

Within the micro perspective, we consider not just the standard behavioral policies that are based on *liberal paternalism* such as *nudging* people to reduce their energy consumption.[3] An important part of nudging people is the dissemination and framing of information in order to persuade and educate people to take pro-environmental actions. We also show how several other behavioral factors can be leveraged for pro-environmental actions. We give particular attention to human behavior in the following two domains within behavioral economics. *Risk preferences* with emphasis on *loss aversion*, and *time preferences* with emphasis on *present-focused preferences*, in influencing pro-environmental actions. We also consider the implications of the *public goods* nature of green investments and a brief consideration of alternative institutions in fostering green investment.

Within our macro perspective, we are interested in situations of *true uncertainty* where it is difficult to even imagine all possible consequences of one's actions and *complex adaptive dynamics* where traditional economic analyses are difficult to apply. Within this framework, we study potential outcomes that might arise from the technology adoption decision by firms to adopt green technologies. This gives rise to complex adaptive dynamics, *chaotic dynamics*, *lock-in effects*, and *punctuated equilibria*.

Section 11.2 gives a brief introduction to traditional economic policy and behavioral public policy for the benefit of interdisciplinary readers with little background in economics. Sections 11.3 and 11.4 constitute the main paper and they discuss, respectively, the implications arising from a micro perspective and a macro perspective. Section 11.5 gives a brief note on the role of behavioral policy in disseminating the best possible science. Section 11.6 concludes.

11.2 A Brief Introduction to the Framework of Behavioral Economic Policy

Neoclassical economics, the currently dominant school within economics, is distinguished by its emphasis on unbounded human rationality. In this view, humans, represented by a mythical entity, the *Econs*, are utility or profit maximizing machines that successfully calculate the mathematically optimal solution in exceedingly complex problems. In so doing, Econs have unbounded calculating ability, information, knowledge, attention, and perfect memory recall that is not hindered by any emotions or moral/ethical values. In particular, there is no merit in this view of educating people

[3] We define or clarify most of the italicized terms in the introduction in the main body of the paper.

on climate change, for if it were to benefit them, they would already have 'rationally' gathered and used this information to their benefit.

Thus, the only way public policy can influence the behavior of Econs is through economic incentives, i.e., by altering the economic benefits and costs of their actions. In economics, this typically takes the form of taxes, subsidies, and regulations on individuals and firms. Public policy that uses such policy instruments is sometimes known as *hard paternalism*. Examples include taxes on polluting firms (carbon taxes), taxes on soft drinks based on their sugar content (sugar taxes); taxes on cigarettes to reduce the incentives to smoke (sin taxes); regulation on the amount of pollutants that can be discharged into the atmosphere by firms backed by a system of tradeable permits (carbon quotas and trading permits).

By contrast, behavioral economics has accumulated an impressive body of evidence to demonstrate that the model of Econs is based on unrealistic assumptions. But even more importantly, it has shown that the predictions of models that rely exclusively on Econs, fail to explain well the relevant empirical evidence (Dhami 2016). By contrast, the evidence from behavioral economics shows that humans are boundedly rational, i.e., they have limited cognitive abilities, calculating power, attention, their memory recall is imperfect, they may make moral/ethical decisions that does not rely adequately on cost benefit decisions, and they often use simple rules of thumb to make important decisions (Dhami and Sunstein 2022).

To be sure, boundedly rational humans try to do the best they can, given their beliefs, but they do not have unlimited rationality, nor the ability to engage effortlessly in complex calculations using mathematics and statistics that are required in neoclassical economics. Furthermore, borrowing from sociology, behavioral economics shows that humans follow social norms as well as internal moral norms that also provide important guides to their actions and behavior. Humans might also be influenced by the way that information is framed to them. By contrast, neoclassical economics assumes that human decisions are framing invariant, i.e., so long as the information content of information does not change, the method of framing has effect on the chosen actions.

In particular, behavioral economics opens up the possibility that real humans can benefit from being persuaded and educated about the pros and cons of environmentally friendly actions. Or that they can be nudged with well framed information, which makes them more likely to take such actions. Its recognition of the importance of social norms can also be leveraged by public policy to undertake group-level programs of persuasion and education about pro-environmental actions.

Behavioral economics does not suggest that incentives are unimportant, although it does highlight many unintended effects of incentives. For instance, Gneezy and Rustichini (2000) considered levying fines on parents who arrived late to pick up their children from day care centres in Haifa. When no fines were levied, there was a relationship of mutual reciprocity between parents and carers in the day centres, which ensured that parents did not turn up excessively late to pick up their children. However, once the fines were introduced, it implemented an agreed upon price for arriving late, which many parents were happy to pay. The opportunity cost of their time was higher than the fine that was introduced. The unexpected result was that the introduction of fines induced the parents to arrive even later to pick up their children. This alerts us to a consideration of the unintended effects of economic incentives that applies potentially to any domain of human behavior, including that which is germane to environmental issues.

There is a strong thread of liberalism that underlies economic policy and neoclassical economists have typically frowned upon paternalistic policies. Providing direct programs of education to the public on issues of climate change might also be considered by some who follow this approach to be paternalistic. In neoclassical economics, since Econs are assumed to take actions that are in their considered best interests, the range of justifiable public policy actions is relatively limited. How can public policy improve individual welfare, if individuals are already taking actions that maximize their welfare? One strong justification for economic policies such as taxes, subsidies, and regulation, which is a form of *hard paternalism*, is economic *externalities*.

Externalities arise when the action taken by one economic entity (e.g., an individual, a firm, or even a country), say, entity A, directly 'benefits' or 'harms' another economic entity, say entity B. If entity B is harmed, then this is a negative externality, and if entity B benefits, then it is a positive externality. Furthermore, if entity B is harmed it receives no direct compensation from entity A for the harm caused. On the hand if entity B is benefitted it make no payment to entity A for the benefits. The classic example of a positive externality is that of two neighbours where one of them keeps bees (entity A) who pollinate the flowers of the other neighbour (entity B), for which B does not pay A any charges. In economics, the classic example of a negative externality is that of an upstream firm (entity A) that discharges harmful pollutants into a river that causes harm to a downstream firm (entity B); and A does not pay any compensation to B for the harm caused to B. Or, consider a firm that discharges harmful toxins into the atmosphere that harm homeowners in the vicinity.

Clearly, in the presence of externalities, private markets fail to produce efficient outcomes. For instance, relative to what is socially optimal, 'too little' action is taken under a positive externality and 'too much' action is taken under a negative externality. For instance, under negative externalities, the upstream firm might discharge too much pollutants into the river relative to what is socially optimal. Hence, there is a gap between what is privately optimal and what is socially optimal. How, then, should we align the actions of economic entities with actions that may be considered socially optimal?

The general solution to the existence of negative externalities is the imposition of taxes on the entity that causes the harmful externality. Such taxes, also known as *Pigouvian taxes*, are levied in such a manner that the marginal (or additional) social costs of a unit of abatement of the externality to entity A (such as discharge of harmful pollutants) equal the marginal (or additional) social benefits of a unit of abatement of the externality to entity B. This does not necessarily imply that it is optimal to completely eliminate the harmful externality. For instance, if the marginal costs of abatement are too high relative to the marginal social benefits of abatement, then partial elimination of the externality might be optimal. Thus, there might be an 'optimal level' of externality reduction that still leaves some remaining externality but balances out the social marginal benefits and marginal costs.

Contrary to a common perception, behavioral economics does not deny the benefits of hard paternalism. However, since humans act with bounded rationality, they might not always be trusted to make the choices that are in their considered best interests that maximize their welfare. Consider some examples.

1. Does a high school student who takes up smoking due to peer pressure or from exposure to passive smoking from a single parent rationally decide to start smoking, taking account of all the benefits and costs of their actions? Should we, in this case, apply the principle of liberalism and respect the choice made by the student, which will likely have long-term consequences? Or should we attempt to understand their choice as arising from peer pressure and poor judgement, and try, at the very minimum, to put warnings on cigarette packs, and even graphic pictures of the dangers of smoking on cigarette packs?
2. Consider a household that accepts the installation of a smart energy meter in their house after a few reminders from their energy company. The household significantly reduces its energy consumption after the installation, and they realize that their previous energy consumption was excessive. Was the extent of energy consumption decision prior to the installation of the smart energy meter in the household a rational choice that needed to be respected by the energy company? If yes, there was no need for the energy company to send reminders to the households for the installation, which eventually persuaded the household to accept the installation, and reduced their energy bills. After all rational households would already have gathered all relevant information to their decision without any reminders.
3. Consider young adults in high school. Should we respect the principle of liberalism and not try to influence their decisions to engage in more environmentally friendly actions, and try to alter their lifetime attitudes? Or should we try to persuade/educate them through a variety of additional methods financed by the public purse? This might include, for instance, offering weekend lectures on climate change, trips to nature reserves, offering additional online resources for them to explore, and offering practical advice on energy consumption and green choices. Indeed, one might wish to encourage group activity, such as school designed social media groups on climate issues that disseminate useful information. The evidence from behavioral economics would support such activity, particularly if it could be demonstrated that the long run benefits exceed the current burden on the public purse from such initiatives. Indeed, this form of intervention might be suggested by a public policy based on 'nudges' that we examine next.

Thaler and Sunstein (2009) suggest a form of *libertarian paternalism* in economic policy, which is sometimes referred to as *soft paternalism*, in comparison to the hard paternalism embodied in taxes, subsidies, and regulation. The basic idea behind libertarian paternalism is that economic policy should minimally distort the behavior of fully rational individuals but "nudge" the behavior of the boundedly rational individuals *in a direction that is in their considered best interests*. For instance, public education campaigns such as graphic labelling or warnings on cigarette packs ought to have little effect on fully rational individuals (they either choose not to smoke, or if they do, they take account of all the relevant costs and benefits anyways); but such warnings might also nudge the boundedly rational type of individuals who have not taken full account of the costs and benefits to either reduce or cease smoking.

The assertion that boundedly rational individuals might be nudged in a "direction that is in their considered best interests" has proved somewhat contentious with the critics (Sugden 2013). The main argument is that how could a third party know what is in the considered best interests of an individual, and any interventions are, therefore, likely to violate liberalism. For instance, how do we know if people like to be nudged, persuaded, or educated about socially beneficial actions? This raises the spectre of a malevolent policymaker who might nudge individuals in a direction that is somehow undesirable either from the point of view of their own best interests or even some social interests. It is, however, quite difficult to come up with actual examples of public policy interventions that are of such a malevolent nature. Nudges have been used to enhance savings rates, take up of pension plans, reduce excessive energy consumption, increase organ donation rates, reduce teenage pregnancy rates, and reduce littering of public spaces; see Dhami and Sunstein (2022) for a discussion. We revisit a related point below.

Nudges can also be used in conjunction with the propensity of humans to follow social norms. For instance, when individuals are informed about the average energy consumption by others in their postcode area, high-energy consumers are likely to reduce their energy consumption. The reason is that the relevant information, in this case, the average energy consumption, alerts consumers to the presence of some sort of social norm of energy consumption that they might also wish to follow. Yet, providing information on average energy consumption is a form of a nudge because it should not influence the behavior of fully rational consumers but possibly alert boundedly rational individuals to the energy consumption norm in their social group.

In general, behavioral public policy can leverage any aspect of human behavior that is identified within behavioral economics to lead to a potential welfare improvement. Designing such policies requires a serious study of the subject matter of behavioral economics and we would argue that this has the potential to lead to welfare improving policy design with respect to issues of climate change.

11.3 A Micro Perspective

In leveraging changes in individual human behavior towards environmentally friendly actions (the micro perspective), the greatest role so far within behavioral economics has been played by "nudges."

In behavioral economics, nudge-type interventions have been shown to be efficacious in successfully reducing energy consumption. These interventions take several possible forms. For instance, include information provision to households that offers social comparisons of their energy usage (Allcott 2011b); observation by the experimenter of energy usage reduces such usage (Schwartz et al. 2013); voluntary installation of smart energy meters has been shown to reduce energy consumption (Rivers 2018); and simply providing information on the energy usage of the household to the household (Allcott 2011a; Jessoe and Rapson 2014). An experimental literature also exists on priming individuals for other-regarding or moral/empathetic preferences and then analyzing their pro-environment behavior; for surveys, see Stern (2000), Heinz and Koessler (2021), Schultz and Zelezny (1999), Dietz et al. (2005).

An 'exclusive' focus on nudge-type interventions in climate change research, despite the undoubted importance of this line of research, can be restrictive. It does not fully leverage insights from the large literature on other aspects of behavioral economics. The problem of climate change is too complex and multifaceted to be solved by simply employing nudges or behavioral interventions alone. It needs the full gamut of public policy instruments, including traditional taxation (e.g., carbon taxes), subsidies (e.g., for the adoption of green technologies), and regulation (e.g., tradeable permits and limits on emissions). Sensible proponents of 'liberal paternalism' will readily concede this.

That said, we are not very sympathetic with the literature that outright dismisses all positive effects of nudges. Nudges are clearly a very important policy instrument; the quantitative effects in several kinds of interventions are often significant and compare favorably with traditional economic instruments, and it is a low-cost alternative. There has been a recent trend critical of nudge-type interventions on the grounds that they have unexpected side effects; for a survey see Dhami (2020, Vol. 7). However, one must note that this is not a criticism that applies just to nudges. It applies to 'any' public policy instrument including all the classical economic policy instruments such as taxes, subsidies, and regulation.

Economics distinguishes between the *partial equilibrium effects* (or the local effects) of an intervention and the *general equilibrium effects* (or the global effects) of an intervention. A policy that is hurried, not well thought out, or not piloted adequately is likely to have some general equilibrium effects that could not have been foreseen by a partial equilibrium analysis. There is nothing new in this argument and it has been well known for decades. It would be more constructive if the critics advocated more caution and piloting of nudge type interventions before they are implemented.

A second objection to nudges is that they are coercive and violate human agency. This sort of criticism has already been dealt by Richard Thaler and Cass Sunstein in various responses (see, e.g., the summary in Dhami and Sunstein 2022). To this, one may add that the alternatives, e.g., the 'contractarian approach,' outlined in Sugden (2018), with prescribes simply giving information to the citizens, are subject to similar criticisms. For instance, if the desire is to reduce smoking by providing information to the citizens, problems similar to those high- lighted by the critics of nudges arise. Who decides when the information is to be given? Who decides how widely the information is to be disseminated? Who decides how frequently the information is to be repeated? Who decides how much information to give? Who decides how the information is to be framed? Who decides on the quality and content of the information to be given? Is there not an element of paternalism involved in these choices? All these questions are related to the nitty gritty of the design of these policies (or *choice architecture*). This brings us back to the point that Thaler-Sunstein often highlight– that whenever we think of public policy, some type of choice architecture, or framing of information and deciding on the menu of choice of alternatives is unavoidable. Even the most extreme liberals must live with that fact.

We believe that the contribution of behavioral economics to issues of climate change is much broader than a simple focus on nudges alone. Several important areas in behavioral economics stand out in issues of human behavior and climate change. Drawing on concurrently progressing research (Dhami et al. 2023), one might identify the following minimum list of important factors.[4] These factors are important in understanding and influencing human behavior through persuasion/education and also important for more effective public policy. In the discussion below, we may refer to investments designed to mitigate the effects of climate change or to abate its harmful effects as 'green investments.'

1. *Present versus future tradeoffs and time preferences*: Green investments are costly in terms of current resources foregone, but the benefits of green investment materialize in the future. This involves a tradeoff between expending either resources on current consumption or on green investments whose uncertain benefits accrue only in the future. Hence, we need evidence on how humans resolve the 'today versus future tradeoffs' in the allocation of resources. Neoclassical economics identifies impatience towards the present, relative to the future, as a key determinant of temporal choices. However, the empirical evidence from behavioral economics shows that, in addition, humans have *present-focused preferences* (e.g., hyperbolic discounting) which leads to a preference for instant gratification that favours current consumption at the cost of making green investments for the future. In this regard, the key question, from the perspective of behavioral economics, is the following: What are the effects of present-focused preferences on the decision to engage in green investment, and how might we mitigate such effects?
2. *Inherently risky outcomes and risk preferences*: Green investments chosen by society to- day, can be mapped into a distribution of risky future climatic outcomes. In particular, the higher the green investments society chooses today, the lower will be the probability of future harmful environmental outcomes, although green investments may not fully eliminate the possibility of such harmful outcomes. Stern (2008) identifies the following channels through which greenhouse gases or GHGs cause climate change. In each of these cases, the higher is the stock of GHGs the higher is the probability of climate change and the effects are experienced with a time delay.

[4] Our list does not exhaust the set of factors that are relevant to the modeling of climate change behavior. For instance, it is useful to make a distinction between flows and stocks of, say greenhouse gases (GHG). Long run climate change is associated with the stock rather than the flow of GHGs (Stern 2008; Dannenberg et al. 2015).

In order to satisfactorily model issues of risk and uncertainty, one needs a rigorous theoretical model. The best available decision theory under risk in behavioral economics is prospect theory (Kahneman and Tversky 1979; Tversky and Kahneman 1992). *Reference dependence* and *loss aversion* are the key concepts underlying prospect theory.[5] By reference dependence is meant that in evaluating their satisfaction from outcomes, people have in mind a reference point, or target outcome, and they evaluate actual outcomes 'relative' to this target outcome. If the actual outcome is greater than the target outcome, they are in the domain of gains and if it is lower, they are in the domain of losses.

Separating the actual outcomes into the domains of gains and losses is not for purely se- mantic reasons. The evidence shows up two differences in behavior in the domain of losses relative to the domain of gains. Both factors are likely to be important in a variety of ways for those who wish to engage in persuasion/education to promote pro-environmental actions.

(i) Loss aversion: Loss aversion implies that losses bite more than equivalent gains. Based on the median estimates from Kahneman and Tversky (1979), and Tversky and Kahneman (1992), the loss aversion parameter is approximately 2.25. Thus, while a gain of £10 may appear as a gain of £10, a loss of £10 is often mentally coded as a loss of £22.5 (10×2.25). In several meta studies carried out since, the loss aversion parameter hovers around a value of 2.
(ii) Risk attitudes: People are generally risk averse (akin to being cautious) in the domain of gains and risk loving (or akin to being reckless) in the domain of losses.[6] For instance, an individual might buy home insurance against potential thefts and burglaries when the house has not yet been burgled (risk averse in gains). Yet the same person might gamble money in a casino and once having lost a bit of money (domain of losses) might gamble even more money to try to recover past losses, but often unsuccessfully (reckless in the domain of losses). The key question here is the following. What are the effects of loss aversion and risk attitudes on the decision to engage in green investment, and if these effects are negative then how might we mitigate them?

3. *Public goods*: A public good is one that is (i) non-rival in consumption (one person's consumption does not reduce another person's consumption) and (ii) non-excludable (it is not possible to exclude others from the consumption of the good). Examples include national defense, law and order, and public broadcasting stations; the classic example is a lighthouse that all ships in the harbor can use equally but none can be excluded. Green investment has the nature of a public good. Thus, any individual's green investment benefits all other individuals in society (e.g., all benefit from a reduction in climate risk). Thus, if

[5] Non-linear probability weighting is another key concept. For an introduction, see Dhami (2019, Vol. 1)

[6] The terms risk averse and risk loving have a precise meaning within economics and readers interested in formal definitions of these terms can consult Dhami (2019, Vol. 1). By contrast, the terms 'cautious' and 'reckless' lack a precise meaning in economics.

individuals are selfish and care only about themselves (self-regarding preferences) they might choose to *free ride*, i.e., not contribute towards the green investment, relying on others to make the green investment. Thus, members of society face a *social dilemma* when they contribute to green investments, in the following sense. It might be that the *privately optimal decision* of each member of society is to free-ride on the costly contributions of others (i.e., make zero contributions) that create public environmental benefits for everyone in the future. However, the *socially optimal decision* of a social planner, whose objective is to maximize the welfare of society, might be to ensure that individuals make positive and high contributions.

In other words, each individual's contributions to the green investment creates a positive externality to society, but in making their contribution decisions, individuals ignore this externality and are likely to contribute too little relative to the social optimal. In other words, we are interested in collective risk social dilemma, a term due to Miliniski et al. (2008). By this is meant that individual and social interests are at odds with each other, and the current actions of individuals lead to delayed and risky outcomes (Nordhaus and Boyer 2000; Stern 2007, 2008; Barrett and Dannenberg 2012). The Nobel Prize winning work of Eleanor Ostrom is also related, even if it is largely conducted in the concrete situation of the problem of the commons and not directly in a climate change context (Ostrom et al. 1992).

4. *Institutions*: The contributions to green investment may be quite different when (i) people are asked to make voluntary contributions, relative to the case where (ii) their contributions are financed through mandatory taxes earmarked for green funds. For instance, there has been a growth in direct referenda on different issues of local and national importance in Europe and in the US; this may be termed as the institution of representative democracy. Members of society might directly vote for higher mandatory green taxes and such taxes provide an institutional mechanism for the provision of green investments. This may override potential problems of the low expectations of contributions of others that might arise in private solutions to the problem. On the other hand, mandatory taxes reduce human agency and autonomy. Ultimately, the relative efficacy of voluntary and mandatory contributions is an empirical question.

Dhami et al. (2023) use the above-mentioned 4 components in the construction of their theoretical model. In their model, green investments made today by a group of individuals (hence, the public goods nature) reduce the risk of a future harmful environmental state from occurring. The authors then measure individual-specific parameters of loss aversion and present focused preferences, which is a challenging and time-consuming task. The data is generated by direct experiments with student subjects in India. The following predictions of their underlying model are verified.

1. Decision makers whose parameter of present-focused preferences is higher (i.e., they have a taste for instant gratification) are more impatient. They divert more resources towards current consumption, reducing green investments and, hence, increasing the probability that a harmful environmental state will arise in the

future. In other words, impatience reduces green investments and the possibility of a future harmful environmental state.
2. An increase in loss aversion has two opposing effects. (i) An increase in green investment today reduces current consumption, creating current loss aversion (a cost). (ii) But at the same time, an increase in green investment today, reduces the probability of a future harmful environmental state. Hence, green investment today, reduces loss to future consumption, thus mitigating future loss aversion (a benefit). However, the beneficial future effect in (ii) is discounted and the harmful environmental state occurs with a probability less than 1.[7] Hence, the future benefits in (ii) appear diminished to people relative to the current costs in (i). For this reason, the current cost effect dominates, and loss aversion is predicted to reduce contributions. In a regression analysis, Dhami et al. (2023) find a statistically significant negative effect of loss aversion on green investments.
3. A postponement of the date at which a future climate change event is likely to occur, by postponing the risk far enough into the future, is predicted to reduce green investment. In the experiments, when the date at which a future harmful effect of climate change recedes into the more distant future, people today invest a smaller amount towards green investment.
4. The effect of institutions, i.e., voluntary contributions to green investment or tax financed contributions towards green investment under representative democracy, is an empirical question. However, Dhami et al. (2023) find that these two institutions made no significant difference in contributions towards green investment. However, the relationship between institutions and green investment is likely to be much richer due to the variety of institutional forms that are relevant to this issue. We need more empirical evidence to provide a more definitive answer.

This discussion does not exhaust the set of behavioral factors that one might leverage to generate more pro-environmental behavior through public policy at the level of individuals. Other behavioral factors whose implications are potentially likely to be worthwhile to study include salience, limited attention, and the leveraging of the typical heuristics used by people in such situations. Indeed, this is an entire research agenda on the subject.

These insights can be leveraged and built into the design of educational/persuasion programs in order to induce people to contribute more towards green investments. Consider, for instance, the finding that loss aversion reduces green contributions for the reasons mentioned above. A driving mechanism for that result is that current contributions create a current cost due to loss aversion relative to future benefits. However, the current cost arises because individuals mentally code a contribution towards green investment from their current income, as a loss. Persuasion/education programs might wish to focus on this aspect and develop, for instance, social norms of contributions from current income towards green investment. The aim of such

[7] The evidence shows that people discount future monnetary values due to impatience towards the current period. For instance, an amount Y received t time periods (e.g., weeks, months, or years) in the future, when discounted back to the current time period is worth only $Y/(1+r)^t$ where r is either the interest rate, or a subjective discount rate, depending on the context.

programs is that such investments do not appear as a cost but as a socially approved mechanism that confers social benefits, even if such benefits are hedonic. This might enhance contributions. Furthermore, we have seen that self-regarding preferences lead to the free rider problem in social dilemma situations. Education can be helpful in developing a regard for others (other-regarding preferences) and an appreciation of the environmental harms caused to others by one's actions. This might mitigate the free rider problem. Finally, education might also stress the immediacy of the environmental threat because the evidence shows that the more the perception of this threat recedes into the future, the lower the green contributions.

11.4 A Macro Perspective

Recall that by a macro perspective, we mean a study of the aggregate properties of a system, conditional on the boundedly rational rules that people might be using. The following three key features provide the basis for a 'non-standard' macro approach that is highlighted in Stern (2022) and Stern and Stiglitz (2022).

1. *Scale economies*: When adopting green technologies, firms experience strong scale economies of adoption and increasing returns. For instance, as the number of people adopting an existing green technology increases, its unit costs of production and assimilation fall.
2. *True uncertainty*: In a situation of *risk*, one is able to assign objective probabilities to all possible consequences of one's actions (for instance, to the future consequences of green investment made today) and these consequences can be imagined, and listed, if required. Under *uncertainty*, even if no objective probabilities are available, one can assign subjective probabilities to all imaginable and listable consequences of one's actions. However, in the real world, a third kind of situation, *true uncertainty* (sometimes referred to as Knightian uncertainty or radical uncertainty) is the pervasive state of affairs (Knight 1921; King and Kay 2020).

Under true uncertainty, it is impossible to even imagine (let alone list) all the possible consequences of one's actions. The typical practice in economics is to choose socially optimal policies by using a social welfare function.[8] Under true uncertainty, we cannot calculate 'optimal choices' with the use of the standard social welfare functions because we simply do not have the data to be used in computing these welfare functions. Furthermore, social welfare functions might even be defined

[8] A social welfare function aggregates the welfare levels of all the members of society to arrive at a measure of aggregate social well-being. Several kinds of social welfare functions are possible. Two commonly used types are as follows. A utilitarian social welfare function takes the sum of the welfare of all members of society as a measure of social welfare. By contrast, a Rawlsian social welfare function takes the welfare of the 'least well off' member of society as a reflection of the measure of the social welfare. Other types of social welfare functions assign different weights to different segments of the population in aggregating social welfare.

under catastrophic climate change risks that involve deaths/displacement of several billion people. In such catastrophic events, the utility function of individuals would tend to minus infinity, precluding any meaningful analysis of risk based on expected utility theory (Weitzman 2009).[9]

In the absence of the standard methods, under true uncertainty, we need a 'guardrail approach.' This is characterized by broad objectives/aspirations/targets, e.g., limiting the increase of global temperatures at 1.5°.[10]

3. Complex adaptive dynamics: Real world economies are complex adaptive systems with the existence of 'tipping points' that lead to large system-wide changes arising from small.

Initial accidents of history. A well-known illustrative example of this is the "butterfly phenomena" whereby a butterfly flapping its wings in one part of the world can cause a natural hazard to occur in another part of the world.

Stern and Stiglitz (2022) lament that none of the three features listed above, individually, or as a group, has been satisfactorily incorporated in the standard economic analysis of the problem of climate change.

Inspired by the framework in Arthur (1989), Dhami (2023) takes account of the three features listed above in the following stylized manner in order to construct a theoretical model. Assume that firms make technology choices between two kinds of technologies. A *green technology* that is pro-environment, and a *brown technology* that is not environmentally friendly. Suppose also that a large number of firms make these sorts of technology choice decisions but make their choices sequentially.

Both technologies offer increasing returns to scale. By this is meant that as a technology is adopted, the costs of adoption to subsequent adoptors is lower (and/or its benefits are higher). For instance, as technologies are adopted, there is 'learning by doing' which reduces technological/organizational costs of production. There can also be an increase in the reputational benefits from adopting the green technology over time and network externality effects that further reduce costs (Katz and Shapiro 1985). Other factors that lead to increasing returns as a technology becomes more widely adopted are common standards that reduce costs, social interaction effects (Young 2009) and strategic complementarities between factors of production. Green

[9] The 'utility function' of an individual captures the welfare or well-being of an individual, as perceived by the individual, from the consumption of goods and services.

[10] It is instructive to quote Stern (2022, p. 1277) on this point: "...the consensus across more than 190 countries embodied in the Paris 2015 UNFCCC agreement did not require full agreement on the utility function to be maximised, the correct damage function, discounting or the probabilities of outcomes. Instead, as it became clear, and broadly accepted, that with temperature rise over 2 °C there was a significant probability of extremely bad outcomes, and that those outcomes could be avoided at moderate costs, there emerged consensus that we should act strongly to try to avoid them. It should be noted that it is this understanding that has led the international community to focus on achieving net zero emissions by around 2050, not the recommendations of economists based on IAMs [integrated assessment models]." Another relevant example is the decision to phase out incandescent bulbs and internal combustion engines even when no viable alternatives were available. Yet, the alternatives were rapidly developed, and were shown to exhibit increasing returns. They were also quickly adopted.

technologies have relatively 'stronger' returns to scale from adoption as compared to the brown technology. Firms that adopt technologies are of two types, G and B. Type G firms are relatively more 'suited' to adopting the green technology and type B firms are relatively more 'suited' to adopting the brown technology. For instance, firms with a higher share of energy costs often benefit more from energy-saving innovations. In other firms, the technological/organizational features within it may reduce the costs of adoption of the green technology. There is good empirical evidence for such differences among firms in the relative benefits and costs of adopting the green and brown technologies (Arvanitis and Ley 2013; Hottenrott et al. 2016; Stucki 2019).

The only uncertainty in the model is that nature picks one of the firms randomly in each time period to make a technology choice. The main interest in Dhami (2023) is in the long-run dynamics of this system. The framework is simple, yet offers powerful insights into the adoption of green technology by firms. Within this framework, Dhami (2023) demonstrates the following results below, which we consider under two main headings.

1. *System dynamics*: As one might expect, the dynamics of the system exhibit features of complex adaptive systems; for surveys, see Arthur (2015), Dhami (2020, Vol. 6), and Hommes (2021).

    (i) On account of the factors listed above (true uncertainty, and complex adaptive dynamics leading to extreme dependence on initial conditions) the long-run outcome is *inherently unpredictable.* Thus, after sufficient time has passed, it is possible that all firms eventually choose the brown technology (system state $S = b$) or that all firms choose the green technology (system state $S = g$). Which state eventually arises is unpredictable, and is, in the language of complexity theory, an *emergent property* of the system. Consider, for instance, starling murmurations that create beautiful and complex patterns in the sky. It appears to an outside observer that there is an intelligent design behind these constantly evolving complex patterns. But each bird is following a simple rule of thumb and following the twists and turns of its immediate neighbour. However, the system-wide property of the emergence of starling murmuration is unpredictable and an emergent property of the complex system.

We may also observe *lock-in effects* into either of the green or the brown technologies after the system arrives at a *tipping point*. Such lock-in phenomena in technology choices are well documented. Indeed, there is no guarantee that the complex system will lock into the more efficient technology. Consider for instance, the choice between QWERTY and the Dvoark keyboards; or the choice between internal combustion engine and an electric engine around the turn of the twentieth century.

(ii) The long-run outcome in terms of which technology gets adopted depends on the history of choices made and the dynamic system is not ergodic. Ergodicity of a dynamic system implies that the eventual long run outcome does not depend on where we start off. So, the initial state of the system is irrelevant. For instance, no matter where we induce a movement in a pendulum, its long-run stable steady state is horizontally downward.

(iii) Small accidents of history, such as which type of a firm made a technology choice at what point, can have large effects, as in models of *chaotic dynamics*. This is often popularized in common discourse as a "butterfly flapping its wings in Mexico and causing an earthquake in Japan." For instance, the solution of logistic equations in the neighborhood of certain parameter values creates very large changes to where the system will end up at, even for extremely minute changes to the initial value of the system. A logistic equation is a basic dynamic representation of a predator–prey model and can be represented by a first order difference equation given by $x_{t+1} = \lambda x_t(1 - x_t)$. Chaotic dynamics then exist for a value of $\lambda = 3.5699$, and at this value, the long run properties of the dynamic system become unpredictable as we change the starting value x_0 even minutely.

(iv) Suppose that firms who get a chance to choose their technology choose their optimal technology with some probability $p \in (0, 1)$, but with the complementary probability $1 - p$, they experiment with alternative technologies. Dhami (2023) applies the framework of Young (1998) to demonstrate that this gives rise to *punctuated equilibria.* By this is meant that the system oscillates between the following two technology states: all firms choose the brown technology $S = b$, and all firms choose the green technology or $S = g$. Under stochastic dynamics, the system never stays in one steady state at all times, which is the case with non-stochastic dynamics; rather, it oscillates between different steady states.

2. *Public policy*: The final outcome under complex adaptive systems is unpredictable. The time path of the dynamic system is unpredictable. Yet prior to any technology adoption decisions by the firm, governments need to formulate and announce the relevant economic policies such as taxes and subsidies for each form of technology, green and brown. If such a policy is not formulated at the beginning, and public policy is rapidly and unpredictably changed over time, it will create uncertainty for the firms, which is undesirable. So let us assume that the government announces all relevant economic policies upfront, on the basis of some guesses or expectations of the long-run properties of the system.

   For the reasons discussed above, the government follows a 'guardrail' approach (Stern 2022; Stern and Stiglitz 2022). For instance, the government might announce that its aim is that in the long-run all firms preferably adopt the green technology.[11]

   There are also political/legal/constitutional/fairness constraints on the choices of public policy such as taxes and subsidies which sets 'feasible bounds'

[11] These objectives are not necessarily based on a cost-benefit approach. Indeed, the exact costs and benefits are often unknown due to true uncertainty. For instance, bans on diesel cars/incandescent light bulbs/ozone- depleting CFCs by a cutoff date in different countries based on some holistic public judgement drawing on the best available science.

on these fiscal instruments. In this world, Dhami (2023) shows the following results.

(i) Implementing optimal Pigouvian taxes[12] on emissions from brown technology might be inconsistent with the objective of ensuring the long-run adoption of green technologies. The reason is that Pigouvian taxes that simultaneously ensure long-run green technology adoption might violate the feasible bounds on fiscal instruments. In that case, the government will need to abandon one of the two objectives (optimal Pigouvian taxes, or the stated objectives on long-term green technology adoption).
(ii) In the presence of increasing returns to scale from adoption, the government can announce a time path of subsides that decreases over time.[13] Once enough firms have adopted the green technology, the subsidies can be completely withdrawn, and yet the dynamic system may lock into the green technology, as discussed above.
(iii) The level of subsidies needed to ensure the adoption of the green technology might be unaffordable for poor countries and lead to the proliferation of global environmental bads. These have the nature of a public good, and all countries suffer harm from these environmental bads. In order to avoid this situation, one may propose the formation of an international fund to subsidize green technology adoption in poor countries.

The unpredictability of outcomes under complex adaptive systems makes it somewhat difficult to identify what sort of education/persuasion is needed when thinking from the macro perspective. In this approach, individuals are using simple rules of thumb that reflect boundedly rational behavior, but the system produces complex, unpredictable outcomes, that are emergent phenomena (recall the example of starling murmurations). There is feedback between the simple rules used by individuals and the aggregate outcomes. Guessing which boundedly rational rules to target through education/persuasion in order to influence the aggregate outcome of the system in a desirable direction is an exceedingly difficult task. However, the possibility of chaotic dynamics alerts us to the possibility that small actions may have large eventual effects (recall the example of 'butterfly flapping its wings in Mexico and causing an earthquake in Japan."). Insofar as public policy is able to identify what these small actions might be that lead to large eventual effects, persuasion/education may have an important role to play in a macro perspective as well. However, it remains exceedingly difficult to offer concrete proposals.

[12] Pigouvian taxes, as described above, are corrective taxes levied on one party that causes harm to another party taking into account the damages caused and the costs of mitigating the harm.

[13] Many examples of such subsidies can be given. In the UK, in 2021, a £166.5 million cash injection was announced for a green industrial revolution. Subsidies towards renewable energy technologies in the US amounted to US$140 billion in 2016. Over the period 2016–2018, the Chinese government spending on ecological/environmental protection was CNY 2.451 trillion. Many of the richer OECD countries have followed similar policies. In some cases, tax rebates on the use of green technology are also used to subsidize its usage.

11.5 Best Available Science

Behavioral interventions need to be informed by the best available science. Hence, the need for an even more interdisciplinary approach than is typically the case. Indeed, it is the use of the best available science that has enabled us to set guardrail targets such as limiting temperature increases to no more than $x\%$ percent by a certain data beyond which we might reach a tipping point of no return. Once such targets are set, then public policy swings into action to try to achieve these targets. Behavioral science might be useful here in guiding the nature/framing of the dissemination of information in the form of education and persuasion of the lay public as well as of politicians and policymakers.

11.6 Conclusion

In this chapter, we distinguished between public policies that can leverage behavioral changes at the level of the individual (micro perspective) and those that affect the dynamic system at large, conditional on the boundedly rational behavior of individuals (macro perspective). We highlighted the key non-standard behavioral insights arising from a micro and a macro perspective. Within a micro setting, this included considerations of risk preferences and time preferences in conjunction with the underlying structure of the problem that is based on public goods, and institutions. Within a macro setting, we considered the system wide dynamics that are influenced by complex adaptive systems that give rise to emergent phenomena, chaotic dynamics, lock-in effects, and punctuated equilibria. Public policy is particularly challenging to apply in this framework due to several reasons. For instance, it is difficult to use social welfare functions under true uncertainty, and complex adaptive dynamics lead the system to meander in all sorts of unpredictable directions, making it difficult to commit ex-ante to a precise policy package.

For each of these two perspectives, micro and macro, we have also offered preliminary insights on what sort of education/persuasion might be required in order to achieve environmentally beneficial outcomes. The nature and the type of education/persuasion are reasonably clear within the micro perspective but somewhat unclear under a macro perspective.

Acknowledgements This paper is an expanded version of the background material provided by Sanjit Dhami as Chair of the panel for a Panel Discussion in the Interdisciplinary Workshop on Behaviour and the Environment on 31 May, 2023 at Oxford University.

References

Allcott H (2011a) Rethinking real-time electricity pricing. Resour Energy Econ 33(4):820–842
Allcott H (2011b) Social norms and energy conservation. J Public Econ 95(9):1082–1095
Arthur WB (1989) Competing technologies, increasing returns, and lock-in by historical events. Econ J 99(1):116–131
Arthur WB (2015) Complexity and the economy. Oxford University Press, Oxford
Arvanitis S, Ley M (2013) Factors determining the adoption of energy-saving technolo- gies in Swiss firms: an analysis based on micro data. Envir Resour Econ 54(3):389–417
Barrett S, Dannenberg A (2012) Climate negotiations under scientific uncertainty. PNAS 109:17372e17376
Dannenberg A, L¨oschel A, Paolacci G, Reif C, Tavoni A (2015) On the provision of public goods with probabilistic and ambiguous thresholds. Envir Resour Econ 61(3):365–383
Dhami S (2016) The foundations of behavioral economic analysis. Oxford University Press, Oxford
Dhami S (2019) The foundations of behavioral economic analysis: Behavioral decision theory, vol 1. Oxford University Press, Oxford
Dhami S (2020a) The foundations of behavioral economic analysis: Models of learning, vol 6. Oxford University Press, Oxford
Dhami S (2020b) The foundations of behavioral economic analysis: Further topics in behavioral economics, vol 7. Oxford University Press, Oxford
Dhami S, Sunstein CR (2022) Bounded rationality: Heuristics, judgement, and public policy. MIT Press, Massachusetts
Dhami S (2023) Green technology adoption, complexity, and the role of public policy: A simple theoretical model. CESifo Working Paper No. 10364, Available at SSRN https://ssrn.com/abstract=4416793 or https://doi.org/10.2139/ssrn.4416793
Dhami S, Hajimoladarvish N, Mamidi P (2023) Climate change risk, and human behavior: Theory and evidence. CESifo Working Paper No. 10678. Available at SSRN https://ssrn.com/abstract=4592965 or https://doi.org/10.2139/ssrn.4592965
Dietz T, Fitzgerald A, Shwom R (2005) Environmental values. Ann Rev of Envir Resour 30(1):335–372
Gneezy U, Rustichini A (2000) A fine is a price. J Legal Stud. 29(1):117
Heinz N, Koessler A-K (2021) Other-regarding preferences and pro-environmental behaviour: An interdisciplinary review of experimental studies. Ecol Econ 184:106987
Hommes C (2021) Behavioral and experimental macroeconomics and policy analysis: A complex systems approach. J Econ Lit 59(1):149–219
Hottenrott H, Rexh¨auser S, Veugelers R (2016) Organisational change and the productivity effects of green technology adoption. Resour Energy Econ 43: 172–194
IPCC (2014) Climate change 2014: Mitigation of climate change. In: Contribution of working group iii to the fifth assessment report of the intergovernmental panel on climate change. Cambridge University Press
Jessoe K, Rapson D (2014) Knowledge is (less) power: Experimental evidence from residential energy use. Am Econ Rev 104(4):1417–1438
Kahneman D, Tversky A (1979) Prospect theory: an analysis of decision under risk. Econometrica 47(2):263–291
Katz ML, Shapiro C (1985) Network externalities, competition and compatibility. Am Econ Rev 75:424–440
King M, Kay J (2020) Radical uncertainty: Decision-making for an unknowable future. W.W. Norton, New York
Knight FH (1921) Risk, uncertainty and profits. Houghton Mifflin Company, New York
Milinski M, Sommerfeld RD, Krambeck H-J, Reed FA, Marotzke J (2008) The collective-risk social dilemma and the prevention of simulated dangerous climate change. P Natl Acad Sci USA 105(7):2291–2294

Nordhaus W (1991) To slow or not to slow: The economics of the greenhouse effect. Econ J 101(407):920–937

Nordhaus W (2017) Social cost of carbon in DICE model. P Natl Acad Sci. 114(7):1518–1523

Nordhaus W (2019) Climate change: The ultimate challenge for economics. Am Econ Rev 109(6):1991–2014

Nordhaus WD, Boyer JG (2000) Warming the World: Economic models of global warming. MIT Press, Cambridge, MA

Ostrom E, Walker J, Gardner R (1992) Covenants with and without a sword: self-governance is possible. Amer Pol Sci Rev. 86(2):404–417

Rivers N (2018) Leveraging the smart grid: The effect of real-time information on consumer decisions. OECD Environment Working Paper No. 127. Available at www.oecd.org/environment/workingpapers.htm.

Schultz W, Zelezny L (1999) Values as predictors of environmental attitudes: evidence for consistency across 14 countries. J Envir Psy. 19(3):255–265

Schwartz D, Fischhoff B, Krishnamurti T, Sowell F (2013) The Hawthorne effect and energy awareness. P Natl Acad Sci USA:110

Stern PC (2000) Toward a coherent theory of environmentally significant behavior. J Soc Issues 56(3):407–424. https://doi.org/10.1111/0022-4537.00175

Stern N (2007) The economics of climate change: The stern review. Cambridge University Press, Cambridge, UK

Stern N (2008) The economics of climate change. Am Econ Rev 98(2):1–37

Stern N (2022) A time for action on climate change and a time for a change in economics. Econ J 132(644):1259–1289

Stern N, Stiglitz JE (2022) The economics of immense risk, urgent action and radical change: Towards new approaches to the economics of climate change. J Econ Method. 29(3):181–216

Stucki T (2019) Which firms benefit from investments in green energy technologies? – The effect of energy costs. Res Pol 48:546–555

Sugden R (2013) The behavioural economist and the social planner: To whom should behavioural welfare economics be addressed? Inquiry 56(5):519–538

Sugden R (2018) The community of advantage: a behavioural economist's defence of the market. Oxford University Press, Oxford

Thaler RH, Benartzi S (2004) Save more tomorrow: Using behavioral economics to increase employee saving. Jo Pol Econ 112(S1):S164–S187

Thaler RH, Sunstein CR (2009) Nudge: improving decisions about health, wealth, and happiness. Penguin books, New Haven

Tversky A, Kahneman D (1992) Advances in prospect theory: cumulative representation of uncertainty. J Risk Uncertainty 5(4):297–323

Weitzman ML (2009) On modelling and interpreting the economics of catastrophic cli- mate change. Rev Econ Stat 91(1):1–19

Young HP (1998) Individual strategy and social structure: an evolutionary theory of Institutions. Princeton University Press, Princeton, NJ

Young HP (2009) Innovation diffusion in heterogeneous populations: contagion, social influence, and social learning. Am Econ Rev 99:1899–1924

Chapter 12
Climate Change Education Through Local Education Programs: A Case Study in Vietnam

Thi Kinh Kieu, Thi My Thi Tong, Ha Thu Nguyen, Thoa Kim Do, and Laura Stamp

Abstract Shifting from a purely scientific approach, climate change education now embraces interdisciplinary knowledge with more hands-on pedagogical approaches. Local education, grounded in students' realities, is a powerful tool for climate change education in Vietnam. Local education program strengthens students' connections with their community while also serving as a seed for spreading climate change throughout the community. These both help to strengthen community education. Vietnam's Ministry of Education and Training implemented the 2018 New General Education Program, mandating the nationwide inclusion of local education programs. Despite this policy shift, a research and practical gap exists regarding the national integration of climate change education and Disaster Risk Reduction Education within Local Education Programs. This chapter contributes to addressing the existing gap by examining climate change education implementation in local education programs across four provinces and cities nationwide. Findings reveal variations in climate change education integration, highlighting areas for improvement. The chapter concludes with recommendations to strengthen climate change education within local education programs, including fostering educator collaboration, diversifying assessment, leveraging community resources, and prioritizing teacher training. These recommendations aim to equip Vietnamese students with

T. K. Kieu (✉)
The University of Danang University of Science and Education, Da Nang, Vietnam
e-mail: ktkinh@ued.udn.vn

T. M. T. Tong
Vietnam Institute of Economic, Vietnam Academy of Social Science, Hanoi, Vietnam

H. T. Nguyen
The Vietnam National Institute of Educational Sciences, Hanoi, Vietnam

T. K. Do
Earth Systems, Hanoi, Vietnam

L. Stamp
Science Teacher in Oak Park Elementary School District 97, Oak Park, Illinois, US

A. Yildiz and R. Shaw (eds.), *Disaster and Climate Risk Education*, Disaster Risk Reduction, https://doi.org/10.1007/978-981-97-5987-3_12

the knowledge, skills, and agency to address climate change and contribute to local environmental sustainability.

Keywords Climate change education · Local education programs · Vietnam · Policies · Strategy

12.1 Introduction

Traditionally, climate change education (CCE) has fallen within the domain of science classes, focusing on the scientific causes and impacts of excess greenhouse gases and disrupted carbon cycles. However, the ramifications of climate change extend far beyond scientific realms, permeating through political, social, and economic dimensions. Research spanning a decade underscores the necessity for a multidisciplinary approach in CCE to effectively address its diverse impacts across various spheres. Understanding scientific principles does not inherently motivate students to adopt behaviors conducive to mitigating climate change; without offering actionable localized solutions, it often leads to feelings of "climate anxiety" or apathy. Moreover, with climate crises increasingly making headlines, the urgency for action has never been more apparent, prompting schools and communities worldwide to transition from mere education to implementing tangible adaptation and mitigation strategies, especially those fit in the local context.

Climate change education for sustainable development (CCESD) embodies a holistic approach that empowers individuals and communities to mitigate and adapt to climate change while fostering environmental sustainability, social justice, and economic equity. By integrating diverse disciplines, engaging students in meaningful experiences, and transforming learning environments, CCESD equips learners with the knowledge, skills, and agency to become active participants in addressing the pressing challenges of climate change (Mochizuki and Bryan 2015). The ESD for 2030 Roadmap (UNESCO) emphasizes transforming learning environments to facilitate meaningful educational experiences aligned with sustainability goals. This involves not only physical changes to school infrastructure but also pedagogical shifts towards project-based and interdisciplinary learning to empower youth along with supporting educators to create transformative experiences. Place-based education and problem-based learning emerge as promising approaches, grounding learning in authentic contexts and fostering connections between students and their communities while teaching critical thinking, communication and collaboration skills (Khadka et al. 2021; Littrell et al. 2020; Mochizuki and Bryan 2015).

Community education is defined as programs and activities that encourage learning and social development within local communities, generally outside of traditional educational institutions such as schools, colleges, and universities (FAU 2024). Community education was also mentioned as a key target for climate change education since it empowers children and young people to take action, develop solutions, change behavior, and work together for a more sustainable future (Rousell and Cutter-Mackenzie-Knowles 2020; Trott 2019). In Vietnam, community-based climate change education is significantly highlighted in the Local Education, a new obligatory subject under Vietnam's 2018 New General Education Program (MOET 2018). Embracing the adage of "think globally, act locally", community based CCE underscores the need for individuals to not only understand global challenges but also take tangible actions within their own communities. CCE, through Local Education Programs (LEPs), goes beyond traditional education by imparting practical knowledge and fostering a deep-rooted commitment to community well-being. By integrating CCE into local curricula, students gain invaluable insights into disaster response, develop a nuanced understanding of local contexts, and actively engage in initiatives that strengthen the bond between schools and their surrounding communities. This synergy between education and local action cultivates a generation of empowered individuals who are equipped to address global challenges while championing the resilience and sustainability of their communities.

Despite the importance of LEPs and their implementation recently, there is a lack of research that has been conducted in Vietnam to evaluate the nationwide integration of Disaster Risk Reduction Education (DRRE) and CCE within LEPs. Thus, this book chapter will look at how CCE is taught in the LEPs of four provinces and cities in Vietnam's north, central, and south regions. Based on the analysis of existing challenges and opportunities, the book chapter provides several strategies to promote CCE in LEPs.

12.2 CCE and Its Integration into Local Education Programs in Countries

Many countries have recognized the importance of integrating CCE into their national curricula and have developed guidelines and resources to support teachers in delivering effective climate change education. Austria's current climate change adaptation strategy emphasizes the need for climate change education to implement the present strategy and develop robust solutions for the future. Many schools in Austria include climate change in their curriculum but fail to acknowledge its interdisciplinary nature. Unfortunately, it is often presented as separate subject lessons, making it difficult for students to grasp its essential concepts in an interdisciplinary manner (Federal Ministry of Agriculture, Forestry, Environment and Water Management 2012; Jackson and Pang 2017). Besides, research by Feldbacher et al. (2023) conducted in 7 Austrian school classes with students from 13 to 17 found that changing the duration of a climate change unit did not necessarily lead to improved pro-environmental behavior or a deeper understanding of the issue. Furthermore, many Austrian textbooks focus on individual behaviors like recycling, which may deflect attention away from more effective solutions and disregard the complex political, social, technological, and economic factors contributing to climate change. As a result, it is seen that climate change education through local education programs is underscored.

In Spanish, incorporating CCE into local education programs can be made clearer by the government taking into account CCE on the regional and local scale, not only for students but also for society, and by implementing more specific measures through engaging learners in hands-on labs, field trips, and expert interactions (Monroe et al. 2019). These experiences enhance students' understanding and catalyze positive environmental behaviors, as stated in the research of the Organization for Economic Co-operation and Development (Mann and Chang 2024). The locally focused approach with CCE was acknowledged in some areas in the United States. White et al.'s research in a rural area of the United States has shown that combining climate change concepts with local agricultural science is an effective way of dispelling climate skepticism among teachers, students, and community members (White et al. 2022). Collaborating with climate scientists and local farmers, teachers gathered local data on climate change, which they then used in the classroom to demonstrate the real-life impacts of climate change within their community. The participation and validation of farmers added further credibility to the study, making it a positive experience for teachers who had previously been hesitant to teach climate change.

CCE has been integrated into local projects in Thailand and Indonesia. In Thailand, teachers and curriculum developers worked together to create a project for students to investigate the impact of human activity on the mangrove habitat of the mudskipper. The mudskipper is considered a bio-indicator of ecosystem health. As part of the project, students collected data in a local mangrove forest. After completing the project, 100% of the participants enthusiastically supported the experience and were able to explain in detail the reasons why the mudskipper's habitat should be protected. In Indonesia, the project involved creating a vegetable production business on school grounds with the help of families and community agribusiness centers. The projects offered an opportunity for the students to actively participate and experiment with real-life problems. This exposure can help them modify their outlook and potentially influence their attitudes and behavior to climate change (Aligaen and Mangao 2012).

12.3 Overall of CCE in Vietnam

12.3.1 General Introduction to the New General Education Program (from 2006 to 2018)

In Vietnam at present, both GEP 2006 and 2018 are being implemented concurrently. GEP 2006 was issued in 2006 under Decision No. 16/2006/QĐ-BGDĐT dated May 5, 2006 by the Minister of Education and Training shortly after Vietnam, along with 195 countries, agreed to the Paris Agreement on climate change—the first global legal document binding the responsibilities of all parties involved in climate change. However, the GEP 2006 document does not include any themes related to climate change. In Vietnam's GEP 2006, Education for Sustainable Development (ESD) is mainly integrated into textbook content as a highly practical approach in Vietnam (Thi 2015).

GEP 2018 was approved by the Ministry of Education and Training under Circular No. 32/2018/TT-BGDĐT dated December 28, 2018 (MOET 2018). According to the capacity development orientation, it was officially implemented in the academic year 2021–2022. GEP 2018 aligns with many principles of Education for Sustainable Development (ESD), such as setting specific goals for the development of students' competencies and qualities. This includes fostering responsibility towards the environment, applying learned knowledge and skills to address issues related to climate change, enhancing experiential learning methods, encouraging teachers to be more creative in their teaching, and promoting local education (MOET 2018). It can be said that the innovation of the GEP 2018 is expected to create a learning environment that can promote Education for Sustainable Development (ESD).

12.3.2 Introduction to the Local Education Program (from 2006 and 2018)

Vietnam's Local Environmental Program (LEP) has undergone a transformation from being optional content developed and taught according to the needs of each locality in the GEP 2006 to becoming a "mandatory" educational content in the GEP 2018, with a duration of 35 periods per year in each class (MOET 2018), p. 11). The main objective of the LEP, as stipulated in the GEP 2018, is to "equip students with knowledge about their living environment to nurture their love for their homeland, raise awareness to understand and apply what they have learned, and contribute to addressing issues in their homeland" (MOET 2018), p. 32). LEPs have been developed with the aim of fostering competencies through practical, relevant, and realistic knowledge of a province and through active engagement of learners. It integrates fundamental issues of nature, geography, history, culture, arts, social welfare, economics, politics, and environmental protection of the province, linking theory with practice and educational content with reality. LEP is designed to flexibly apply positive teaching methods and forms to maximize the students' autonomy, positivity, and creativity; enhance practical activities, apply what has been learned to identify and solve practical problems in local life, using primarily educational methods such as organizing experiential theme topics; organizing learning projects; organizing according to the STEM model. Therefore, LEP in localities expects to integrate at least one theme related to Education for Sustainable Development (ESD) and Environmental Education (EE), and the number could be higher depending on the localities.

12.4 Status of Integrated CCE Through LEP

This study evaluates the content, objectives, and methods of CCE in the LEPs of four provinces and cities representing the northern region (Ha Giang, Hanoi), central region (Quang Nam), and southern region (Ho Chi Minh City). These regions were chosen because they are heavily impacted by typical effects of climate change and various natural disasters, including landslides, flash floods, storms, floods, severe cold spells, heatwaves, droughts, and saline intrusion.

For the objectives and content of Education for Sustainable Development (ESD), we examined the presence of ESD themes in the learning content and the duration of their teaching in the curriculum. Regarding pedagogical methods: we assessed which teaching approaches are used in the integrated ESD themes in the LEP. This evaluation strategy can provide a deep insight into how LEPs align with ESD principles and identify necessary changes to better integrate ESD and address climate change issues in LEPs (Table 12.1).

The analysis of the LEPs of the four provinces and cities reveals that the integration of CCE content varies in different ways. This includes full integration equivalent to a duration of 4 periods (out of a total of 35 teaching periods) dedicated solely to CCE themes, as well as partial integration with less than 1 period embedded sporadically within various other content. Out of the total of 12 searched and analyzed CCE themes, only 3 themes (accounting for 33.3%) were fully integrated in the LEPs in Ho Chi Minh City (Grade 8), Quang Nam (Grade 9, Grade 12).

The content progression ranges from awareness to building daily habits in adaptation to climate change and communication about climate change. The impacts and coping actions related to climate change are divided into two main themes: climate change with the natural environment and climate change with humans. These are allocated to the curriculum in Grade 9, where students have previously learned about climate change in other subjects. For Grade 12, climate change is taught towards the end of the curriculum with the aim of guiding students' careers and actions as adults to adapt to climate change.

Notably, Quang Nam fully integrated two climate change themes into the LEP, followed by Ho Chi Minh City with one fully integrated theme and four partially integrated themes. Meanwhile, in Hanoi and Ha Giang, we did not find any fully integrated themes. Hanoi only integrated four themes related to environmental education and sustainable development, while Ha Giang only integrated one theme related to biodiversity. Partially integrated climate change themes are only briefly mentioned in lessons.

Regarding content, the analysis of the LEPs shows the compatibility and alignment of CCE content with the natural characteristics, climate, and types of natural disasters in each locality. Ha Giang is considered a frontier province in the northern mountainous region of the country. With its location in the highland region characterized by a tropical monsoon climate with cold features, complex geological terrain, and high altitudes, Ha Giang faces significant climate and disaster risks, including

Table 12.1 CC themes integrated into local education programs of provinces/cities in Vietnam

LEP	Grade	Objective	Content	Integrated duration
Ha Giang	Grade 7	List some solutions for conserving rare species of animals and plants in Ha Giang	Topic 10. Biodiversity in Ha Giang	15–45 min
Ha Noi	Grade 6	Identify the causes and consequences of environmental pollution in Hanoi, including ozone depletion and the phenomenon of acid rain	Topic 8. Environmental pollution and its effects on the lives and health of Hanoi people	15–45 min
	Grade 10	Outline solutions to CCE, including solutions to protect Hanoi's ecological diversity	Topic 7. Hanoi's diverse ecosystem	15–45 min
	Grade 11	Explain the negative impacts of production and consumption on Hanoi's environment Take specific and practical actions to protect the living environment in Hanoi	Topic 8. Hanoi's "*production—consumption*" policy aims at sustainable development	15–45 min
	Grade 12	State solutions to minimize the harmful effects of climate change Have awareness and appropriate actions to adapt to climate change	Topic 8. Hanoi students participate in protecting a green—clean—beautiful environment	15–45 min
Quang Nam	Grade 9	State the effects of climate change on the environment in Quang Nam Present the causes and some measures to limit the impact of climate change in Quang Nam Be aware and take appropriate actions to contribute to limiting the impact of climate change	Topic: Impact of climate change on the environment in Quang Nam	4 h
	Grade 12	Recognize the impact of climate change on economic, social activities and people's lives in Quang Nam Outline solutions to adapt to climate change in Quang Nam Have awareness and appropriate actions to adapt to climate change	Topic: Adaptation to climate change and sustainable development in Quang Nam	4 h

(continued)

Table 12.1 (continued)

LEP	Grade	Objective	Content	Integrated duration
Ho Chi Minh city	Grade 4	State the manifestations of climate change in Ho Chi Minh City such as frequent water shortages due to long dry seasons, environmental pollution problems and the speed of industrialization causing the natural environment to degrade	Topic 1. Nature and people of Ho Chi Minh City	15 min - 45 min
	Grade 6	State the climate characteristics of Ho Chi Minh City	Topic 1. Nature and people of Ho Chi Minh City	15–45 min
	Grade 7	Learn about the effects of the greenhouse effect on life on Earth; Present some common negative manifestations of the environment (greenhouse effect, global warming, rising sea levels,…); Point out the impact of climate change on human health; State the basic concepts of responding to climate change; Propagate and mobilize relatives and friends to consciously carry out actions to mitigate climate change; Implement communication strategies to reduce the greenhouse effect in different forms	Topic 7. Communication on climate change prevention and control in Ho Chi Minh City	4 h
	Grade 8	Identify some issues about natural resources and the environment in Ho Chi Minh City, including air, water, and air pollution issues	Topic 7. Protect natural resources and prevent environmental pollution in Ho Chi Minh City	15–45 min
	Grade 10	Present the current situation of environmental pollution in Ho Chi Minh City; State the impact of humans on nature; State the solution and purpose of participating in environmental education activities; Proposal to participate in environmental pollution prevention activities in Ho Chi Minh City	Topic 7. Environmental pollution in Ho Chi Minh City	15–45 min
Total		**12**		

flash floods, inundation, and freezing weather.[1] Therefore, the issue of "Biodiversity in Ha Giang" is considered a pressing matter here.

Hanoi is the capital of Vietnam with the highest urbanization rate and most developed economy. Nationwide, Hanoi has increased production activities, coupled with environmental pollution and large greenhouse gas emissions, then resulting in extreme weather phenomena (such as prolonged heatwaves, sudden mutations, heavy rainfall, and concentrated rain) causing flooding, traffic congestion, and significant impacts on people's health.[2] In this context, CCE is integrated into LEP through themes related to environmental pollution and sustainable production and consumption. For example, in Quang Nam, a province in Central Vietnam facing significant climate-related risks such as floods and forest fires due to factors such as excessive rainfall, storms, dam and dike breaches, and where agriculture plays a predominant role in the economy, CCE has been integrated into the LEP through themes related to the impact of climate change on agriculture and solutions to climate change adaptation in the agricultural sector. Similarly, in Ho Chi Minh City, the most populous city in the country, accounting for 23% of the total Gross Domestic Product (GDP) and 20% of the total foreign direct investment in Vietnam, CCE is integrated into the LEP through various themes related to urban sustainability, resource management, and economic development.[3] With its characteristic of being situated close to sea level and experiencing continuous population growth, Ho Chi Minh City is greatly affected by natural disasters such as thunderstorms, storms, coastal flooding during storms, and tidal surges. The themes related to CCE in Ho Chi Minh City are often associated with greenhouse gas effects, the negative impacts of climate change on people's health, and proposing solutions in the climate change communication strategy. Thus, the integrated content of CCE has helped elevate students' awareness in the Local Environmental Programs (LEPs) of various provinces regarding practical issues related to climate change at the local level. This aligns with the objective of LEP outlined in the GEP 2018, which is to "equip students with knowledge about their living environment,…to raise awareness to understand, apply, and solve issues related to their homeland." (MOET 2018), p. 11).

Pedagogical approaches: Generally, a wide range of were capture in 4 LEPs. The frequency of CCE approaches was given in Table 12.2.

These methods include project-based learning (50%) and experiential learning (33.3%), followed by drawing, presentations, and reports (25%). Notably, mind mapping and situation handling techniques are also mentioned for conducting Education for Sustainable Development (ESD) in the Local Environmental Programs (LEPs). Particularly in the curriculum, schools are encouraged to actively organize a practical working session for students to develop actions to reduce the impact of climate change on the environment around them, along with creating mind maps, drawing, and action plans to address climate change. Thus, the pedagogical methods used in LEPs focus mainly on active methods such as real-life experiences, local issue

[1] https://dangcongsan.vn/xa-hoi/ha-giang-chu-dong-ung-pho-voi-bien-doi-khi-hau-545832.html.

[2] https://accgroup.vn/dan-so-ha-noi.

[3] Report of Climate Change Adaptation of Ho Chi Minh (funded by ADB), 2010.

Table 12.2 The frequency of CCE approaches in 4 LEPs

Pedagogies	Curriculum utilizing activities (N = 12) (%)	
Project based learning	6	50,0
Experiential learning	4	33,3
Drawing	3	25,0
Presenting	3	25,0
Group discussion	2	16,6
Problem solving	2	16,6
Communication	2	16,6
Real-life connection	1	8,3
Mind map	1	8,3
Field work	1	8,3

investigation methods, and project-based methods. These methods are considered appropriate and have a strong impact on changing individual motivation, aligning with the recommendations of UNESCO, 2017 (Monroe et al. 2019).

Although there is an emphasis on using active and experiential teaching methods, the integration of CCE into current Local Environmental Programs (LEPs) varies significantly across different localities. The design of LEPs still heavily relies on local capacities, with some places having limited local and teacher capacities (such as Ha Giang and Quang Nam). In these cases, the integrated content may not adequately provide the necessary knowledge for students, nor provide enough resources for students to have experiential learning opportunities and practical experiences. This is considered a barrier to integrating CCE into LEPs in Vietnam.

12.5 Implications and Recommendations

CCE in Vietnam has undergone positive changes compared to the 2010 period when CCE was only taught through projects. By 2018, CCE had been integrated into compulsory subjects and taught in the mainstream curriculum of general education, including local education subjects. This approach is entirely consistent with the global trend of CCE, which is place-based education (Hernandez et al. 2022; Meya and Eisenack 2018). An analysis of four LEPs reveals that CCE content has been substantiated as pertinent to the distinct climatic and environmental features of each locality. This is regarded as a salient advantage of the new LEP. Furthermore, the coding results of the LEPs implemented in various localities underscore the utilization of a multitude of active pedagogical methodologies, which are deemed indispensable for fostering the development of learners' action competence (Monroe et al. 2019).

CC is a multifaceted issue that necessitates an interdisciplinary approach. However, the majority of the current teaching workforce in Vietnam (excluding preschool and primary education) receives training within a single discipline. Consequently, the integration of CCE into LEPs will facilitate a transformative shift in the approach to CCE instruction, transitioning from integration within discrete subjects such as chemistry, geography, and physics to a single-disciplinary framework. Moreover, LEPs furnish opportunities for educators to implement project-based learning and bolster experiential learning at the local level.

This approach can potentially yield benefits for engagement and community, empowering learners to comprehend field and place-based issues. Additionally, its implementation outside the classroom setting enables learners to grasp real-world challenges, engage with the local community, and tackle local concerns. This approach is believed to be aligned with CCE principles and possesses a significant impact on the transformation of individual motivation.

On the other side, not only do students and schools raise awareness and take action in response to climate change and natural catastrophes, but the community is also empowered with knowledge and skills to deal with climate change. Participating in CCE events for students is one approach to reinforce and assist the local community in reviewing and advancing their climate change responses. School students and teachers might propose a variety of solutions or initiatives to help the community cope with climate change and natural disasters.

Indeed, community-based climate change education has multiple advantages for both schools and communities, creating a symbiotic link between the two. Working together, they can create innovative solutions to climate change concerns, such as establishing sustainable habits, pushing for policy changes, and starting community projects. Schools and communities can make a long-term difference by continuing to educate and involve people, ensuring that future generations are prepared to face the problems of climate change.

Nevertheless, action-oriented learning is constrained by the pedagogical approach, which features a remarkably limited number of topics that incorporate out-of-classroom activities such as fieldwork and real-life connections (as illustrated in Table 12.2). Notably, project-based learning, a quintessential field-based learning method that has been substantiated as being most pertinent to CCE within the LEP framework, has yet to be effectively implemented in Vietnam.

In practice, while teachers still conduct experiential learning or project-based learning on CC in local education lessons, there appears to be a dearth of student participation in monitoring the CC process and its impact on the local environment and population. Students are still predominantly "informed" rather than being actively encouraged to explore information independently, as is customary in classes in the United States, Thailand, and Indonesia. Furthermore, place-based education and inviting local community members (such as farmers, scientists, etc.) to share their stories of coping with climate change in classrooms seem to be relatively uncommon in Vietnam.

This can be attributed to the fact that the LEP is regarded as a mandatory program in the GEP 2018. Prior to this, the LEP was an elective program that was developed and taught in accordance with the specific requirements of each locality within the GEP 2006.

Of the four LEPs that were analyzed, the program implemented in Quang Nam province is considered to be relatively advanced for the following reasons:

i. CCE content is comprehensively integrated into the curriculum for grade 9, at a stage when students have already acquired a foundational understanding of CC through their studies in previous years.
ii. Consequently, students are empowered to proactively undertake self-directed learning and research activities, while simultaneously applying an interdisciplinary approach under the guidance of their teachers
iii. Furthermore, the LEP of Quang Nam province demonstrates notable development in terms of both content and student capacity, with a strong emphasis on fostering connectivity between different grade levels.

Building upon the preceding analyses and discussions, the following recommendations are proposed for Climate Change Education (CCE) within Local Education Programs (LEPs) in Vietnam:

1. *Strengthening Experience and Practice Sharing*: Given the recent implementation of GEP 2018, further forums are necessary for educators across various localities to share lesson plans and specific CCE teaching experiences within LEPs. This collaborative approach will foster mutual learning among educators, enabling them to progressively refine and elevate their teaching methodologies.
2. *Diversifying Student Competency Assessment Methods*: Concurrently with investments in CCE teaching activities, local Departments of Education and Training (DOETs) should provide guidance on more innovative methods for evaluating student competency. This could involve assessments based on student outputs such as project deliverables, innovative solutions, or proposed models for addressing climate change.
3. *Leveraging Community Resources*: Organizing experiential learning activities for students often demands significant resources in terms of cost, time, and design effort. Consequently, each locality should mobilize community resources (e.g., farmers, fishermen) to share indigenous knowledge and real-life stories with students. Additionally, local authorities should map potential locations that can serve as learning sites for students exploring climate change issues.
4. *Additional Considerations*: Prioritizing teacher capacity building regarding CCE instruction within LEPs remains crucial; developing a system of teaching materials and learning resources aligned with the new curriculum is essential; strengthening collaboration between schools, families, and communities is vital for effective CCE of students.

Through the collective efforts of all stakeholders, CCE instruction within Vietnamese LEPs can be demonstrably enhanced. This will ultimately contribute to fostering greater awareness and action among students in environmental protection and climate change adaptation.

Acknowledgements Thi Kinh Kieu acknowledged the Sumitomo Foundation Small Grant to support her in data collection and analysis.

Tong Thi My Thi acknowledged the Vietnam National Foundation for Science and Technology Development (NAFOSTED) under grant number 507.01-2019.302.

Nguyen Thu Ha acknowledged the PhD Scholarship Programme of Vingroup Innovation Foundation (VINIF), code. VINIF.2023.TS028

References

Aligaen JC, Mangao DD (2012) Climate change integrated education: a model for lifelong learning system. In: International conference on educating the future we want: a green Southeast Asia

Federal Ministry of Agriculture, Forestry, Environment and Water Management (2012) The Austrian strategy for adaptation to climate change. https://climate-adapt.eea.europa.eu/en/metadata/publications/national-adaptation-strategy-austria/austrianadaptationstrategy_context_final_25092013_v02_online.pdf

Florida Atlantic University (2024) FAU | What is Community Education? Retrieved April 8, 2024, from https://www.fau.edu/education/centersandprograms/csmcc/whatiscommunityeducation/

Feldbacher E, Waberer M, Campostrini L, Weigelhofer G (2023) Identifying gaps in climate change education—a case study in Austrian schools. Int Res Geogr Environ Educ 1–16.https://doi.org/10.1080/10382046.2023.2214042

Florida Atlantic University (n.d.) FAU|What is community education? Retrieved April 8, 2024, from https://www.fau.edu/education/centersandprograms/csmcc/whatiscommunityeducation/

Hernandez J, Scherr R, German M, Horowitz R (2022) Place-based education in high school science: situating energy and climate change in students' communities. Sustain Climate Change 15(1):58–67. https://doi.org/10.1089/scc.2021.0058

Jackson L, Pang M-F (2017) Secondary school students' views of climate change in Hong Kong. Int Res Geogr Environ Educ 26(3):180–192. https://doi.org/10.1080/10382046.2017.1330036

Khadka A, Li CJ, Stanis SW, Morgan M (2021) Unpacking the power of place-based education in climate change communication. Appl Environ Educ Commun 20(1):77–91. https://doi.org/10.1080/1533015X.2020.1719238

Littrell MK, Tayne K, Okochi C, Leckey E, Gold AU, Lynds S (2020) Student perspectives on climate change through place-based filmmaking. Environ Educ Res 26(4):594–610. https://doi.org/10.1080/13504622.2020.1736516

Mann A, Chang Y (2024) Enhancing green career guidance systems for sustainable futures. https://www.oecd-ilibrary.org/education/enhancing-green-career-guidance-systems-for-sustainablefutures_e6ad2d9c-en?crawler=true&mimetype=application%2Fpdf

Meya JN, Eisenack K (2018) Effectiveness of gaming for communicating and teaching climate change. Clim Change 149(3–4):319–333. https://doi.org/10.1007/s10584-018-2254-7

Mochizuki Y, Bryan A (2015) Climate change education in the context of education for sustainable development: rationale and principles. J Educ Sustain Dev 9(1):4–26. https://doi.org/10.1177/0973408215569109

MOET (2018) General Education Program 2018. https://moet.gov.vn/tintuc/Pages/CT-GDPT-MOI.aspx?ItemID=8421

Monroe MC, Plate RR, Oxarart A, Bowers A, Chaves WA (2019) Identifying effective climate change education strategies: a systematic review of the research. Environ Educ Res 25(6):791–812. https://doi.org/10.1080/13504622.2017.1360842

Rousell D, Cutter-Mackenzie-Knowles A (2020) A systematic review of climate change education: giving children and young people a 'voice' and a 'hand' in redressing climate change. Children's Geographies 18(2):191–208. https://doi.org/10.1080/14733285.2019.1614532

Thi TTM (2015) Integrating climate change education into teaching biology at high school in Vietnam. UED J Soc Sci Human Educ 5(4B):109–113. https://doi.org/10.47393/jshe.v5i4B.622
Trott CD (2019) Reshaping our world: collaborating with children for community-based climate change action. Action Res 17(1):42–62. https://doi.org/10.1177/1476750319829209
White PT, Wolf KJ, Johnson-Maynard J (2022) Changes in teacher attitudes relating to climate science. Natural Sci Educ 51(2):e20086. https://doi.org/10.1002/nse2.20086

Part III
Governance

Chapter 13
Advancing Climate Risk Governance and Education Through UNESCO

Sadaf Ismail, Faduma Ali, and Soichiro Yasukawa

Abstract As climate change and disasters increasingly threaten education systems globally, there is an urgent need for improved governance from the local school level to international frameworks to enhance resilience across this vital sector. UNESCO, the leading UN agency on education, science, and culture, plays an active role in supporting governments, educational institutions, and communities in reducing risks and adapting to intensifying geo-hazards and climate-exacerbated hazards such as earthquakes, floods, landslides, and wildfires that endanger school infrastructure and disrupt learning. Key strategies focus on protecting school facilities, learners, and education personnel while also building longer-term resilience through strengthened climate change adaptation and mitigation policies, implementation of disaster preparedness programs, and promotion of sustainable environmental practices within the education process. Central to its approach, UNESCO engages in multi-stakeholder collaboration, involving education officials, school administrators, teachers, parents, communities, and crucially, learners themselves. By engaging all education stakeholders in a transformative process and actively collaborating with partners, UNESCO is working toward achieving these agendas at a global level through initiatives aimed at promoting disaster risk reduction and education. This collaborative effort aims to transform teaching and learning methods to equip future generations with the skills to navigate escalating risks and build a more sustainable and equitable future. UNESCO collaborates closely with partners to advance risk governance and climate education, including the Global Alliance for Disaster Risk Reduction and Resilience in the Education Sector (GADRRRES) and the Greening Education Partnership (GEP), launched in 2022 in the aftermath of the United Nations Transforming Education Summit. Furthermore, UNESCO has developed and implemented various projects and activities to advance climate disaster risk reduction and education, focusing on providing quality education to all, from schoolchildren to professionals in the field, to strengthen their capacity in the built environment.

S. Ismail · F. Ali (✉) · S. Yasukawa
Scientific and Cultural Organization (UNESCO), United Nations Educational, Paris, France
e-mail: f.ali@unesco.org

S. Ismail
Graduate School of Advanced Integrated Studies in Human Survivability, Kyoto University, Kyoto, Japan

A. Yildiz and R. Shaw (eds.), *Disaster and Climate Risk Education*, Disaster Risk Reduction, https://doi.org/10.1007/978-981-97-5987-3_13

Keywords UNESCO · Global initiatives · Climate change · Disaster risk · Resilience · Education · Global Alliance for DRR and GADRRRES · Greening Education Partnership (GEP)

13.1 Introduction

The worsening effects of climate change and natural disasters clearly show how important it is to have proper management and education about these risks all over the world. As the top UN agency for education, science, and culture, UNESCO is a key player in promoting resilience and climate action through its global educational programs and policies. The rising frequency and severity of disasters pose major obstacles for the education sector, especially in maintaining consistent, inclusive, quality, and safe education opportunities for vulnerable communities.

Education, as stated by UNESCO (2011), is crucial for mitigating the destruction caused by natural hazards. It strengthens both individual and community capacity to understand hazards and exercise practical skills to minimize damage, mortality rates, and disturbance to livelihoods and culture. The integration of disaster preparation into schools improves readiness and ensures continued instruction post-disaster (Shaw 2014; UNESCO 2011). Furthermore, education equips individuals and communities with knowledge and skills to reduce the impact of hazards, thereby lessening fatalities, limiting damages, and decreasing disruption to economic, social, and cultural activities.

Guided by frameworks such as the Sendai Framework for Disaster Risk Reduction, the Paris Agreement, the 2023 Agenda for Sustainable Development, and Africa 2063: For the Africa We Want, UNESCO supports Disaster Risk Reduction (DRR) and promotes school safety and risk governance. UNESCO pursues its disaster risk reduction and school safety efforts through collaborative multi-stakeholder platforms like the Global Alliance for DRR and Resilience in the Education Sector (GADRRRES), which comprises United Nations agencies, international non-governmental organizations, humanitarian, and development bodies, as well as private sector entities. In 2022, GADRRRES introduced an updated Comprehensive School Safety Framework (CSSF) aimed at bolstering the resilience of education systems against disasters and climate change impacts. UNESCO is also involved in the recently established Climate Smart Initiative pilot program, which delivers targeted country-level support for climate change education and policy reforms in disaster-prone regions in Africa, Asia, the Pacific, Small Island Developing States (SIDS), and Latin America and the Caribbean (LAC).

At the program level, UNESCO employs a diverse range of modalities to advance disaster and climate education as well as risk governance, demonstrating a comprehensive commitment to fostering resilience. Through strategic workshops and collaborative initiatives, UNESCO engages with various countries, facilitating curriculum review and validation workshops to seamlessly integrate risk reduction principles and risk governance frameworks.

In addition to these practical initiatives, UNESCO develops authoritative resources to serve as guidelines for the global integration of disaster and climate education. The organization recently published a new guide in 2023, offering valuable assistance to countries in incorporating commitments from key international frameworks, such as the Paris Agreement and the Sendai Framework, into their educational policies and curricula (UNESCO 2023). UNESCO's commitment extends to the local level through its field offices, which actively promote context-specific solutions and facilitate knowledge exchange, ensuring that the global strategies resonate with the unique challenges and needs of individual regions. UNESCO also forges partnerships with a diverse array of stakeholders to drive collaborative action, particularly in areas related to school safety and the establishment of climate-resilient education systems.

UNESCO's intersectoral efforts underscore the pivotal role of education in empowering citizens to comprehend, adapt to, devise solutions, and effectively govern risks in a world undergoing transformation due to climate change. By leveraging expertise across various domains, UNESCO recognizes the interdisciplinary nature of the challenges at hand related to climate change adaptation, mitigation, and risk governance. However, sustained commitment and ongoing innovation are imperative to fully unlock education's potential and harness its transformative power in strengthening societal resilience and developing effective risk governance frameworks. The continuous evolution of strategies and approaches is essential to meet the dynamic and evolving needs posed by the changing climate landscape and associated disaster risks that require robust governance mechanisms.

Realizing the vision of advancing disaster and climate education as well as risk governance globally requires sustained efforts by UNESCO and its partners worldwide. It necessitates embedding climate action within flexible, relevant, and participatory learning accessible to all. Teachers will need support to facilitate engaging learning experiences on risk reduction, resilience, preparedness, sustainability, and proposing policies for effective risk governance. UNESCO plays a vital convening role in this endeavour, guided by key frameworks, and propelled by escalating risks.

Its diverse efforts empower learners, build capacities, and promote risk reduction as well as risk governance strategies, laying the foundation for more sustainable, just, and resilient futures. However, transforming education systems and policies to respond to the climate crisis and escalating disaster risks demands renewed collective action. More initiatives of UNESCO about advancing climate risk governance and education will be discussed in further sections of this book chapter, providing a deeper understanding of UNESCO's role and its diverse efforts in this critical area. Overall, UNESCO's plays an important role in advancing disaster and climate education and risk governance policies globally.

13.2 Disaster Risk Governance

Effective disaster and risk governance is crucial for enhancing global resilience and reducing vulnerability. Various factors contribute to risk, including poverty, inequality, environmental degradation, rapid urbanization, weak infrastructure, and population growth in hazard-prone areas. However, progress in risk reduction is impeded by challenges such as legal and policy gaps, issues with institutional coordination, and financial constraints. It is vital to address these barriers to strengthen governance capacities and mitigate escalating risks exacerbated by climate change.

Global frameworks like the Sendai Framework, the Paris Agreement, Agenda 2030, and Africa's Agenda 2063 provide guidance, but their implementation requires context-specific strategies and resources. There is no uniform solution across diverse nations with varying resource levels and geography-dependent risks. Understanding local customs, beliefs, and practices is essential for effective localized action. Strategies must align with available resources and recognize this diversity. Countries can translate commitments into action through engaged leadership, strategic investment, inclusive participation, and evidence-based planning, thereby enhancing governance foundations to manage complex current risks and build sustainable long-term capacity to address future threats.

One such example is European Union (EU) Member States; despite being highly industrialized and technocratically governed, they face substantial disaster management issues. Albris et al. (2020) identified three major science-policy gaps undermining disaster risk reduction in EU Member States: an epistemological gap, an institutional gap, and a strategic gap.

The epistemological gap refers to differences in the way scientific knowledge and policy-making processes operate. According to the International Federation of Red Cross and Red Crescent Societies (IFRC), legal gaps can undermine community resilience. In disaster risk reduction, there is often a disconnect between scientific expertise and policy decisions. While scientists generate valuable knowledge, policymakers may not fully understand or effectively utilize it. Bridging this gap involves improving communication channels, translating complex scientific findings into actionable policy recommendations, and fostering mutual understanding between scientists and policymakers. Facilitating these necessitates significant governance improvements.

The institutional gap pertains to the structures and organizations responsible for disaster risk management. Institutions play a crucial role in implementing policies and coordinating disaster response. A study (Berchtold et al. 2020) conducted across seven EU member states (Germany, Netherlands, United Kingdom, Ireland, Italy, Austria, and Greece) highlighted the importance of interorganizational collaboration in disaster response. The study was based on expert interviews with representatives of different types of disaster response organizations (health services, police services, fire services, and other crisis management organizations). The findings emphasized the need for effective interorganizational crisis management to address new and complex threats. However, gaps exist in terms of capacity, coordination, and alignment between different institutions. A lack of robust emergency preparedness protocols and response plans can result in critical tasks being overlooked and officials becoming overburdened when crises occur. For instance, there might be fragmentation across agencies, a lack of standardized procedures, or insufficient resources. Addressing this gap involves strengthening institutional frameworks, enhancing collaboration, and ensuring that disaster risk reduction efforts are well-coordinated across various bodies.

Additionally, the strategic gap relates to long-term planning and vision for disaster risk reduction. Often, disaster management focuses on short-term responses rather than proactive, strategic planning, as highlighted in the article "Advancing the Field of Disaster Response Management: Toward a Design Science Approach" (Frykmer et al. 2021) published in the International Journal of Disaster Risk Science. The paper argues that if emergency and disaster response management interventions are developed and implemented without being empirically tested, this might worsen rather than improve response, emphasizing the importance of testing interventions and their effectiveness before implementation. The Disaster Management Cycle discussed by University of Central Florida (UCF) Online outlines five crucial stages of disaster management, which include preparedness, response, recovery, and mitigation, but the challenge lies in the need to be prepared for a wide range of contingencies (UCF Online, n.d.). Policymakers may prioritize immediate crises over long-term resilience-building measures. To bridge this gap, strategic thinking is essential. It involves developing comprehensive risk reduction strategies, investing in preventive measures, and considering the socio-economic and environmental dimensions of disaster resilience. Poor coordination between institutions and insufficient allocation of resources may compound these issues. To effectively integrate policies across organizations, strategic alignment of goals and close collaboration are essential.

The IFRC Handbook on Law and DRR (IFRC 2015) offers practical guidance for lawmakers, officials, and practitioners seeking to review and strengthen laws and policies to prioritize DRR. Its primary aim is integrating DRR across all sectors and levels of government, a key tenet of the Sendai Framework for Disaster Risk Reduction. The handbook stresses the importance of clear institutional responsibilities and accountability mechanisms. It is designed for use alongside the IFRC's Checklist for Law and DRR (IFRC and UNDP 2015) to ensure holistic incorporation of DRR principles. As part of the IFRC's broader efforts to elevate humanitarian standards, promote a culture of peace, and enhance societal resilience in the face of crises, this handbook serves as an invaluable tool for all disaster risk management stakeholders and legislators.

The IFRC (2015) identifies gaps in frameworks related to natural hazards and stakeholder roles, emphasizing the need for clearly defined guidelines for cross-sectoral and multi-stakeholder interventions in disaster risk management. A key governance challenge globally is deficiencies in legal frameworks and policies for disaster risk management. Assessments indicate gaps in comprehensive disaster risk management legislation and alignment with international frameworks like the Sendai Framework for DRR in many countries.

Insufficient institutional capacities for risk monitoring and analysis also constrain evidence-based policymaking. Limited data collection and risk assessment capacities are common in disaster-prone countries, undermining the establishment of risk baselines to inform planning and decision-making. Even where policies exist, implementation often lags. For example, as of 2021, only 81 countries out of all the 193 UN member states had adequate systems to report on Sendai Framework global indicators. This illustrates the challenges faced by many nations in establishing robust DRR strategies. Significant efforts are still required in most nations to systematically gather data and conduct research to characterize risks.

Climate change acts as a risk accelerator, intensifying the frequency and severity of extreme events worldwide. Comprehensive strategies are vital for managing the interconnected risks of disasters, climate change, and unstable development. However, reviews show that climate adaptation and disaster prevention initiatives often remain uncoordinated in many nations. This fragmentation stems from several causes, including viewing climate change as a global concern without local or national accountability, and the absence of systemic approaches. Though integrating strategies is widely acknowledged as crucial for handling disaster, climate, and development threats in an all-encompassing manner, putting such unified tactics into practice can prove difficult, leading to disconnected efforts on the ground.

The education sector faces governance challenges related to disaster risk management and resilience globally. Assessments indicate that DRR is not sufficiently prioritized or mainstreamed in national education policies, plans, and budgets. Specific gaps highlighted by UNESCO include inadequate risk assessments, insufficient inclusion of marginalized groups, limited integration of risk awareness into curricula, and resource constraints. While frameworks like the CSSF provide guidance, successful implementation requires high-level commitment, allocating adequate resources, conducting comprehensive risk assessments, including marginalized groups in planning and implementation, and integrating risk awareness into curricula.

UNESCO has implemented impactful and inclusive initiatives strengthening disaster risk governance. The Risk Governance and Social Resilience initiative emphasizes holistic DRR involving youth, women, indigenous groups, and civil society. It focuses on the Sendai Framework guidelines to build resilience by enhancing governance. The Local and Indigenous Knowledge Systems (LINKS) programme promotes including such knowledge in climate science and policy so local insights inform issues like biodiversity, climate change, disaster preparedness, and sustainable development. Recognizing their vital roles, UNESCO's Youth Engagement aims to incorporate the perspectives and needs of youth, women, and indigenous groups for greater societal resilience. The Managing Disaster Risks for World Heritage project (UNESCO 2010) focuses on building disaster risk management capacity for African site managers and facilitating peer exchange to strengthen management in East Africa. Environment and Ecosystem-Based DRR includes reducing ocean hazards, using nature to reduce hazard impacts, and planting sea grass to reduce coastal erosion. These diverse initiatives demonstrate UNESCO's whole-of-society approach to strengthening disaster risk governance by incorporating insights from marginalized groups and promoting accountability, transparency, and partnerships. Frameworks such as the Sendai Framework promote collaborative action aligned on implementation, data gathering, and cross-country peer learning to strengthen governance.

UNESCO has undertaken efforts to bolster governments' data capabilities, which focuses on filling data gaps and increasing gender equality in water through applying gender-disaggregated data and evidence-based decision-making. Additionally, the STEM and Gender Advancement (SAGA) project builds member state's capacity for collecting gender-disaggregated Science, Technology and Information (STI) data, identifies policy gaps, and advances gender-related STI policies. Moreover, by working across governments, civil society, research institutions, and international organizations, UNESCO has driven interdisciplinary cooperation on diverse disaster and climate risk reduction aspects. It also endorses the Sendai Framework's preventative and risk-reduction priorities.

Initiatives such as Green Solutions for DRR (UNU 2019) and Climate Change Adaptation (UN, n.d.), as well as the International Conference on Climate Risk, Vulnerability and Resilience Building (April 2023), showcase such cross-sectoral partnerships. Green Solutions advances DRR and climate adaptation through ecosystem services and nature-based solutions. For instance, UNESCO participates in the Partnership for Environment and Disaster Risk Reduction, promoting ecosystem based DRR implementation. The International Conference on Climate Risk, Vulnerability and Resilience Building also bridges science, policy, and decision-making to support effective resilience building by tackling challenges in scaling innovative adaptations, climate funding, and emerging citizen participation and open science roles. Through diverse partnerships and conferences, UNESCO bolsters disaster and climate risk governance efforts by providing resources, expertise, and multi-stakeholder support.

Adequate sustained investment and financing are critical global priorities to realize disaster and climate risk governance commitments. However, assessments show ex-ante DRR receives inadequate resourcing compared to post-disaster response in many countries. There is an urgent need to increase budget allocations and establish predictable, accessible contingency financing. International assistance can catalyse governance improvements, especially for lower-income countries facing constraints. Climate change adaptation funds are increasingly available but require capacities to access them. Further efforts are vital to ensure inclusive, equitable financing that empowers local stakeholders and vulnerable groups through risk-informed investments that account for short and long-term cost-effectiveness.

13.3 Disaster Risk Education

Disaster risk education is a critical tool in strengthening global disaster and climate risk reduction capabilities. It serves as a shield for communities against the escalating threats posed by climate change and disasters. Disaster risk education involves the integration of DRR principles across all forms of education and age groups, adopting an all-hazards approach. The increasing intensity, frequency, and impacts of disasters on access to continuous, inclusive, quality, and safe education make them a major concern for education. In this context, disaster risk education plays a vital role in preparing communities and building resilient societies. It provides tools to integrate DRR across educational settings and train educators on risk education.

Children are particularly impacted, with over 1 billion children worldwide affected by climate-related disasters. This is a result of high enrolment rates (UNICEF 2021), as more children enrolled in schools equates to more young lives exposed to the risks associated with climate-related disasters. This exposure is especially significant in regions where schools may not be equipped to handle such events. Annually, disasters disrupt the education of more than 175 million children, disproportionately affecting the most marginalized populations (GADRRRES 2023). These communities often have less access to resources needed for recovery and resilience.

The link between effective disaster and climate risk governance and education systems is critical. They jointly build resilience through comprehensive risk reduction strategies. However, disruptions to these systems can hinder the development of crucial skills and knowledge. UNESCO advocates for disaster risk education through initiatives like the GADRRRES. GADRRRES promotes comprehensive safety in education through the CSSF 2022–2030, shifting emphasis from reactive to preventative disaster risk management. Several UNESCO institutes and chairs, like the International Institute for Educational Planning (IIEP), also contribute to disaster risk education by strengthening crisis-sensitive planning and management capacities.

Despite the crucial role of education, there are major systemic gaps in the comprehensive integration of disaster risk education globally. Shockingly, only 27% of countries report the inclusion of risk reduction across their curricula, revealing a significant deficiency (GADRRRES 2023). This shortfall is primarily attributed to insufficient coordination between government agencies overseeing climate change and education agendas. DRR tends to be overlooked in donor education strategies and national education budgets, underscoring the need for improved prioritization. This improvement is achieved by recognizing the importance of DRR in education and taking active steps to integrate it into curricula through effective coordination and dialogue.

UNESCO assists Member States to build resilience against natural hazards by harnessing knowledge and technology and furthering education to improve preparedness and reduce vulnerability. Positive initiatives, such as the Comprehensive School Safety Initiative (CSSI), set exemplary standards by bringing together disaster management and education stakeholders for constructive dialogue and effective coordination (UNDRR 2022). These initiatives ensure that DRR is not overlooked and is instead integrated into education strategies, thereby addressing the systemic gaps in disaster risk education globally. Despite progress, major systemic gaps exist globally in risk education, like insufficient coordination across government agencies overseeing climate and education agendas. This causes issues like overlooking risk reduction in budgets and strategies.

Another substantial challenge lies in the focus of risk education primarily on basic awareness, rather than skill development for disaster prevention and preparedness. The limitations in teacher capacities are evident, with fewer than 50% feeling fully prepared to teach on risks (GADRRRES 2023). At the school level, a notable gap exists in the alignment of safety plans with national disaster risk management strategies, emphasizing issues in decentralized disaster education governance (UNDRR 2022).

Positive strides in enhancing school safety are also noticeable in certain regions, particularly in the Caribbean, where several countries have successfully aligned their school safety plans with national strategies, following the CSSF model (UNDRR 2022). While this is an encouraging development, there is a pressing need for substantial efforts to scale and replicate similar governance coordination initiatives globally. At the international level, UNESCO plays a crucial role in facilitating the integration of DRR into educational policies, curricula development, and school planning through initiatives like the CSSF (UNESCO 2015). The organization also co-leads the GADRRRES, focusing on the implementation of the CSSF's updated framework through effective coordination.

UNESCO's commitment to fostering resilience extends to supporting extensive curriculum reviews, as evidenced by its collaborative project in The Gambia in 2021, which aimed to integrate DRR into the national curricula. This project serves as a noteworthy example of comprehensive efforts that bridge global guidance with localized governance processes, demonstrating the importance of replicating similar localized governance coordination efforts for systemic progress. A holistic and coordinated approach across various levels is indispensable to not only bolster resilience but also achieve universal access to quality education and effectively manage the escalating challenges posed by climate change and natural hazards.

Partnerships are instrumental in promoting governance coordination across sectors and stakeholders in disaster risk education. The CSSI serves as an exemplary model of multi-sectoral collaboration, bridging ministries of education, disaster management offices, and various partners (UNDRR 2022). Networks like the CSSI facilitate the exchange of information, development of localized curricula, enhancement of teacher capabilities, and the formulation of integrated planning. Despite the significant efforts required to coordinate and resource comprehensive risk education through integrated governance systems, strategic investments hold the potential to cultivate widespread societal resilience in the face of escalating climate and disaster risks.

In disaster risk education, partnerships still offer possibilities to promote information exchange and multi-sectoral collaboration. CSSI, not only enable the development of context-specific curricula but also facilitate teacher training and integrated planning. Despite the substantial effort required to coordinate and resource localized disaster risk education through integrated governance systems, the potential impact on societal resilience makes these strategic investments essential.

13.4 UNESCO's Initiatives in Disaster Risk Governance and Education

Recognizing the urgency of disasters and climate change, UNESCO's role in disaster and climate risk governance is significant. The programs undertaken by UNESCO are designed not only to impart knowledge but also to instil values and develop skills, enabling individuals to make meaningful contributions to sustainability. UNESCO takes the position that climate governance is achieved through education in the classroom, in the field with professionals, and ensuring the safety of educational buildings through resilient infrastructure.

In its many programmes, UNESCO aims to safeguard learning to contribute to a more inclusive world, for example in Education for Sustainable Development (ESD) programme aims to empower learners of all ages with the knowledge, skills, values, and attitudes for a peaceful world through its components focused on climate change education and disaster risk reduction. The Man and the Biosphere (MAB) Programme combines natural and social sciences, economics, and education to establish a scientific basis for enhancing the relationship between people and their environments while improving human livelihoods and safeguarding ecosystems. Whereas the LINKS programme recognizes the crucial role of indigenous knowledge in managing and adapting to environmental change and thus integrates this knowledge into broader policies, programs, and practices. The UNESCO Global Geoparks program manages sites and landscapes of international geological significance with a holistic concept of protection, education, and sustainable development.

In 2022, GADRRRES released an update outlining educational resilience priority. In disaster-prone nations, UNESCO has collaborated to embed risk reduction in teaching informed by local contexts. At the programme level, UNESCO employs diverse modalities to advance disaster and climate education, demonstrating a comprehensive commitment to resilience. Through collaborative initiatives, UNESCO facilitates curriculum reviews to integrate risk principles. Notably, UNESCO strongly emphasizes participatory pedagogies that actively engage youth as change-makers in building resilient societies, imparting not just knowledge but values and skills for sustainability.

UNESCO plays a major role in both disaster and climate risk governance and education. Its 2022–2029 strategy prioritizes supporting Member States to strengthen prevention, preparedness, management, response, and recovery, including open-access knowledge exchange (UNESCO 2022). An index also connects people to existing open data platforms on priority areas like disasters, climate, biodiversity, and geosciences to enable cross-disciplinary knowledge sharing. The strategy prioritizes marginalized groups and promotes inclusion through partnerships. Additionally,

UNESCO assists science and technology for resilience, early warning, school safety, risk reduction for heritage sites, and ecosystem-based risk reduction.

Alongside practical initiatives, UNESCO develops authoritative resources to guide the global integration of disaster and climate education. A 2023 guide offers valuable assistance for countries to incorporate frameworks like the Paris Agreement and the Sendai framework into policies and curricula (UNESCO 2023). The commitment extends to regional offices tailoring context-specific solutions and enabling local knowledge exchange. Diverse partnerships, particularly regarding school safety and climate-resilient education systems, also catalyse collaborative action. Intersectoral efforts underscore education's pivotal role in empowering citizens to comprehend, adapt and devise solutions amid climate change.

This has been shown through the Capacity Building for DRR in the Built Environment in Latin America and the Caribbean (BERLAC) project lead by UNESCO. It aims to enhance the built environment's resilience against natural hazards, particularly earthquakes, in the region. It comprises four crucial components: Safer New Buildings for Small-Scale Houses, School Facilities' Safety, Strengthen Existing Buildings Against Seismic Risk, and Risk-Informed Policy Making. The BERLAC initiative's comprehensive approach, incorporating capacity building, technical expertise, and policy development, represents a concerted effort to strengthen the built environment's resilience in Latin America and the Caribbean, safeguarding communities and ensuring their preparedness for the challenges posed by natural disasters.

The "Safer New Buildings for Small-Scale Houses" component focuses on improving the safety of new small-scale houses, ensuring they can withstand various natural hazards prevalent in the area. The "School Facilities' Safety" component is dedicated to enhancing the safety of school facilities, which often serve as community centres and shelters during disasters. The "Strengthen Existing Buildings Against Seismic Risk" component, conducted by the Japan Building Disaster Prevention Association, aims to strengthen existing buildings against seismic risks. The association organized a committee of Japanese engineers who evaluated the seismic resilience of an existing building in Cuba and proposed technical improvements and retrofitting methods. The "Risk-Informed Policy Making" component seeks to develop and revise earthquake-resistant policies in Latin America and the Caribbean, addressing the challenges posed by resource constraints, rapid urbanization, and the prevalence of self-construction that disregards building codes.

Through its efforts in Haiti, the BERLAC project has facilitated the implementation of crucial measures to strengthen the safety of small-scale buildings. This includes training 200 local engineers, with 40% being women, in the use of 3D printers for engineering classes at universities. Additionally, the project contributed to the training of 800 masons and foremen, enhancing the resilience of construction in Haiti against seismic risks and other natural hazards, ensuring safer infrastructure. While in the Dominican Republic, the project has developed and deployed an artificial intelligence (AI) tool to optimize investment plans for retrofitting critical infrastructure, such as school buildings and bridges. By leveraging AI, the project aims to minimize the disruption of education in case of natural hazards, such as floods and earthquakes, in Santo Pedro de Marcos, a city with a population of 200,000.

A 2023 UNESCO-UNFCCC conversation series on climate education and social transformation led up to COP28. As a key priority of the ESD for 2030 framework, UNESCO mainstreams education within the climate response. The first Greening Education Hub with the UAE Ministry of Education promoted concrete solutions through over 200 sessions.

In collaboration with the University of Udine, UNESCO developed a multi-hazard school safety assessment methodology called Visual Inspection to define the Safety Upgrading Strategies (VISUS) and empower policymakers to direct their attention towards effective disaster risk reduction efforts. This rapid visual triage methodology (Grimaz and Malisan 2016) enables the identification of seismic vulnerabilities in school facilities to inform regional risk reduction policies. Notably, UNESCO has successfully implemented this comprehensive methodology in diverse contexts, including the Dominican Republic, Bosnia and Herzegovina, Mozambique, Haiti, Indonesia, and Cambodia (UNESCO 2018).

Recent examples of UNESCO's work include aiding Nepal's school safety initiative to teach disaster preparedness to millions (WVIN 2020); promoting integration of climate change into Pacific Island nations' education procedures and teachers (Frontiers in Marine Science 2019), launching the 2022 GEP to accelerate climate learning and response; updating a framework outlining educational resilience goals; and mapping seismic risks across global structures (UNESCO 2018). Beyond national initiatives, UNESCO champions regional collaboration. In the Pacific Islands, they partnered to embed climate adaptation into policies and practices. This entailed developing teacher programs, revising curricula, and creating localized resources, better preparing students for a transforming climate (Frontiers in Marine Science 2019; PLOS Climate 2022; UNESCO 2012).

In addition to these initiatives, UNESCO's Social and Human Sciences sector also plays a crucial role in addressing health emergencies. They have published many resources and guidelines that provide insights into managing such crises in education. Moreover, UNESCO organizes workshops to facilitate experience sharing among educators and administrators. For instance, the "Experience Sharing Workshop for More Effective Management of Teacher Allocation" held in Dakar in 2016 provided a platform for participants to exchange valuable experiences and perspectives. These workshops serve as a valuable tool for capacity building and promoting innovative practices in education.

One of UNESCO's core strategies is fostering a culture of safety and resilience through education. This involves integrating DRR and climate change adaptation (CCA) concepts into school curricula, teacher training, and community outreach programs (UNESCO 2023). The Education for Sustainable Development (ESD) framework promotes DRR and CCA education as integral components of sustainable development, equipping learners with the knowledge, skills, and values to navigate future challenges.

CSSF emphasizes four key pillars: Safe Learning Facilities, School Disaster Management, DRR and Resilience Education, and School Safety. GADRRRES has become a key platform for collaborative action on school safety. The CSSF extends beyond merely ensuring structurally sound school buildings. It enables policies and systems for effective implementation (UNISDR 2022); makes safer facilities by improving infrastructure and taking risk reduction measures (UNESCO 2023); emergency preparedness plans, drills, and response mechanisms to ensure education continues even during disruptions (GADRRRES 2023).

GADRRRES plays a vital role in shaping a future where education systems are prepared to mitigate disaster risks and build resilience. GADRRRES has contributed to raising global awareness and political commitment to disaster and climate risk reduction education. Through training programs and resource provision, GADRRRES has empowered educators, policymakers, and communities with the knowledge and skills to implement effective education measures for disaster and climate risk reduction. Many countries have reported improved school infrastructure and disaster preparedness plans thanks to GADRRRES' work. Schools equipped with disaster and climate risk reduction knowledge and plans have shown better responsiveness and recovery during disasters.

GADRRRES and the CSSF have had a significant impact on communities and individuals worldwide, including increased safety and preparedness, enhanced disaster response, improved educational continuity, empowered individuals, and promotion of gender equality. In the Philippines, after Typhoon Yolanda in 2013, GADRRRES supported the Philippines in developing a national school safety policy and training teachers in DRR (UNESCO 2023). In Nepal, following the 2015 earthquake, GADRRRES assisted in rebuilding safer schools and integrating DRR into the curriculum, contributing to Nepal's improved response to subsequent disasters.

The ESD for SDGs initiative aims to integrate education for sustainable development into curricula and teaching practices worldwide. Its targets specifically address issues like climate action, responsible consumption and production, and life below water, all linked to disaster and climate risk reduction (UNESCO 2017). It equips learners with knowledge and skills on sustainability, and the ESD for SDGs initiative fosters responsible behaviours and contributes to building resilient communities.

GEP's collaborative approach has multifaceted impacts, as it has strengthened education systems by improving the capacity of education systems to address disaster and climate risks. It has increased access to quality education by enabling vulnerable communities to access knowledge and skills for resilience. GEP improved livelihoods by empowering individuals with education to adapt to changing environments and secure sustainable livelihoods. GEP also enhanced community collaboration through multi-stakeholder partnerships fostered by GEP, leading to community-driven solutions for resilience. It supported Pakistan in developing a national DRR framework for education, reaching over 40 million students (UNGEI 2018). It also partnered with the Ethiopian government to integrate climate change education into the national curriculum, benefiting millions of students.

UNESCO's support has extended beyond curriculum reform to include teacher training and the provision of materials to enable effective implementation. For example, in the case of The Gambia, a validation workshop was held in December

2021 to review teacher toolkits developed through the DRR curriculum integration initiative. Teachers and education officials provided feedback to finalize the toolkits containing sample learning activities, assessment guidance, and pedagogical resources aligned with the revised curriculum. Such efforts to equip teachers with contextualized materials and build their capacities are crucial to translating curriculum revisions into enhanced learning outcomes.

UNESCO is well-positioned to leverage its unique mandate and expertise to further scale up disaster and climate risk reduction education. As the lead UN agency for the Education 2030 Agenda, it can advocate the integration of risk reduction across educational policies, planning, and monitoring frameworks.

13.5 Recommendations

Inclusive strategies are essential to enhance disaster risk governance globally. This entails acknowledging local needs, insights, and capabilities by actively engaging communities, marginalized groups, civil society organizations, indigenous peoples, and the private sector. Their participation ensures their perspectives are considered and needs addressed, since local communities often profoundly comprehend their distinctive risks. Facilitating participatory dialogue and integrating local knowledge into policies, as UNESCO's youth networks showcase, can empower stakeholders. Human rights-based approaches promoting openness, accountability, and equitable partnerships further empower policies by building social capital.

Governments require assistance to develop robust systems that routinely gather quality disaggregated data to identify and monitor vulnerable groups and risks. This is indispensable for targeted policies but necessitates financial commitment and coordination. Concurrently, with disasters and climate impacts transcending borders, global cooperation and knowledge sharing are vital to address escalating governance gaps.

To realize disaster and climate risk governance commitments, adequate sustained investment and financing are critical global priorities. However, many countries allocate insufficient funding to disaster risk reduction measures implemented before a disaster occurs, compared to the response after a disaster. This highlights an urgent need to increase budget allocations and establish predictable, accessible contingency financing. Assistance from other countries can make it easier for nations to improve governance and systems. This is mainly true for poorer nations that face challenges. However, getting money from funds meant for adapting to climate change requires specific capacities. Therefore, capacity building in these countries is crucial. Further efforts are vital to ensure inclusive, equitable financing. This involves empowering local stakeholders and vulnerable groups through risk-informed investments. These investments should account for both short and long-term cost-effectiveness, ensuring that funds are used efficiently and effectively. This approach will contribute to a more resilient and sustainable future.

Monitoring, evaluation, and transparency are integral for effective governance. Feedback loops revealing shortfalls enable timely improvements. However, assessments show countries lacking adequate aligned monitoring systems. Strengthening statistical capacities and standardizing methodologies can address these gaps. Modern technologies like satellite imagery and climate modelling can support risk analysis when integrated. Regularly reviewing spending, outcomes, and sharing results publicly enhances accountability. Inclusive participation in monitoring provides vital community insights. Independent oversight bodies and audits can also identify deficiencies. Evidence-based observation, explicitness, and participation can elevate policies and service delivery.

Advancing governance necessitates strong leadership and political commitment. Without high-level champions providing momentum, laws, policies, and institutions may languish. Effective risk communication and framing by leaders influence public discourse and behaviours. Visionary leadership at all levels drives innovations and empowers institutions with appropriate mandates and capacities. While systemic changes take time, inspirational leaders can catalyse cultural shifts towards proactive, participatory governance, conquering inertia, and resistance. Entities like disaster management agencies rely on robust guidance from the top. Ultimately, global leaders must prioritize inclusive resilience through words, policies, and actions to bolster foundations.

The expanding burden of disasters underscores the pressing need to strengthen capacities. Despite substantial obstacles, enhancing governance is morally and practically imperative to safeguard communities. Collective action grounded in equity, science and solidarity can transform acceptance of disaster risk as inevitable into strengthening resilience and protecting the vulnerable as a communal duty. All stakeholders have roles in this shift towards more effective governance, improving well-being and security.

Frameworks like the CSSF provide solutions through coordinated efforts, but alignment across governance levels still requires urgent systemic action. Disaster risk education initiatives are often hindered by a lack of localized materials. Furthermore, most countries have limited monitoring and evaluation of outcomes, preventing comprehensive assessments of teaching effectiveness. Therefore, an improved monitoring and evaluation procedure can lead to improved overall impact.

A significant gap exists in aligning school safety plans with national strategies, highlighting decentralized governance issues. This necessitates urgent systemic action to fulfil education's role in building resilience amid 21st-century risks, preparing students to understand, adapt to and mitigate various challenges, including those posed by hazards while integrating risk reduction across curricula. The CSSF provides a roadmap prioritizing this integration, emphasizing monitoring student learning, advocating teacher capacity building, and linking school planning with national strategies. This dual role is increasingly vital in today's rapidly changing world, where adaptability is ever more important.

A paradigm shift is imperative, balancing reactive disaster responses with greater emphasis on prevention and preparedness. Education emerges as pivotal in enabling this shift, but realizing its full potential necessitates a profound governance transformation through comprehensive multi-stakeholder coordination. Achieving this shift requires a balance between national policies and strengthened local governance, allowing for contextualized and effective actions. Greater community participation and youth engagement is essential to support localized priorities and relevant curricula. Incorporating indigenous knowledge and gender perspectives can create culturally appropriate materials and localized content. Exploring technology innovations and networks can facilitate scalable teacher training, accessible student resources, and shared risk monitoring.

The COVID-19 pandemic has underscored the critical need for education systems to be resilient and responsive to emerging threats. UNESCO is well-positioned to incorporate pandemic preparedness and response into its disaster and climate risk governance initiatives. Entry points include developing guidance for public health emergency planning, integrating health literacy into curricula, and leveraging networks to provide distance learning during school closures. UNESCO can also facilitate experience sharing between countries to identify good practices and lessons learned during the pandemic. Moreover, UNESCO can advocate for increased investments in education as a crucial element of national resilience strategies. By broadening its efforts to encompass health emergencies while sustaining momentum across existing priorities, UNESCO can drive holistic risk governance to safeguard education access and quality in an increasingly complex and uncertain world.

The goal is to integrate disaster resilience education into society, leveraging education to empower citizens with awareness, anticipation skills, and resources to adapt to accelerating climate impacts. However, this demands strategic governance and collective action alongside fundamental policies, enhanced capacities, and cultural shifts for education to strengthen resilience effectively. As UNESCO continues advocating reform, it must also invest in grassroots implementation, collaborating with educators to deliver engaging risk reduction education and shape resilient behaviours.

UNESCO's multifaceted expertise positions it ideally to catalyse transformative change. Leveraging its education and science-based approach, the organization can devise solutions to enhance learning outcomes, while its cultural prowess enables it to foster behavioural shifts vital for resilience. Significantly, UNESCO facilitates cross-border knowledge sharing, accelerating the adoption of best practices globally. Through its cross-sectoral collaborations and engagement with diverse stakeholders, it can champion a holistic, society-wide approach to resilience-building. However, despite progress, the looming climate crisis demands that UNESCO take rapid and decisive action to unlock education's potential in empowering learners and communities to rebuild more robustly in the aftermath of disasters. While past initiatives have laid a solid groundwork, sustained commitment to innovation is now essential to achieve a truly transformative global impact.

13.6 Conclusion

Strengthening disaster and climate risk governance is imperative worldwide as hazards intensify. It is vital to address these barriers to strengthen governance capacities and mitigate escalating risks exacerbated by climate change. Education plays a vital role in building risk resilience by empowering learners with knowledge and skills for preparedness, response, and recovery. UNESCO has made significant contributions through diverse programs that promote evidence-based policies, data-driven risk monitoring, capacity building of educators, and localized curricula. These initiatives shape national policies and education sector plans, build the capacities of teachers and administrators, and provide resources for climate change education, empowering learners at the community level with knowledge and skills for resilience and facilitating participatory risk reduction.

UNESCO brings together diverse stakeholders, offers research-based guidance, and cultivates scientific collaborations across borders to support disaster and climate risk governance. Regional platforms further cooperate by customizing assistance and enabling exchanges based on local contexts. Recent technologies like crowdsourced risk mapping, online-enabled data collection, and capacity building between countries also aid these efforts. Though national governments hold key accountability, partnerships spanning the UN, civil society, academia, and industry sectors seek to bolster initiatives and tackle ongoing deficiencies in data strategies, capacity building, technology, financing, regulations, and policies.

Ultimately, resilience must be built across every level—from individual learners to schools, communities, education systems and policies. UNESCO can continue providing vital technical guidance while enabling collaborative action worldwide. All countries need risk-informed education policies, disaster-resilient schools, and learners equipped with knowledge, skills, and values for a sustainable future. But strengthening disaster risk education is not just a policy priority—it is an ethical imperative to protect lives and safeguard our shared future.

UNESCO has a key role to play in steering integrated progress between humanitarian response, sustainable development, and climate action. With strategic governance and education, we can empower generations to build back better from disasters and shape an inclusive, just, and risk-resilient world. Specifically, resilience against rising threats should be fostered at multiple levels, from preparing individual students with knowledge and capacity-building to help communities rebound after hazards through policies, partnerships, and on-ground teaching interventions. UNESCO can keep generating critical technical inputs on education systems even as it galvanizes coordinated action by public and private actors globally. UNESCO is uniquely placed to drive unified progress across disaster relief, sustainability efforts and climate action by guiding education programs that allow younger generations to recover from disasters and advance inclusive, and just societies.

References

Albris K, Lauta KC, Raju E (2020) Disaster knowledge gaps: exploring the interface between science and policy for disaster risk reduction in Europe. Int J Disaster Risk Sci 11(1):1–12

Berchtold C, Vollmer M, Sendrowski P, Neisser F, Müller L, Grigoleit S (2020) Barriers and facilitators in interorganizational disaster response: Identifying examples across Europe. Int J Disaster Risk Sci 11:46–58

Frontiers in Marine Science (2019) Lessons from the Pacific Islands—adapting to climate change by supporting social and ecological resilience. https://www.frontiersin.org/articles/https://doi.org/10.3389/fmars.2019.00289/full

Frykmer T, Tehler H, Uhr C, Wester M (2021) Advancing the field of disaster response management: toward a design science approach. Int J Disaster Risk Sci 12:220–231. https://link.springer.com/article/https://doi.org/10.1007/s13753-021-00330-0

Global Alliance for Disaster Risk Reduction and Resilience in the Education Sector (GADRRRES) (2023) Comprehensive school safety framework (CSSF) 2022–2030 for child rights and resilience in the education sector. https://gadrrres.net/

Grimaz S, Malisan P (2016). VISUS: a pragmatic expert-based methodology for the seismic safety triage of school facilities. Bollettino di Geofisica Teorica ed Applicata 57(2)

International Federation of Red Cross and Red Crescent Societies (IFRC) (2015) Handbook on law and disaster risk reduction. https://www.ifrc.org/document/handbook-law-and-disaster-risk-reduction

IFRC and United Nations Development Programme (UNDP) (2015) The handbook on law and disaster risk reduction

IFRC & United Nations Development Programme (UNDP) (2015) The checklist on law and disaster risk reduction. https://www.ifrc.org/sites/default/files/2022-02/Checklist_law_drr_2015.pdf

PLOS Climate (2022) Lessons for adaptation pathways in the Pacific Islands. https://journals.plos.org/climate/article?id=https://doi.org/10.1371/journal.pclm.0000011

Shaw R (2014) Educational governance in disaster risk reduction. In: Risk governance: the articulation of hazard, politics and ecology. Springer Netherlands, Dordrecht, pp 167–184. https://link.springer.com/chapter/https://doi.org/10.1007/978-94-017-9328-5_10

University of Central Florida (UCF) Online (n.d.) The disaster management cycle: 5 key stages & how leaders can help prepare. https://www.ucf.edu/online/leadership-management/news/the-disaster-management-cycle/.

United Nations (UN) (n.d.) Climate adaptation. https://www.un.org/en/climatechange/climate-adaptation

UNESCO (2010) Managing disaster risks for world heritage. https://whc.unesco.org/en/managing-disaster-risks

UNESCO (2011) Disaster Risk Reduction (DRR) in education: an imperative for education policymakers. https://unesdoc.unesco.org/ark:/48223/pf0000213925

UNESCO (2012) Disaster risk reduction in school curricula: case studies from thirty countries. https://unesdoc.unesco.org/ark:/48223/pf0000217036

UNESCO (2015) Disaster risk reduction: UNESCO's contribution to a global challenge. https://unesdoc.unesco.org/ark:/48223/pf0000233348

UNESCO (2017) Changing minds, not the climate: the role of education. https://unesdoc.unesco.org/ark:/48223/pf0000266203

UNESCO (2018) UNESCO 2017 (ERI-2018/WS/1 REV.1). UNESCO. https://unesdoc.unesco.org/ark:/48223/pf0000261558

UNESCO (2022) Medium-term strategy for 2022–2029 (41 C/4). United Nations educational, scientific and cultural organization. https://unesdoc.unesco.org/ark:/48223/pf0000378083

UNESCO (2023) Transforming education for a more inclusive society by building a culture of resilience to climate change: UNESCO Country and Liaison Strategy (2022–2025). https://unesdoc.unesco.org/ark:/48223/pf0000386112

United Nations International Children's Emergency Fund (UNICEF) (2021) Children's Climate Risk Index (CCRI). https://data.unicef.org/resources/childrens-climate-risk-index-report

United Nations Office for Disaster Risk Reduction (UNDRR) (2022) Caribbean safe school initiative (CSSI): thematic case view. https://sendaiframework-mtr.undrr.org/publication/caribbean-safe-school-initiative-cssi-thematic-case-view

United Nations Girls' Education Initiative (UNGEI) (2018) A whole school approach to prevent school-related gender-based violence: minimum standards and monitoring framework. https://healtheducationresources.unesco.org/library/documents/whole-school-approach-prevent-school-related-gender-based-violence-minimum

United Nations University (UNU) (2019) Green solutions for disaster risk reduction and climate change adaptation. https://ehs.unu.edu/news/news/green-solutions-for-disaster-risk-reduction.html

World Vision International Nepal (WVIN) (2020) Annual Report 2019. https://www.wvi.org/publications/annual-report/nepal/annual-report-2019-world-vision-international-nepal

Chapter 14
Disaster and Climate Change Education: Importance of Local Governance

Yukihiko Oikawa

Abstract Since the twenty-first century, climate change has been becoming more serious year by year due to human economic and social activities. It causes many losses and damages in not only environmental but also social and economic aspects, especially as disaster. Therefore, SDGs set challenges of climate change as Goal 13, and ESD emphasizes Climate Change Education (CCE) as top priority on "Berlin Declaration on ESD" in 2021. In this context, UNESCO has launched "Green Education Partnership" focused on CCE. To implement CCE effectively at each nation and local level, governance should take crucial role at all level. In this chapter, how policies and strategies of local, national and global governments promote and empower the efforts of CCE including disaster education in Japanese case will be discussed.

Keywords Climate change education (CCE) · Disaster education · Sustainable development goals (SDGs) · Education for sustainable development (ESD) · Governance

14.1 Disaster and Climate Change in the Sustainable Development Goals (SDGs)

14.1.1 Disaster and Climate Change in the Context of SDGs

Since the end of the twentieth century, on a global scale, humanity has been facing a significant number of global issues such as (i) persistent conflicts such as religious conflicts and ethnic conflicts; (ii) emerging environmental issues such as climate risk, biodiversity, ocean issues, increasing severe natural disasters such as typhoons, flood, drought, earthquakes, tsunamis, etc.; (iii) globalization including the influx of refugees and foreign workers, on the other hand; (iv) the rise of protectionism and nationalism; (v) widening economic disparity with a few wealthy people controlling

Y. Oikawa (✉)
Nara University of Education, Nara, Japan
e-mail: oikawa.yukihiko.rj@cc.nara-edu.ac.jp

A. Yildiz and R. Shaw (eds.), *Disaster and Climate Risk Education*, Disaster Risk Reduction, https://doi.org/10.1007/978-981-97-5987-3_14

most of the world's wealth in the world; (vi) advances in science and technology such as the rise of AI and changes in a social structure called "Society" 5.0, which was proposed in the 5th Science and Technology Basic Plan in Japan as a future society that Japan should aspire to and it is a human-cantered society that achieves both economic development and the resolution of social issues through a system that highly integrates cyberspace (virtual space) and physical space (real space); (vii) population disparities such as population explosion in developing countries and labour shortages in developed countries; and (viii) the pandemic of the new coronavirus infection, and so on.

Currently, these global problems are on the rise and are becoming more complicated and unsustainable; those have become apparent in environmental, social, and economic aspects. Therefore, efforts towards the achievement sustainable societies is taking on an important role throughout the world.

In light of that, the UN Sustainable Development Summit was held to solve the global critical issues and to realize a sustainable society in September 2015. Leaders from more than 150 countries gathered at UN Headquarters to formally adopt an ambitious new Sustainable Development Agenda that is named "Transforming our world: the 2030 Agenda for Sustainable Development". The agenda aims for a world where "No one will be left behind" and has set 17 goals and 169 targets as action goals shared by humanity in the world. These are the Sustainable Development Goals (SDGs). The SDGs officially came into effect on January 1, 2016 (UN 2015).

As for climate change, SDGs set goal 13 as "Climate Action: Climate change is a global challenge that affects everyone and everywhere". Climate risk is a serious and urgent challenge facing humanity.

Compared to prior to the Industrial Revolution, the average global temperature in 2020 had risen by about 1.2 °C due to the emission of greenhouse gases including carbon dioxide caused by the consumption of fossil fuels in human economic and social activities up to the present. In particular, the number of automobiles has more than tenfold to 1.4 billion units in 60 years. Electricity consumption from fossil fuels will increase more than 25 times in 70 years. According to a 2013 report by the Intergovernmental Panel on Climate Change (IPCC), the average temperature of the atmosphere is projected to rise by up to 4.8 °C by the end of the twenty-first century if countries around the world do not take action against global warming (IPCC 2013).

The acceleration of climate change will have a critical impact on both the global environment and human life. Greenhouse gases in the atmosphere is continuing to increase, global warming will melt ice sheets such as Greenland, and rising sea levels will give significant impacts on island nations and coastal regions.

On the other hand, climate change is causing disasters to occur more frequently and severely around the world. Due to the increasing number of disasters, such as frequent droughts, extreme temperatures, desertification, super typhoons, and floods, more and more people are forced to evacuate and are exposed to severe food crises. In addition, forest fires and droughts are increasing, leading to the extinction of flora and fauna and the loss of biodiversity. During 2019–20 Australian bushfires killed more than one billion animals, causing health hazards in California in 2020. The area of forest fires in the world in 2021 was approximately 9,300,000 ha, which is

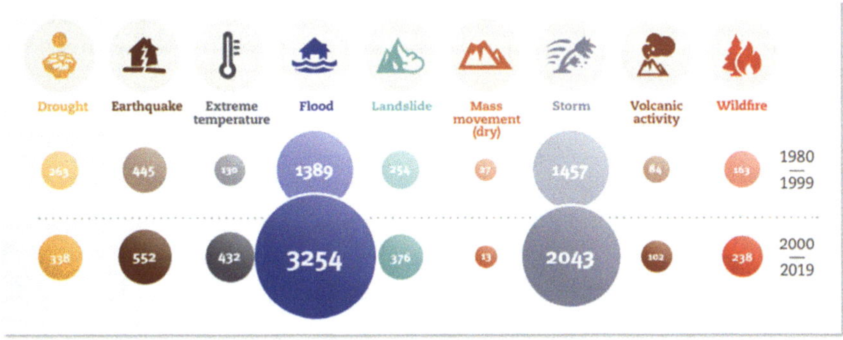

Fig. 14.1 Total disaster events by type: 1980–1999 versus 2000–2019. *Source* UNDRR human cost of disasters: an overview of last 20 years (2020)

equivalent to two times in 2001. In addition, the thawing of permafrost in the Arctic has led to the emergence of unknown viruses and the risk of new infectious disease epidemics.

Disaster and climate change are two sides of the same coin. Climate change will have an impact on human life and society and bring about many types of disasters. From the end of the twentieth century to the beginning of the twenty-first century, Disaster caused by climate change, such as floods, storms, extreme temperatures, droughts and wildfires, have increased drastically (Fig. 14.1). Therefore, disaster education and Climate Change Education should be integrated to the learning and implemented from both perspectives.

14.1.2 Disaster and Climate Change Education Multi-linked to the SDGs

It should be considered that the approach of Disaster and Climate Change Education is multilinked and interconnected to the SDGs. When the Climate Change Education will be promoted, it should be expected that Climate Change Education will develop and contribute to the entire Goals of SDGs in a coordinated and comprehensive measure.

For example, by merging Goal 13's "climate change" to Goal 4's "education," students can learn about basic countermeasures of climate change issue. That is a key component of Climate Change Education in a narrow sense.

On the other hand, in the course of this research, it will also face the intensification of seawater temperature and rising sea level due to climate change (Goal 14), the resulting intensification of meteorological disasters (Goal 11, etc.), and the impact on terrestrial ecosystems (Goal 15). In addition, in the course of this research, it will face the intensification of seawater temperature due to climate change (Goal 14), the resulting intensification of meteorological disasters (Goal 11, etc.), and the impact

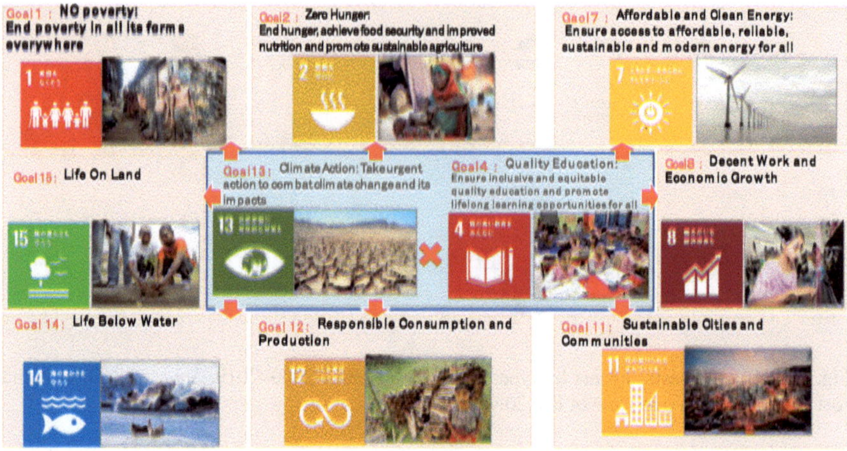

Fig. 14.2 Climate change education multi-linked to SDGs. *Source* Author

on terrestrial ecosystems (Goal 15). Furthermore, as a social aspect, it is connected to food shortages due to climate change (Goal 2), the shift from fossil fuels to renewable energy (Goal 7), responsible consumption and production (Goal 12), poverty (Goal 1) and employment (Goal 8) problems caused by climate change.

In this way, Disaster and Climate Change Education plays a critical role in connecting each goal of SDGs and strengthens the visualization of these linkages and interconnected efforts. In other words, this is the integrated and interdisciplinary approach to Climate Change Education in a broad sense or perspective (Fig. 14.2).

14.1.3 Importance of Local Governance Towards Achieving the SDGs

The local initiative of SDGs, including the adaptation and mitigation for disaster and climate change, is very important to achieve all SDGs. Especially, the role of local governance is crucial to tackle and solve issues based on local situations practically as the bottom-up approach. Since the SDGs has launched in 2015, a lot of local governments in Japan have been promoting the projects of SDGs positively tackling local issues those they are facing. It is locally called "Local SDGs" in Japan.

In this context, Japanese government has been promoting "SDGs Future City" project since 2018. "SDGs Future Cities" are selected as cities and regions that have a particularly high potential to realize sustainable development through the creation of new value in the three aspects of economy, society, and environment from among the cities and regions that are trying to promote basic and comprehensive initiatives in line with the philosophy of SDGs. Japanese government has designated 182 local governments as the SDGs Future City from 2018 to 2023. Moreover, during same term, the Japanese government has designated 60 local governments as "Municipal SDGs Model Projects" that are leading the way in the SDGs Future Cities initiative and are expected to create a self-sustaining virtuous cycle in their communities through collaboration with diverse stakeholders (CAO 2023). Each SDGs Future City as well as Municipal SDGs Model Project has been implementing diverse and characteristic strategies and initiatives based on the local issues towards achieving the SDGs in Japan.

Thus, it is recognized that the local governance is taking a critical role of contribution to the SDGs in Japan as the promotion of Local SDGs. Therefore, the initiatives of local governance also accelerate Disaster and Climate Change Education, which is the approach to SDGs.

14.2 Disaster and Climate Change Education in the Context of Education for Sustainable Development (ESD)

14.2.1 Disaster and Climate Change Education in the Evolution of ESD

In Japan, Disaster and Climate Change Education has been promoted in the context of ESD. As UNESCO described, ESD is the education that fosters abilities and attitudes needed to create sustainable societies and future, solving global issues such as environmental destruction, poverty, conflict or war, as well as the climate change and the natural disaster and so on. ESD educates people to perceive global problems to be their own problems and to take action to think globally, act locally, and refers to learning and activities that aim to transform to new values and behaviours that will help solve those issues. As a result, it is able to create sustainable societies and future. Taking that into consideration, disaster and climate change education are parts of the ESD approaches, and it is an urgent and important issue for the survival of human beings. Currently, its priority is getting high year by year in the world.

First, it should be confirmed the progress of ESD has been about 30 years so far. In 1992, Earth Summit in Rio de Janeiro, Brazil, emphasized the importance of education in order to build a sustainable society in chapter 36 of Agenda 21. Based on this message, in 2002, "United Natation Decade of Education for Sustainable Development (UNDESD)" was proposed at World Summit on Sustainable Development in Johannesburg, South Africa. It was launched in 2005 until 2014 by UNESCO as lead agency. At present, ESD is facing new stage after the SDGs was launched in UN Sustainable Development Summit in 2015.

Japanese government has been committed and contributing to all the process of ESD Evolution. Especially, Japan initiated to propose UNDESD with co-proposing countries at World Summit on Sustainable Development in 2002. Flowing that, at the end of UNDESD, UNESCO conducted the World Conference on Education for

Sustainable Development cosponsored with Japan in Aichi City, Nagoya Prefecture, Japan, and "Global Action Programme on Education for Sustainable Development (GAP)" and "Aichi Declaration" was launched as follow-up of DESD at the World Conference (Oikawa 2014a).

The GAP was officially adopted in UN assembly in 2014, which is the final year of UNDESD, and it was implemented from 2015 to 2019.

Moreover, at the new phase of ESD after GAP, Japan is one of the main countries related to the launch of "ESD for 2030". Its official title is "Education for Sustainable Development: Towards achieving the SDGs (ESD for 2030)" (Table 14.1).

Table 14.1 Evolution of the concept of education for sustainable development

Year	Events	Place	Agenda/Declaration/Output
1992	Earth summit (United Nations Conference on Environment and Development: UNCED)	Rio de Janeiro, Brazil	The role of education in achieving sustainable development was emphasized in Chapter 36 of Agenda 21
2002	World Summit on Sustainable Development	Johannesburg South Africa	"UN Decade of Education for Sustainable Development (DESD) (2005–2014) was proposed by Japan
2009	UNESCO World Conference on ESD	Bonn, Germany	Bonn Declaration was issued at mid-term of DESD for the second half of DESD Emphasized the disaster education in ESD
2012	UN Conference on SD (Rio + 20)	Rio de Janeiro, Brazil	"The future we want" was addressed in, 20 years later from Earth Summit was held in Brazil, 1992
2014	UNESCO World Conference on ESD	Aichi-Nagoya/ Okayama, Japan	At the end of DESD, held in Japan, and launched Global Action Programme(GAP) and Aichi Declaration
2015	Sustainable Development Summit 2015	New York, USA	"Transforming Our World: 2030 Agenda for Sustainable Development" including SDGs was adopted
2019	40th UNESCO General Conference	Paris, France	"Education for Sustainable Development: Towards achieving the SDGs (ESD for 2030)" was adopted

Source Author

14.2.2 Disaster and Climate Change Education in the Context of the ESD for 2030

The overall goal of "ESD for 2030" is aiming to achieve a more just and sustainable world by contributing to the achievement of the all goals of SDGs through strengthening ESD. The objective of ESD for 2030 is to integrate the concept and practice of ESD and SDGs into learning environments, the capacity building of educators, the empowerment and mobilization of young people, policies and local level action. Therefore, "ESD for 2030" insists to strengthen the contribution of ESD to all SDGs including addressing climate change and disaster. In other words, Education is at the key enabler of the 2030 Agenda for Sustainable Development essential for the achievement of 17 goals of SDGs (UNESCO 2023).

ESD for 2030 is suggesting three dimensions of ESD learning as follows (Fig. 14.3).

① Dimension 1: Cognitive learning—Learners will be expected to understand sustainability challenges and their complex interlinkages, and to explore disruptive ideas and alternative solution.
② Dimension 2: Social and emotional learning—Learners will be expected to build core values and attitudes for sustainability, to cultivate empathy and compassion for other people and the planet, and to motivate to lead the change.
③ Dimension 3: Behavioural learning—Learners will be expected to take practical action for sustainable transformations in the personal, societal and political fields.

These three dimensions of ESD learning could be adapted to the disaster and climate education. Through those learnings, Disaster and Climate Change Education could contribute to the achievement of SDGs. As to achieve Goal 11, "Sustainable Cities and Communities" including Disaster Risk Reduction (DRR) could support creation the cities and communities as places of meaningful transformative actions. As towards achieving Goal 13, "Climate Action" is to tackle climate risk by improving education, raising awareness as well as fostering human and institutional capacity. And as the approach towards achieving Goal 4 "Quality Education" is for educators to develop quality learning contents and methods which can contribute to the survival and prosperity of humanity. In order to accelerate Disaster and Climate Change Education, it is crucial to merge these dimensions or approaches to its learning.

Fig. 14.3 ESD toward achieving all SDGs. *Source* UNESCO

14.3 Greening Education Partnership for Promoting Climate Change Education

14.3.1 UNESCO Greening Education Partnership for Climate Change Education

In 2021, UNESCO conducted World Conference on ESD in Berlin as the kick-off of "ESD for 2030", and adopted Berlin Declaration on ESD. An urgent action is crucial to address the drastic interrelated crisis under the recognition that the world is facing the climate risk in particular. Its commitment insists that ESD for 2030 is the foundational element of its educational systems at all levels with Climate Change Education as a core curriculum component.

In this context, UNESCO has launched "UNESCO Greening Education Partnership" as an initiative of Climate Change Education at international level in 2022.

Mr. Antonio Guterres, the Secretary-General of UN, calls the climate crisis "a battle for our lives" in the situation which human beings is still struggling to transform societies to reach the 1.5°C-degree path recommended by the Paris Agreement. The Greening Education partnership was launched as a global initiative to encourage member countries in accelerating the implementation of Climate Change Education in response to the climate risk and crisis; that concept was emerging from the 2022 UN Secretary General's Transforming Education Summit; also, it was urged by the Youth Declaration.

However, according to the survey on current status of Climate Change Education by UNESCO below, sad to say, the promotion status of education related to climate change at global level is not so active.

- It was revealed that about half of the 100 countries reviewed had no mention regarding to climate change (education) in their national curriculum.
- 95% of teachers at surveyed primary and secondary schools were conscious that teaching climate change is crucial. On the other hand, less than 30% in those school response a readiness to teach it.
- About 75% of youth are frightened to their future due to climate change. Young people demanded practical strategies and action, and they called upon governments to change education systems those are fit to address the challenges of climate change at COP26 in Glasgow and the TES pre-Summit in Paris in June 2022.

Therefore, the partnership of stakeholders of Greening Education should be enforced beyond national boundaries and sectors. UNESCO launched "Greening Education Partnership" titled "Getting every learner climate-ready" in 2022.

The Greening Education Partnership is a global initiative by UNESCO, that should be a whole-of-system approach to empower member countries in addressing the climate change by utilizing the important role of education. As a collaborative platform for governments and other stakeholders, including inter-governmental organizations, civil society, youth, academia, and the private sector, Greening Education Partnership aims to deliver strong, coordinated and comprehensive action to prepare every learner to acquire the knowledge, skill, value, and attitude to tackle climate change and to promote sustainable development.

The practices monitored by members of the Partnership adopt an interdisciplinary approach across all subjects, not only looking at the scientific facts of climate change but addressing the social, economic and behavioural facets of climate justice, eco-anxiety, action-based solutions and ensuring that it is contextualized to learners' real-life experience and aspirations.

14.3.2 Pillars of Greening Education Partnership

The Greening Education Partnership has 4 pillars of action areas as follows (Fig. 14.4). Member countries and organizations are encouraged to participate in the Greening Education Partnership, expressing their interest in at least one of the four action areas.

① Pillar 1: Greening Schools

Greening Schools should engage to ensure that all schools achieve green school accreditation, from early childhood through adult education including teacher training and higher education institution.

② Pillar 2: Greening Curriculum

Greening Curriculum should try to embrace a life-long learning approach that integrates Climate Change Education into curriculum, teaching materials, pedagogy and assessment as well as technical and vocational education, workplace skills development.

③ Pillar 3: Greening Teacher Training and Education Systems' Capacity

Teacher Training and Education Systems' Capacity is to strengthen the systems of education and school to be climate-smart with appropriate measures, and guarantee the training of teachers, policymakers and so on.

④ Pillar 4: Greening Communities

Greening Communities is to engage communities to climate action by integrating Climate Change Education into life-long learning through community learning centres and learning cites in particular.

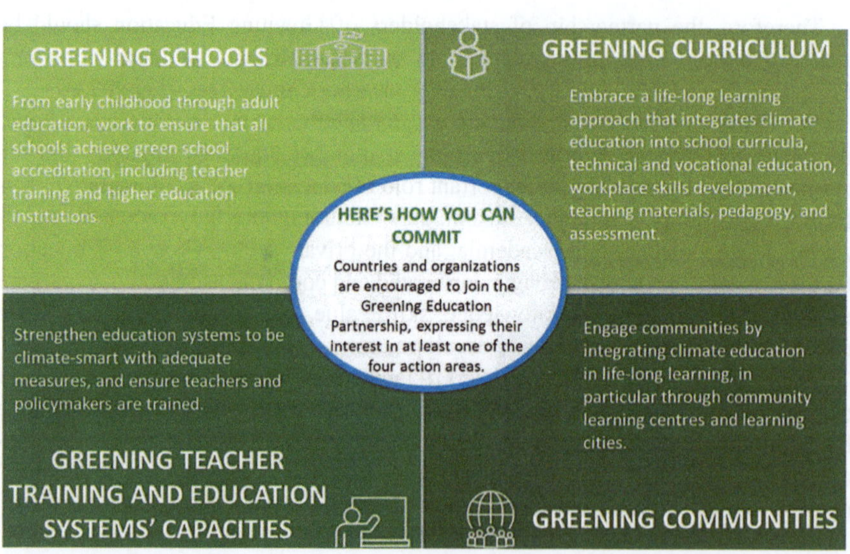

Fig. 14.4 Pillars of UNESCO greening education partnership. *Source* UNESCO

UNESCO has defined the stakeholders of Greening Education Partnership such as member states, organizations and institutions including civil society organizations, UN agencies, research institutions, youth organizations and international development partners as an open and inclusive community.

The Greening Education Partnership inspires the climate action from countries to empower learners with the skills required for inclusive and sustainable development in the context of the transition toward digital and green economies, drawing on ESD's holistic approach to learning (UNESCO 2022).

Thus, the Greening Education Partnership can be expected to develop and promote Climate Change Education at global level and it opens the door towards a new frontier of Climate Education. In order to accelerate a climate education and action around the world, the initiative of global governance like UN agencies such as UNESCO will take a critical role through the harnessing international partnerships and collaborations.

14.4 Policies for ESD Including Disaster and Climate Change Education in Japan

14.4.1 Renewing Law for Promotion of Environmental Education

At first, the Japanese government, mainly the Ministry of Environment, renewed the "Act on the Promotion of Environmental Conservation Initiatives through Environmental Education" in the context of ESD in 2011 after the Great East Japan Earthquake and Tsunami. It describes the improvement of environmental education including climate change in school education and the reinforcement of environmental education. The main principles of this renewal are as follows (MOE 2011).

① As enhancement of the basic philosophy, valuing on collaborative efforts, respect for life, integrated the development with the economy and society, formation of a recycling-oriented society.
② Forming the framework for practical promotion by local governments.
③ Innovating the environmental education in school education.
④ Reinforcing of systems for environmental education.
⑤ Introducing the opportunities for providing nature experiences.
⑥ Promoting the participation of private sectors in environmental administration and collaborative efforts.

14.4.2 National Curriculum Guidelines and Basic Plan for the Promotion of Education

Especially in formal education, Japan government has been promoting ESD at all school levels from elementary schools to high school since UNDESD started in 2005. The Ministry of Education, Culture, Sports, Science and Technology (MEXT) reformed the National Curriculum Guidelines based on the concept of ESD in 2017. It mentions in the preamble, "All schools from now will seek to enable each and every child (student) to recognize his or her own worth and potential, and while respecting the value and worth of all others and engaging with a diverse range of people, to overcome various social transformations, open up a rich life for themselves". And as the goal of school education, it sets for students to become creators of sustainable societies (MEXT 2017). Moreover, in General Principles of the national curriculum guideline, it mentions "aiming to equip children, who are expected to become creators of sustainable societies, with rich creativity and foster in them a zest for life, the curriculum will clarify the qualities and capabilities that schools will seek to foster through the entire school education as well as through each subject". This means that Japanese School Education is based on the ESD (Table 14.2).

The third Basic Plan for Promotion of Education approved in 2018 as well as the fourth plan in 2023 insists to foster creators of the sustainable society forecasting society after 2040 and to improve well-being based on Japanese society (MEXT 2023).

In this way, in Japan, fostering the "Creators of a Sustainable Society" which is the purpose of ESD is incorporated into the new National Curriculum Guidelines as well as the Basic Plan for Promotion of Education as a crucial concept. This means that the foundation of ESD which every school all over Japan can implement ESD in formal education and curriculum has established. This is the most important role of national government to promote ESD including Disaster and Climate Change Education in national wide. Therefore, Japanese schools have a potential to implement systematic and practical Disaster and Climate Change Education in formal education according to the national curriculum guideline, developing the curriculum or lesson plan for them at each school. It is the advantage of Japanese educational governance for the promotion of Disaster and Climate Change Education in the context of ESD.

14.4.3 Japanese National Implementation Plan for "ESD for 2030"

Japan government has also established new National Implementation Plan on ESD for "ESD for 2030" by the ESD Inter-Ministerial Meeting as multi-stakeholder approach in 2021. The plan aims to promote ESD as an all-Japan effort and to initiate ESD in the world as an ESD advocate country. It involves various stakeholders to put into

Table 14.2 Educational policy for ESD promotion in Japan

New National Curriculum Guidelines	Third Basic Plan for the Promotion of Education
(Issued in March 2017) **【Preamble】** All schools from now will seek to enable each and every child (student) to recognize his or her own worth and potential, and while respecting the value and worth of all others and engaging with a diverse range of people, to overcome various social transformations, open up a rich life for themselves, and **become creators of sustainable societies** **【Chapter 1 General Principles】** 1. Fundamentals of elementary (junior high school) education and role of curriculum 1–3. ...in aiming to equip children, who **are expected to become creators of sustainable societies**, with rich creativity and foster in them a zest for life, the curriculum will clarify the qualities and capabilities that schools will seek to foster through the entire school education as well as through each subject, moral education class...period for integrated study, special activities...and will enhance the education activities	(Approved by the Cabinet in June 2018) Section 14.2 **Aims and Measures for Education Policy in the Coming Five Years** **<Mainly at primary and secondary education stages>** MEXT will enhance the activities of **UNESCO Associated Schools, which are considered as the bases for ESD in Japan**, and will communicate and spread their successful examples throughout the country. Moreover, MEXT will promote **the implementation** and **diffusion of ESD** and exchanges among schools, as well as deepen the reach of ESD. Through these initiatives, leaders who will **create sustainable societies will be fostered** **<Mainly at higher education stage>** MEXT **will facilitate the implementation of ESD** through collaboration among a variety of regional stakeholders (schools, boards of education, universities, companies, NPOs, community centers, etc.), as well as **deepen the reach of ESD** so as to contribute to the realization of the SDGs through various interdisciplinary and other initiatives. Through these efforts, students will be encouraged to view global problems as their own, develop a mindset to take action nearby, and **become leaders for the creation of sustainable societies**

practice as throughout Japan. Furthermore, it describes specific initiatives for each stakeholder as follows(Interministerial Meeting on ESD 2021).

① Involving various sectors and stakeholders to put into practice

- Forming the multi-layered networks connecting diverse stakeholders such as schools, local research institutions, governments, companies, NGOs/NPOs, etc.
- Strengthening advocacy to disseminate information not only in Japan but also in international community, and to promote mutual collaboration.

② Describing priority initiatives for specific stakeholders defined in each of 5 Priority Action Areas for promoting "ESD for 2030" below:

- Priority 1: Promotion (Advance) of policy support
  - National and local governments (including boards of education).

- Priority 2: Transforming the learning environments
  - schools, national and local governments (including boards of education), other educational institutions (social education facilities, universities, institutions that practice education through ESD), citizen groups including NGOs/NPOs and public interest corporations, private companies, local communities, etc.
- Priority 3: Building the capacity of educators
  - Educators who practice education through ESD, such as school teachers, national and local governments (including boards of education), other educational institutions (social education facilities, universities, institutions that practice education through ESD), citizen groups including NGOs/NPOs and public interest corporations, private companies, etc.
- Priority 4: Encouraging youth empowerment and participation.
  - Youth organizations, youth support organizations, civil organizations including NGOs/NPOs and public interest corporations, national and local governments (including boards of education), educational institutions (schools, social education facilities, universities, private companies, etc., institutions that practice education through ESD), and private enterprises.
- Priority 5: Accelerating activities and action at the local level.
  - National and local governments (including boards of education), citizen organizations including NGOs/NPOs and public interest corporations, educational institutions (schools, social education facilities, universities, private companies, etc.) that practice education through ESD institutions), private companies, and all members of the local community.

14.4.4 Guidebook for ESD Promotion

The Ministry of Education (MEXT) also published and renewed "Guidebook for ESD promotion for teachers" to accelerate ESD mainly in school education in 2021. The guidebook shows how to design and manage ESD curriculum. It also gives the guidance how to establish and promote ESD as the whole school approach, and to formulate the partnership and collaboration for ESD promotion (MEXT 2021).

- Objective of the Guidebook:
  "Guidebook for ESD promotion" for educators teachers was issued in 2016 and renewed in 2021 as version 3 to accelerate ESD practice at school.
- Pillars of the Guidebook:

  ① Method of the design and management of ESD curriculum in school education to implement ESD at each school
  ② Process of promoting whole institution (school) approach for ESD promotion

③ Strategy for building the linkage and collaboration between schools and communities or institutions to spread and deepen ESD.

14.4.5 *UNESCO Associated School (ASPnet School) in Japan*

In Japan, Climate Change Education has been implemented in the context of ESD. Especially, UNESCO Associated School network (ASPnet School) is the hub to promote ESD including Disaster and Climate Change Education in each region. The number of ASPnet Schools has been increasing dramatically since 2005, when DESD started. It is locally called as "UNESCO School" in Japan. At present, Japan has more than 1100 ASPnet schools all over Japan from kindergarten to university (Fig. 14.5). This number equates to 10% of UNESCO Associated Schools in the world.

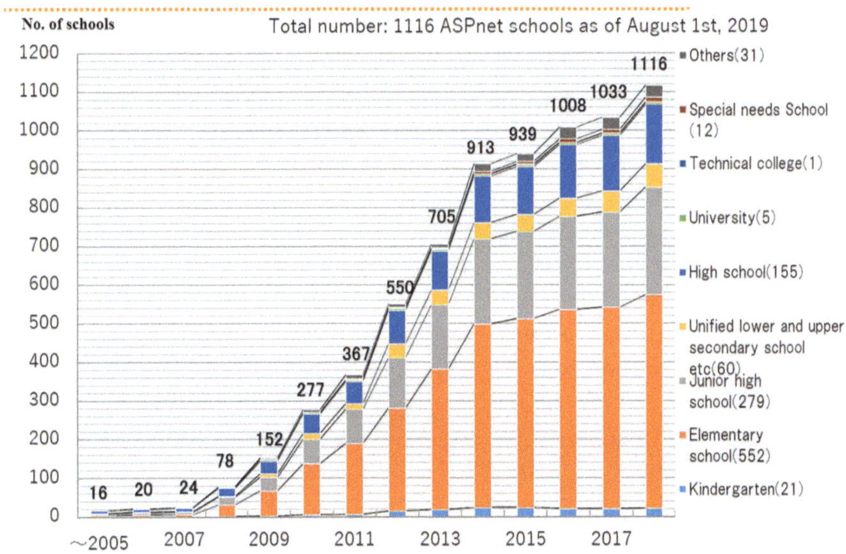

Fig. 14.5 Transition of UNESCO associated schools in Japan from 2005 to 2019. *Source* MEXT

14.5 Policy for Accelerating of Disaster and Climate Change Education in Japan

14.5.1 Notice for Promoting Global Environmental Education Including Climate Change

However, according to the survey of Japanese National Commission for UNESCO (JNCU), ASPnet Schools that tackle Climate Change education are only 20%, Sad to say, the priority of Climate Change Education has not been so high in Japanese School, similar to other countries, although Environment education is over 80% (Fig. 14.6).

Therefore, the Ministry of Education, Culture, Sports, Science and Technology (MEXT) and the Ministry of the Environment jointly issued a notice to schools and Board of Education all over Japan. The summary of the notice is as follows:

- Essentiality of efforts by all actors in order to realize a decarbonized society.
- Importance to shift citizen's lifestyle to a decarbonized one.
- Improvement of the global environment education to foster the creators of a sustainable society, which will deepen the understanding of global environmental issues

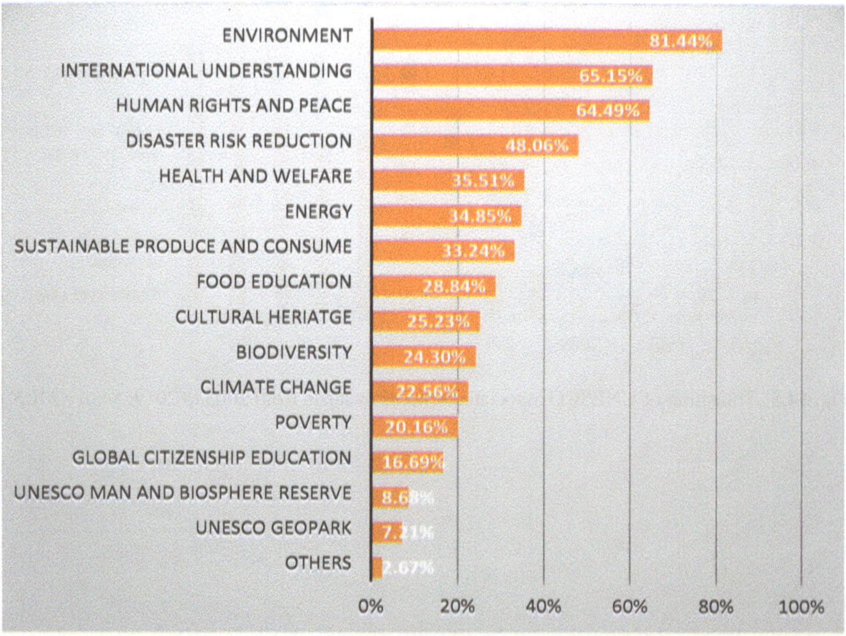

Fig. 14.6 Focusing areas of Japanese ASPnet schools. *Source* National Survey on ASPnet schools, Jan.2018, Japanese National Commission for UNESCO (JNCU)

- Taking actions to protect and conserve the environment
- (Notice from MEXT and MOE to the board of education and schools, 2021).

It aims to accelerate Global Environmental Education, including Climate Change Education, to shift our lifestyle to a decarbonized one(MEXT and MOE 2021).

Following this notification, The Ministry of Environment uploaded the teaching material on their website for students to learn Climate Change Education (CCE), titled "Change the future of the earth together to create a decarbonized society" The contents are below (MOE 2021).

① CCE teaching materials for decarbonization
② Learning map for environmental education including CCE
③ Guidance including good practices and practice examples in order to utilize the teaching materials

The materials adjust the learning contents depend on developing stage of students from primary to junior high school.

14.5.2 Promotion of Climate Change Education by ESD Resource Center of Japan

The Ministry of Environment and MEXT jointly established ESD Resource Center in 2016. It aims to form ESD Promotion Network in order to promote and support ESD practices at local and national level linking multi-stakeholders' efforts.

The center is currently initiating the Climate Change Education in Japan. It tries to expand the possibilities of Climate Change Education from the perspective of ESD through visualizing implementation at schools and social education (ESD Resource Center 2023).

The mission and objects of this initiative are as follows.

- Mission

  Accelerating the activity of Climate Change Education (CCE) from the perspective of ESD through demonstrating and visualizing practices at school and social education.
- Outcome objectives (by 2025)

  ① Mainstreaming of CCE in the context of ESD
  ② Expanding of CCE bases and networks
  ③ Building of the system and mechanism for promotion.

Its concept is to contribute to all SDGs from Climate Change Education. To integrate goal 13's "Climate Change" to goal 4's "Quality Education" is opening the door to contribution to other goals of SDGs. This means the integrated and interdisciplinary approach to climate change education in a broad sense or perspective (Fig. 14.7).

Fig. 14.7 CCE through ESD in Japan. *Source* ESD resource center

14.6 Local Governance and Practice of Disaster and Climate Change Education in Japan

14.6.1 The Case of Kesennuma City on Disaster and Climate Change Education

It will be introduced in two good practices of Disaster and Climate Change Education in school education and local governance in Japan.

The first case is Kesennuma City, Miyagi Prefecture, Tohoku Area in Japan. Kesennuma City is implementing ESD as whole city approach. Currently, entire elementary and junior high schools as well as some kindergartens and high schools were acknowledged as the UNESCO Associated School (ASPnet School) by UNESCO, and each school is implementing characteristic ESD practice which is approach of Environmental Education, Ocean Education, DRR Education, Cultural Education, International Understanding Education and so on, making the best use of rich nature, history, traditional culture as well as some local issues. The practice of ESD in Kesennuma is implemented systematically connected from kindergarten to high school (K to 12) depend on the developing stage of students in long term.

Kesennuma City is also designated as part of "Greater Sendai RCE". RCE is named "Regional Centres of Expertise" launched by United Nation University in 2005 when UNDESD started and started to ESD model project, it was RCE. Greater

Sendai RCE is one of the first group RCE in the world. It is called "Initial 7 RCE". RCE promotes ESD linking with formal education, non-formal education and informal education sectors such as schools, community learning centres, kind of museums, nature centres, city office, private company, NPO/NGO, specialized institution, and so on. Therefore, Kesennuma City, mainly Kesennuma City Board of Education, accelerated to establish the partnership with diverse sectors and institutions for promoting ESD as whole city approach. In this way, Kesennuma City has been promoting ESD as one of the pioneers and model not only in Japan but also in the world since 2002 (Oikawa 2012).

In 2011, Kesennuma City was affected with a mega disaster of Great East Japan Earthquake and Tsunami and it suffered huge earthquake, tsunami and fire. Most of the city was destroyed by this catastrophe and had fallen into an unsustainable situation. Under the local governance, Kesennuma City Board of Education tackled school restart and educational recovery with diverse strategies from ESD perspectives according to recovery processes (Oikawa 2013).

In this process of recovery in school education from the disaster, Kesennuma City had made the most use of ESD practice and network so far. Especially, Kesennuma City Board of Education improved disaster (DRR) education developing curriculum guidebook and renewing the disaster manual (Oikawa 2014b).

Under the initiative of Kesennuma City Board of Education, Shishiori Elementary School tackles Climate Change Education linking with ocean education and in the context of ESD. This school is an ASPnet School located in Kesennuma City.

At first, students are learning the mechanism of climate change through experiments and lectures by experts from universities. They learn the linkage between climate change and ocean events based on experiments such as warm and cold ocean currents, typhoon generation, annual comparison of seawater temperature, etc. (Fig. 14.8).

Flowing that, students exchange acquired knowledge and experience with peers abroad such as Republic of Kiribati, which is located in Oceania, through online conference. They think about the impact by climate change to small islands and its residents, sharing their thoughts and awareness beyond national boundaries. Furthermore, they discuss how to mitigate and adapt climate crisis to take actions for protecting sustainable society with each other. Through this international learning, students could share not only loss and damage of climate change in both countries, but also mutual sympathy to peers abroad and action for addressing climate change (Fig. 14.9).

14.6.2 The Case of Omuta City on Disaster and Climate Change Education

The second case is Omuta City, which is located in the south of Fukuoka Prefecture, Kyushu Area in Japan. Omuta City is also promoting ESD as whole city approach.

Fig. 14.8 Experiment of ocean currents

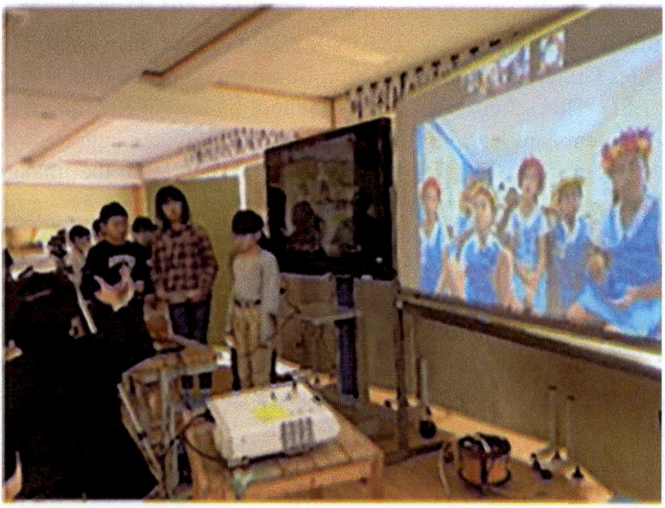

Fig. 14.9 Learning exchange by online. *Source* Kesennuma City Board of Education

Under the leadership of Omuta City Board of Education, entire school in Omuta City such as elementary school, junior high schools and special needs education school were acknowledged as ASPnet School simultaneously in 2012. That is an unprecedented achievement in the world, so that it is called "Omuta City that is UNESCO School City". Omuta City was also designated as "SDGs Future City in 2019 by Japanese government and got the "Special Prize of Japan SDGs Award" in

2019. Moreover, the city was acknowledged "Regional Centres on Expertise on ESD (RCE)" by United Nation University (UNU) in 2020 (Shaw et al. 2021).

In light of these strategies for ESD/SDGs, Omuta City has been accelerating practices of ESD and SDGs and its progress was outstanding as a leading city of ESD in Japan.

However, in July, 2020, Omuta City suffered the severe disaster that flooded the whole street of the city. Minato Elementary School especially, which is located in the urban area of the city was inundated by more than 1 m in just over 2 h due to the heavy rain, something they had never experienced before (Fig. 14.10). The first floor and school yard of the school was flooded soon, so that most of students evacuated to upper floor and many students had to stay overnight in the school building. The next morning, all students evacuated to upper floor of the school building and were safely rescued by the fire department and self-defense force using life rafts (Fig. 14.11).

Based on this terrible experience, Minato Elementary School students started to promote climate change education integrating with disaster education by inquired-based learning utilizing the integrated study period. Students researched the mechanisms and impacts of the disaster caused by climate changes, inviting experts from universities and institutions as well as involving local residents. Students tried to conduct "Town Watching" to find damages, impacts and vulnerabilities of the flooding disaster by communicating with community members. Moreover, they disseminated their work and awareness to the local community by making the DRR map and leaflets. Finally, the students suggested better skills for evacuation from the disaster and lifestyle for mitigation of climate change and they also proposed Build Back Better for the sustainable future of the community and city.

Fig. 14.10 Inundated Minato elementary school

Fig. 14.11 Students rescued by boat. *Source* Omuta city board of education

Thus, both case of Disaster and Climate Change Education in Kesennuma City and Omuta City were under the supervision of both city board of educations. Shishiori Elementary School and Minato Elementary school including other schools in each region have been implementing Disaster and Climate Change Education by receiving appropriate advices and supports by the local government, especially the board of education of each city. Therefore, these cases are not only good practices of schools but also the evidences of local governances in Disaster and Climate Change Education.

14.7 Disaster and Climate Change Education Toward Sustainable Development

In conclusion, in light of governances and practices at global, national, local and school level, those that are described above should be considered the strategy to promoted Disaster and Climate Change Education as the process of sustainable development. The learning process of Disaster and Climate Change Education can be divided into four steps to realize sustainable society achieving the SDGs. By thinking in this way, the process of Disaster and Climate Change Education should be clear and developed as follows (Table 14.3).

Table 14.3 Learning process of disaster and climate change education towards SD

Step	Step 1 mechanism of disaster and climate change	Step 2 impact on society and human life	Step 3 adaptation and mitigation	Step 4 build back better (creative recovery)
Capacity	Awareness and Knowledge for Mechanism of disaster, and climate change	Recognition of Linkages between disaster, climate change and society	Skill for Adaptation and Mitigation to disaster and climate change	Imagination, Creativity and Collaboration for Build Back Better (BBB)
Focus	Understanding Mechanism of disaster and climate change scientifically and critically	Recognizing Influences and Impacts of climate change to society, community and life	Learning how to reduce the risk of disaster and climate change and to implement measures	Learning Perspective and Contribution to Creative Recovery as a process for Sustainable Development
ESD Relations	**Scientific Knowledge on**: Factors and Mechanism of Environmental, Social, Economic and Cultural fields	**System Thinking about**: The Loss and Damage linking Environmental, Economic, Social and Cultural aspects	**Judgment & Action of**: Risk Management, Network Building, Using knowledge, wisdom and skills	**Participatory Contribution to**: Build Back Better as the process for rebuilding Sustainable Society from disaster and climate crisis

Source Author

① 1st Step is "Mechanism of Climate Change" that means the knowledge and awareness for the mechanism of disaster and climate change. At first, it is important for students and learners to understand Mechanism of climate change both scientifically and critically.

② The second step is "Impact on Societies and Lives", which means the recognition of linkages or connections between disaster / climate change and society. Students and learners should recognize the influences or impacts of disaster and climate change on society, community and lives.

③ Following that the 3rd Step is "Action for Mitigation and Adaptation", which means skills for mitigation and adaptation to address disaster and climate change. Students and learners have to learn how to reduce the risk of climate change and take action by implementing measures.

④ Last but not least, the 4th Step is "Recovery and Reconstruction", which means "Build Back Better" and needs the imagination, creativity and collaboration to achieve it. It is essential for leaners to learn about the educational process, perspective and contribution to creative recovery and reconstruction from the loss and damage by disaster and climate change.

To realize this undertaking, it is needed to foster educators and stakeholders who can manage and implement the program and curriculum of the Disaster and Climate

Change Education. Therefore, the governance of the Disaster and Climate Change Education at all level such as UNESCO, MEXT, Ministry of Environment, Local Government, Board of Education and schools should commit to each step of this learning process and should enhance to the learnings and practices for building sustainable society.

In Japanese school, ESD curriculum including Disaster and Climate Change Education would be developed by each school and teacher utilizing Integrated Learning Period advised and supervised by the board of education. On the other hand, disaster evacuation drill at the community or city level is often implemented conducted by local governments to raise residents" awareness and skills for DRR.

Thus, in formal education, nonformal education as well as informal education, the local governance will take a key role to accelerate Disaster and Climate Change Education. The local governments of each region and each level have to innovate their practice of Disaster and Climate Change Education taking 4 steps above into consideration.

References

CAO (2023) SDGsMiraitoshi/JititaiSDGsmodrujigyo Senteitoshi Ichiran (List of SDGs Future Cities/Municipal SDGs Model Project Selection Cities). https://www.chisou.go.jp/tiiki/kankyo/miraitoshi.html. Accessed 1 Mar 2024

(Changing the future of the earth together: To create a decarbonized society). http://eco.env.go.jp/lib/env/cn_education/page/top02.html. Accessed 1 Mar 2024

Chapter 1 General Provisions. https://www.mext.go.jp/content/20201008-mxt_kyoiku02-000005241_1.pdf. Accessed 1 Mar 2024

ESD Resource Center (2023) ESD/Kikouhendokyoiku no Suisin nituite (Climate Change Education for Sustainable Development). https://esdcenter.jp/wp-content/uploads/2023/03/846e01bb201c280e3474ed32d7ef90f4.pdf. Accessed 1 Mar 2024

Intergovernmental Panel on Climate Change (IPCC) (2013) Working Group I Contribution to the IPCC Fifth Assessment Report (AR5), Climate Change 2013: The Physical Science Basis, Final Draft Underlying Scientific-Technical Assessment. https://www.climatechange2013.org/images/uploads/WGIAR5_WGI-12Doc2b_FinalDraft_Chapter09.pdf. Accessed 1 Mar 2024

Interministerial Meeting on ESD (2021), Wagakuni niokeru "Jizokukanouna Kaihatsu notameno Kyoiku (ESD) nikansuru Jisikeikaku (Dai 2 ki ESD Kokunai Jishikeikaku) (Japanese National Implementation Plan on Education for Sustainable Development (ESD)): Second National ESD Implementation Plan. https://www.mext.go.jp/content/20210528-mxt_koktou01-000015385_2.pdf. Accessed 1 Mar 2024

MEXT and MOE (2021) Kikouhendo mondai wo Hajimetosita Chikyukannkyo nikansuru Kyoiku mo Jujitu nituite (Tsuchi) (Notification regarding the Enhancement of Education of Global Environmental Issues including Climate Change Issues). http://eco.env.go.jp/files/material05_r30602.pdf. Accessed 1 Mar 2024

MEXT (2021) Jizokukanona Kaihatsu notameno Kyoiku (ESD) Suishin no Tebiki (Guidebook for the promotion of Education for Sustainable Development (ESD)). https://www.mext.go.jp/content/20210528-mxt_koktou01-000015385_3.pdf. Accessed 1 Mar 2024

MEXT (2023) Kyouiku Shinko Kihon Keikaku (Fourth Basic Plan for Promotion of Education). https://www.mext.go.jp/content/20230615-mxt_soseisk02-100000597_01.pdf. Accessed 1 Mar 2024

Ministry of Environment (MOE) (2011) Kankyokyoiku touniyoru Kankyohozenno Torikumi no Sokushin nikansuru Houritsu (Act on Promotion of Environmental Conservation Efforts through Environmental Education, etc.) https://www.env.go.jp/content/900495299.pdf (Japanese). Accessed 1 Mar 2024

MOE (2021) Minnade Kaeru Chikyu no Mirai: Datsutannso Shakai wo Tukurutameni

Ministry of Education, Culture, Sports, Science and Technology (MEXT) (2017) National Curriculum Guideline of Elementary School

Oikawa Y (2013) Institutional response in education sector in Kesennuma City. In: Shaw R (ed) Disaster recovery: used or misused development opportunity. Springer, Japan, pp 89–113

Oikawa Y (2014a) Education for sustainable development: trends and practices. In: Shaw R, Oikawa Y (eds) Education for sustainable development and disaster risk reduction. Springer, Japan, pp 15–36

Oikawa Y (2014b) City level response: linking ESD and DRR in Kesennuma. In: Shaw R, Oikawa Y (eds) Education for sustainable development and disaster risk reduction. Springer, pp p155-176

Oikawa Y (2012) Education for Sustainable Development and Its Implications to Recovery Process in Kesennuma. In: Shaw R, Takeuchi Y (eds) East Japan earthquake and tsunami. Research Publishing. Singapore, pp 165–186

Preamble. https://www.mext.go.jp/content/20220603-mxt_kyoiku02-000005241_001.pdf. Accessed 1 Mar 2024

Shaw R, Sakurai A, Oikawa Y (2021) New realization of disaster risk reduction education in the context of a global pandemic: lessons from Japan. Int J Disaster Risk Sci 12:568–580. https://doi.org/10.1007/s13753-021-00337-7.Accessed1Mar2024

Cabinet Office of Japanese Government (CAO) (2022) Society 5.0. https://www8.cao.go.jp/cstp/english/society5_0/index.html. Accessed 1 Mar 2024

UNESCO (2022) Greening education partnership. https://www.unesco.org/en/sustainable-developmen/education/greening-future. Accessed 1 Mar 2024

UNESCO (2023) Education for sustainable development: a roadmap #ESD for 2030. https://unesdoc.unesco.org/ark:/48223/pf0000374802/PDF/374802eng.pdf.multi. Accessed 1 Mar 2024

United Nation (UN) (2015) Transforming our world: the 2030 Agenda for Sustainable Development. https://sdgs.un.org/2030agenda. Accessed 1 Mar 2024

Chapter 15
Disaster Risk Management and Climate Change Impact Actions in Türkiye

B. Burcak Basbug

Abstract This chapter briefly touches on the general definition and causes of climate change. Then, it discusses the climate change adaptation situation in Türkiye, focusing mainly on the education sector. It mentions the vital need for education and disaster and emergency risk management, including climate change, in the curricula. Education is the tool from an early age until the end of a university degree to create climatic and disaster awareness. The main sources of information in the chapter are secondary data and a scan of the available literature and documentation. The chapter ends with an overall glance at the actions to be taken in the Turkish education sector.

Keywords Climate · Development · Disaster · Education · Policy

15.1 Introduction

Climate, seasons, and weather structures do not behave as they used to in the past few years in Türkiye. Some signs of global warming and its effects are very visible. Even an ordinary citizen in the community can observe it with the naked eye without being a scientist. For instance, a flash flood in a downtown can suddenly occur even in metropolitan cities such as Ankara, Istanbul, and Izmir, which are thought to have strong infrastructure. In the winter of 2023–2024, no snowfall was observed in Ankara, which is located in mid-Anatolia and, under normal circumstances, has a harsh climate. Türkiye is normally famous for its four-season climate. One could enjoy the summertime on the beach in the west or south of the country, whereas, at the same time, one could enjoy the rain and skiing in the north and eastern parts of the country.

Things changed and will change over time. The community has a discussion about the fact that there will be no more four seasons, but just winter with cold weather and summer with the heat. The mid-seasons, that is, spring and fall, will soon totally disappear. It is not only the community that feels this change, not even being called

B. B. Basbug (✉)
Department of Statistics, Middle East Technical University, Ankara, Türkiye
e-mail: bbasbug@gmail.com

A. Yildiz and R. Shaw (eds.), *Disaster and Climate Risk Education*, Disaster Risk Reduction, https://doi.org/10.1007/978-981-97-5987-3_15

slow-onset anymore; on the contrary, it is very sudden-onset; the policymakers also accept the fact. Scientists have had warnings for years and years in Türkiye, as in other parts of the world. It was stated by TEMA (Türkiye Erosion Response Foundation) that the main Anatolia would be a desert in the next forty-fifty years. That report was published a few years ago, and the speed of the climatic and meteorologic change seems to shorten the estimated period of 40–50 years.

Preparedness and mitigation are the key points to be prepared for climate and disaster risk and related emergencies and crises. This chapter will discuss how existing mitigation strategies for disaster preparedness during ex-ante times link with actions taken before it gets worse by the impact of climate change. It will explain the link between disaster risk and climate risk and how it has been dealt with in the country in terms of education.

In research perspective, this study benefits from secondary data. Academic papers, institutional reports, media news, multilateral agency documents are examples of sources that have been consulted. Research showed that there is an increasing awareness and demand in understanding climatic risks and mitigation against those in the community as well as the policymakers and scientists. The unification of all these is very important to achieve the goals.

15.2 Disaster Risk Reduction/Management and Climate Change Adaptation: Türkiye

Due to its geological location, Türkiye is prone to multi-hazards, where man-made, natural and technology-originated events are observed frequently and also severely in recent years with the changing impact of the climate. Earthquakes, floods and landslides stake the first three ranks when named as the most observed disasters. Actually, a natural phenomenon does not necessarily turn into a disaster, where we expect social, economic, environmental and cultural losses, if there is no human intervention. Disaster scientists work on the three phases of disaster risk management; that is before, during and after times. Investment in capacity building in ex-ante (before) events is proven to be helpful during the event times with quick response and with quick and robust reconstruction and recovery at the ex-post (after) event period.

(Balaban 2022) explains the type of disaster Türkiye faces every year in link with the triggering climate change impact. Floods, forest fires, snow and landslides are observed frequently in different parts of the country and can cause excessive damage and losses. She also mentions that Istanbul, Ankara, and Izmir, the top three most populated cities in Türkiye, have large carbon footprints in Europe and are responsible for high levels of carbon emissions. The economic loss of the climate change impact will be in billions of US Dollars at the 2030 threshold and beyond in 2100. It is important to pay attention to the years of the main frameworks in the World policy.

SDGs target 2030, Sendai Framework for Disaster Risk Reduction covers 2015–2030. Paris Agreement of the Intergovernmental Panel on Climate Change (IPCC), Sendai and SDGs have all been declared post-2015.

Adaptation means making arrangements to have less affected natural and human systems by the consequences of climate change. Also, adaptation is explained as 'the process of strengthening, developing and implementing strategies in this area in order to combat the effects of climate events (of risks), provide benefits and manage the effects' (Küresel Denge Derneği 2016). If adaptation is adapted to climate change, it needs flexibility and a combination of strategies to reduce the impact.

Efficient stakeholder management in all three phases plays a crucial role in achieving success in disaster risk reduction (DRR) and disaster risk management (DRM) in competence with climate change adaptation (CCA) strategies. These all lead to contributing to the Sustainable Development Goals (SDGs) 2023 Agenda. SDG 13 Climate Action, SDG 11 Sustainable Cities and Communities and SGD 17 Partnership for Goals go hand-in-hand to support the development of resilient individuals, communities, nations and countries. In addition to these, SDG 4 is Quality Education. It is a clear fact that education, quality and good education are key to helping with DRR, DRM and all related activities for capacity building and development. It is one of the main sectors which needs to be updated and calibrated according to the changing dynamics and needs of the World.

In a research paper published by Basbug (2023), it has been emphasized that stakeholders, especially national and local governments, define the main areas of connection between climate change adaptation (CCA) and disaster risk reduction (DRR). These are:

Strategic: Goals and priorities in CCA and DRR in local and national development plans,

Operational: Policy frameworks and institutional arrangements to support CCA and DRR,

Technical: Strengthen technical capacities to assess risk and opportunities, identify and prioritise DRR and CCA measures, and finance them.

For this reason, there have been various national and international education projects in Türkiye to enhance DRR/DRM and CCA plans and actions. The country needs strong planning, adequate resources, and commitment to mitigate disaster risks and the impact of climate change at the same time. The plans need to be carefully designed, covering the parameters such as the poor in the community, who is more vulnerable and the worst hit; different geographical locations with different characteristics, and different vulnerabilities (physical, social, economic, cultural, and environmental).

In her research, Basbug (2023) states that climate change impact can affect disaster risks in twofold: first, through the likely increase in weather and climate hazards and second, with the increase in vulnerability of communities to different hazards. The challenge of increasing risks due to climate change makes it harder to implement and achieve Sustainable Development Goals (SDGs). Post-2015, the declaration of the Sendai Framework for Disaster Risk Reduction (SFDRR), the Paris Agreement on

Fig. 15.1 The link between sustainable development goals and climate change (open source UNDP 2016)

climate change and the 2030 Sustainable Development Agenda became inseparable documents and resources for scientists and policymakers. Figure 15.1 picturises the integration and coherence of the Sendai targets, SDGs and climate action.

15.3 Education Perspective

The following section uses the methodology of reading secondary data sources and briefing the outcomes from those. Education is a sector, which is affected by conflict, by refugee situations, by earthquakes, by extreme heat and extreme cold, pandemics and more. For instance, if there is a forced migration, continuous child education is an issue to pay special attention. If it is the case where there are extreme weather conditions plus forced migration plus gender violence, how a scientist or a policy maker will solve these problems? That is why and how the education sector needs strong advance planning, drills as practice, and adaptable and flexible designs in mitigation for DRR/DRM and CCA.

Proceeding in the context of the education sector, variables such as the structure of the school building, the location, the community it serves and its access to resources (e.g., water, energy, health) are some key elements to consider in planning. Increasing awareness of disaster and climate change topics in the communities, among the teachers, families, school administrations and pupils themselves, is not an easy job but surely doable. Education has direct (e.g. problem-solving skills, knowledge) and indirect (e.g. reducing poverty) effects, which have a role in mitigation action to reduce any existing vulnerabilities at the individual, local, national and even global levels.

In the research article by Barak (2020), it is stated that Climate Change Education (CCE) was born at the intersection of environment, sustainable development and disaster risk reduction education. Barak (2020) also explains the initialisation of covering CCE in the national education curriculum in different countries by the call from UNESCO. Countries such as Germany, Canada, Spain and Türkiye use a multidisciplinary approach, Sweden has a disciplinary approach, and the United Kingdom and Finland use both when CCE is included in the educational programs at schools.

The Ministry of National Education (MoNE) is the main responsible line ministry for preparing and implementing education policies of the Türkiye. Disaster and Emergency Management Authority (AFAD) is the main coordinating institution by Law for all disaster and emergency-related policies before, during and after disasters in the country. There is a Ministry of Environment, Urbanisation and Climate Change Adaptation, in which several departments climate change adaptation policies are activated. These three main institutions, with the contribution of other stakeholders such as the Ministry of Health, the Turkish Red Crescent, and universities, work on how to prepare, respond and recover in case of climate and disaster-related hazards.

A research article by Barak (2020) mentions that in Türkiye, the 5th, 6th, 7th, and 8th grades in the Turkish curricula have CCE covered. It mainly has components in the Social Sciences module in the 5th, 6th and 7th grades. The science and Technology module covers the CCE in the 5th, 6th, 7th, and 8th grades. In 2024, MoNE has been working on developments in the national curriculum to teach disaster and climate risks, starting from pre-school ages and lasting until university age. The core element of such activities is to be aware of disasters and climate risks for future generations. It is the responsibility of MoNE to have safe schools in structures in seismic hazard zones, to have safe infrastructure in flood-hazardous locations, and to provide equal opportunities to every enrolled student in the system. In that way, if all work together, it might be possible to have disaster and climate change resilient communities with all units. Ministry of Energy and Ministry of Agriculture are other line ministries which need to support the education system to teach children to save energy and to save soil. Non-governmental organizations (NGOs), academia and the private sector are other stakeholders to include in the system for better disaster and climate preparedness.

MoNE has been working on developments in the national curriculum to teach disaster and climate risks, starting from preschool ages and lasting until university age. The core element of such activities is having disaster and climate risk awareness for future generations. It is the responsibility of MoNE to have safe schools in structures in seismic hazard zones, to have safe infrastructure in flood-hazardous locations, and to provide equal opportunities to every enrolled student in the system. In that way, if all work together, it might be possible to have disaster and climate change resilient communities with all units. The Ministry of Energy and the Ministry of Agriculture are other line ministries that need to support the education system to teach children to save energy and soil. Non-governmental organizations (NGOs), academia, and the private sector are other stakeholders that should be included in the system for better disaster and climate preparedness.

The Ministry of Environment, Urbanization and Climate Change Adaptation in Türkiye have a strong will to ensure a balance between the policy and implementation at the local, national and international levels for the needs of the community and country. Policy preparations and implementation of climate change adaptation in the educational sector and country-level multi-sectoral development are indeed multi-actor systems. As in the DRR/DRM frameworks, there is a coherent and committed need for all the stakeholders, such as the national government, municipalities representing local government, NGOs, academicians, the private sector and volunteers from the community. All of the stakeholders should come into action the short-term, medium-term and long-term impacts of global warming and resulting climate change adaptation, especially for slow-onset and sudden disasters caused by global warming. Universities, with their research and teaching facilities, can support policies and their implementation with the use of scientific data on technical, social, cultural and environmental problems in link with climate change and disaster risk reduction. CCA is not just about changing temperatures; it is about legal systems, livelihoods, mitigation, and protection of vulnerable groups and ecosystems, which are cross-cutting points in DRR/DRM policies. One and each of these sectors need to have an education sector involvement starting from the preschool curriculum, going after and beyond university degrees. It is a lifetime learning process to adapt. It requires transparency, accountability, sustainability, equity and dignity, which are the lifelines of CCA and DRR/DRM together.

A report published by TESEV (2022) explains the detailed work of some main metropolitan provinces in the country and their efforts on different components of climate change adaptation with different capacities of each location with the eye of a local government. Although it is very useful to have, it is also very technical, with almost no educational component. In the very soon future, it will be a 100% must for every province to have resilience against climate change. And the core work for this road passes from education. Some provinces such as Adana, Ankara, Bursa, Denizli, Eskisehir, Hatay, Istanbul, Izmir, Kahramanmaras, Kocaeli, Manisa, Samsun, Tekirdag, Trabzon and Van have 'Action Plans for City xxx Compatible with Climate Change', which covers important topics like transportation, agriculture, communication, zero-waste, infrastructure, biodiversity, waste management, renewable energy, animal-friendly environment. It is important to pay attention that most of the cities in the country sit on the fault line or are located very near to an active fault. Bursa, Istanbul, Kahramanmaras, Hatay, Kocaeli, and Manisa are some big cities both with disaster and climate risks. Educational policies in such specific cities need to be extra carefully planned and prepared by including all the stakeholders. Especially at the local level, these plans need to be integrated into the citizen's daily lives, where there can be positive and useful outcomes for all.

To increase awareness at both the policymaker side and the community side, some suggestions can be provided as increasing in-service training for the employees of the municipalities, NGOs and the public. The top administration from the related Ministry to the municipality should also be educated about expected hazards, disasters and climate risks. The Union of Municipalities in Türkiye can be a hub to coordinate all these activities.

Ozer (2023) states that the most effective tool to cope with the social and economic trauma caused by disasters is education. It is also a fundamental component for humanitarian perspective and human rights as well as mitigating climate risk. All countries are responsible for sustainable and quality education. SDG 4 is the 'Quality Education' and needs to be accessible and available to all with equity.

Following the devastating impacts of the 6 February 2023 Kahramanmaras Earthquake, the Ministry of National Education implemented the post-disaster education services per as in (UNESCO 2010). This requires the collaboration of all the stakeholders. Experienced and trained teachers in the context of disaster and climate risk are required to take a vital role in rebuilding the education system in disaster affected areas. As a mitigation tool, there is a need for a database where students, teachers, parents, and other stakeholders can access trustworthy data that could be used in the education sector.

15.4 Conclusion

Rapid urbanisation, industrialisation, misuse of agricultural land, extreme consumption of fossil fuel and similar hazards mainly rooted in human action increase the climate change impact. MoNE (2023) clearly defines that Türkiye is one of the countries which will be hit the worst by climate change. It is one of the signatory countries of the United Nations Intergovernmental Panel on Climate Change, Kyoto Protocol and the European Green Deal. In line with that, the Eleventh National Development Plan covers relevant sections for mitigation strategies for climate change adaptation, such as the protection of water resources, food security, and increasing environmental protection awareness and education. Education is the only tool at every level to generate 'climate literacy', which is the key phrase that enables the communities, from individuals to the top decision-makers, to have resilience against climate change effects and adapt to the changes as quickly as possible.

In the workshop outcome, MoNe (2023) lists the actions to be taken in the Turkish education system to fight against the impact of climate change. These are:

- To update the national curriculum, from preschool to university level, in line with the Sustainable Development Goals (SDGs) 2030,
- To develop and design educational and learning tools in line with the update of the curricula,

- To increase public awareness of climate risk,
- To increase the infrastructural capacity of the school buildings.

Campaigns such as Zero Waste, Net Zero by 2053 and the Green Development Pact are consulted and used in updating the Turkish national education curriculum to produce effective climate change policies. It requires strong and coherent stakeholder management, where students, families, teachers, school principals, local and national education managers, municipalities, the private sector, academia and volunteers. A sustainable and safer future can only be achieved if all goes hand in hand with determination. And the key to this was education, is education and will always be education.

References

Balaban AY, Akman K (2022) Climate change adaptation policies of metropolitan municipalities in Turkey (Türkiye'de Büyükşehir Belediyelerinin İklim Değişikliğine Uyum Politikaları). Süleyman Demirel University Visionary J 13(36):1132–1149

Barak B, Gonencgil B (2020) A comparison of secondary school curricula in terms of climate change education in the world and Turkey. J Geogr 40:1–16. https://doi.org/10.26650/JGEOG2019-0039

Basbug BB (2023) Climate change and disaster management strategies in the book titled 'Climate Crisis Research Within the Framework of Disaster Preparedness' Livre De Lyon

International Network for Education in Emergencies-INEE (2010) Minimum standards for education: preparedness, response, recovery. https://spherestandards.org/wp-content/uploads/INEE-EN.pdf

Küresel Denge Derneği (2016) https://kureseldenge.org/wp-content/uploads/2016/08/TBMM_ve_iklim_degisikligi_raporu.pdf

MoNE (2023) Ministry of national education: Turkiye 'İklim Değişikliği Farkındalığının Temel Eğitime Entegrasyonu Çalıştayı (Workshop on Integration Climate Change Awareness into Basic Education)'

Ozer M (2023) Education policy actions by the ministry of national education after the earthquake disaster on February 6, 2023 in Türkiye. Bartın Univ J Faculty Educ BUEFAD 12(2):219–232

TESEV (2022) Environmental protection and climate change budgets of metropolitan municipalities: an assessment for 2021 https://www.tesev.org.tr/en/research/environmental-protection-and-climate-change-budgets-of-metropolitan-municipalities-an-assessment-for-2021/

UNDP (2016) UNDP and climate change scaling up climate action to achieve the sustainable development goals

UNESCO (United Nations Educational, Scientific and Cultural Organization) and International Institute for Educational Planning-IIEP (2010) Guidebook for planning education in emergencies and reconstruction. https://www.iiep.unesco.org/en/publication/guidebook-planning-education-emergencies-andreconstruction

Chapter 16
Empowerment of Women and Girls in DRR Policy and Implication to Resilience

Gul Ergun and Salih Taskin

Abstract Global warming has been a global public health concern, especially for vulnerable and underserved communities. Women and girls are often disproportionately impacted by global warming due to gender-specific inequities. Although the Paris Agreement on Climate Change seeks to limit the global temperature rise to 2 °C (and strives towards a rise of 1.5 °C), some communities have already been impacted by global warming and climate change. Similarly, the Sendai Framework has adopted a disaster risk management approach that aims to broadly strengthen resilience and adaptive capacity to climate-related hazards and natural disasters, emphasizing the need for dovetailing climate change adaptation and disaster risk reduction (DRR) efforts. Women and girls, as agents of positive change, must be included in all efforts to combat gender inequality, climate change. This chapter will highlight some of the key issues with examples from Turkiye.

Keywords Woman · Girl · Turkiye · Disaster · Resilience · Climate change

16.1 Introduction

Global warming has become a major problem for the whole world. Climate change, stemming from various factors, have different effects on individuals and societies globally. In addition to the geographical location of countries, their level of development can also be decisive in the effects of climate change (Sırma 2023). In a comprehensive study examining the adaptation of developed countries to climate

G. Ergun (✉)
Burdur Mehmet Akif Ersoy University, Faculty of Health Sciences, Department of Nursing, Burdur, Türkiye
e-mail: ergun@mehmetakif.edu.tr

S. Taskin
Burdur Mehmet Akif Ersoy University, Institute of Social Sciences, Family Studies Master's Program with Thesis, Burdur, Türkiye
e-mail: psksalihtaskin@gmail.com

A. Yildiz and R. Shaw (eds.), *Disaster and Climate Risk Education*, Disaster Risk Reduction, https://doi.org/10.1007/978-981-97-5987-3_16

change, Belgium, the United Kingdom, the United States, Greece and Belgium were found to be effective in the adaptation process. These countries were followed by the member countries of the European Union (Gagnon-Lebrun and Agrawala 2006). On the other hand, it has been found that undeveloped countries, which are affected by climate change very much, are very affected by climate change in many areas and are inadequate in taking precautions. In East Africa, the yield of wheat, the most sensitive crop, will decrease by 72% due to the change in precipitation characteristics, costs will increase in agricultural products in Tanzania, and incomes will decrease significantly (Adhikari et al. 2015). Turkiye, which is one of the developing countries that strives to combat the effects of climate change, is struggling with sustainable development and energy problems (Kaygusuz and Toklu 2012). Turkiye, which makes decisions in the fight against climate change and endeavors to regulate the laws, has problems in the implementation of these decisions in the field (Gürçam 2022).

Recent research findings show that temperatures affect human health and physiology (Tam et al. 2020; Ebi et al. 2021; Karahasan and Pınar 2023). For this reason, heat waves have caused deaths in Europe in recent years (Rossati 2017). As the body temperature is affected by the environment, it causes more diseases, heart diseases, problems in pregnancy and newborns, and therefore deaths. For instance, 1 °C increases the number of deaths by 0.2–5.5% on average (Basu et al. 2010).

According to the estimates made in the projects on climate change and its effects on human health involving European countries, it is estimated that the temperature will increase by approximately 3 degrees Celsius in 2071–2100 and an extra 86,000 deaths will occur every year (Pink and Pink 2018). The consequences of climate change include epidemics, financial and economic crises, extreme weather changes. Floods, landslides, frost, sudden cold or hot weather changes, etc. It is thought that the reason why natural events have been irregular since the industrial revolution is related to climate change. In recent studies, it is thought that if global warming exceeds 1.5 °C between 2030 and 2052, the greenhouse gas effect will continue (Talu 2016).

The main environmental problems caused by climate change include floods, erosion, forest fires, air pollution, extreme heat/cold, drought and thirst. These extreme events are becoming more problematic especially in developing countries such as Turkiye. Women are particularly underrepresented in the earth sciences, which are most relevant to climate science (Akbaş and Taner 2021; Lee et al. 2020). This chapter focusses on one of the vulnerable group, women and girls, to empower them in disaster risk reduction policy and implications to resilience.

16.2 Types of Disasters in Turkiye

16.2.1 Earthquakes

Turkiye is one of the countries experincing the effects of climate change and disasters caused by the natural hazards frequently. Especially in recent years, large earthquakes along with fires, floods and landslides due to the climate crisis have become the biggest problem (Çamaş and Turan 2023). The 7.7 and 7.6 magnitude Kahramanmaraş Earthquakes that occurred on February 6, 2023 in Turkiye heavily affected 11 provinces. In these earthquakes, more than 50 thousand people lost their lives and many settlements were destroyed. The earthquake particularly affected women and girls who were left behind in the normal life cycle to participate The earthquake particularly affected women and girls who, in normal life, are not involved in the production level of society and are economically dependent on men (Tamers et al. 2020).

Studies have found that depression and suicide attempts are extremely high in women after disasters compared to men (Sohrabizadeh et al. 2016). They had to cope with many problems such as physical, spiritual, moral and economic. Findings showed that violence, especially against women, increased in families and even in the civilian population who could not cope with these problems after the disaster. Bozkurt (2023) conducted interviews with earthquake survivors of the earthquake centered in Kahramanmaraş and examined the difficulties they experienced from a social perspective; the stories of cutting the arms of earthquake survivors to get their bracelets were also told by some interviewees. In the same study, it was emphasized that there were many girls who never left their tents due to the trauma they experienced during the earthquake. According to a recent study conducted in Turkiye, especially girls were subjected to repetitive trauma during the earthquake, when journalistic ethics regarding disaster reporting and interviewing victims were ignored (Bozkurt 2023). In the same study, three common themes emerged: “Miracles and hope,” “patriarchal state” and “sensational/dramatic elements.“ As a result of the research, it was shown that the media representations of earthquake survivor children in Türkiye aestheticize their pain and suffering and reduce them to objects of pity or spectacle (Atalay and Muratoğlu Pehlivan 2023). The United Nations Population Fund (UNFPA) reported that there were 226,000 pregnant women of reproductive age in the eleven provinces affected by the earthquake who survived the devastating effects of the earthquake, and an estimated 25,000 of these women gave birth within a month. One pregnant woman gave birth under the rubble. Many pregnant women did not survive the rubble (Kipay 2023). Women’s most basic needs such as access to bathrooms and toilets, underwear and sanitary napkins were ignored. After the earthquake, many women did not have access to sanitary napkins, all their clothes were buried under the rubble, they had to wear their dirty clothes for weeks, and they could not bathe for two months. Women are more affected by natural disasters than men. Women are vulnerable to the effects of natural disasters. Especially their psychological resilience and economic power are not sufficient before a natural disaster. Therefore, women have higher rates of injury and even death than men (Thurstone et al. 2021; Castañeda et al. 2020; Fatema et al. 2019; Topgül 2012).

16.2.2 Fires

In a study examining the natural disasters experienced in Turkiye between 1970 and 2012, it was determined that our country has a great risk for fires. In the same study, it was stated that the highest number of fires across the country occurred in Antalya province (269 fires) (Özşahin 2013). With the effect of global warming in Turkiye, many people suffered material and moral losses in the fires that occurred in the summer months, especially in the Aegean and Mediterranean regions. The most common headlines in the news reports are that individuals went to the fire zone as volunteers and actually participated in the efforts to extinguish the fire. Among the activities reflected in the news, it came to the forefront that young people, including girls, coordinated and helped the fire brigade team. Women established workshops and tried to reduce financial loss by selling handmade products (Kolukırık et al. 2022). This reveals that women and girls can actively participate in the post-disaster recovery process and the importance of empowering them in terms of psychological, skills and self-confidence in pre-disaster preparedness is highly important. In the fire affected areas, there were power and water cuts in order to reduce the impact of the fire. In the hot weather, especially the hygiene needs of women and girls could not be met to a great extent.

Women and girls may face problems such as lack of access to resources in disaster settings. They are vulnerable groups due to the possibility of being caught unprepared for disasters. It is stated that there is a significant correlation between the number of deaths and illnesses of women and especially girls as a result of disasters. In this sense, it is extremely important to support women to receive assistance before and after disasters. Furthermore, climate change is a major risk factor for overall health, including reproductive health and well-being. Rising temperatures, greenhouse gases accumulating in the atmosphere, forest fires, hurricanes, hurricanes, floods, stagnant water causing vector dispersal have serious negative impacts on human health. These disproportionately affect disadvantaged and vulnerable groups, including women at special life stages, pregnant women, young children, the elderly, the poor and people with disabilities. Women's reproductive functioning affected by climate change can result in an unhealthy newborn. Although it seems to concern only women's health in particular, it actually directly affects public health (Yalçın et al. 2022).

For women and girls, the goal of post-disaster and post-emergency activities should be to return to normal life as soon as possible. It is to quickly heal their wounds and prepare the conditions they are used to again. For both groups, the disaster preparedness process is to ensure that health services planned by administrative units are also accessible in case of a disaster. In Türkiye, especially due to the nature of the patriarchal structure, men are the individuals who represent the family more often in receiving and providing outside services. Therefore, in the event of a disaster, if the man loses his life or takes on different duty responsibilities, especially girls and women will have difficulty in accessing services. At this point, local people can be asked for help to access the services needed and to benefit from administrative and health institutions. These activities can be coordinated through volunteering activities. Thanks to the volunteering network, a supervision system will be created for disaster victims who are alone and need someone else, and elderly individuals who are under the responsibility of volunteers can be prevented from being alone during the disaster. A sustainable volunteering work can also be done for post-disaster care (Sırma 2023; Glasser 2019).

16.2.3 Floods, Droughts and Storms

Floods, droughts and storms caused by climate change affect people in different parts of the world. Looking at the natural disasters worldwide between 2000 and 2019, 1.65 billion people were affected by floods and 1.43 billion by droughts (Aydın and Genç 2023). Women and girls are particularly vulnerable to these disasters (Güneş 2023). Women and girls also have to cope with acts of violence against them in these difficult conditions. Since women and girls are left with the task of taking care of the house and cooking for the household, they need access to water sources. The long journeys to access water are quite challenging. In addition, when there is a water or food crisis, girls are provided with less food and water than boys (Okur 2023). For these reasons, girls cannot ensure their adequate development.

It is seen that women cannot meet their needs in water management systems created by patriarchal culture (Bilgili 2017). Women with low levels of education or socioeconomic status may not be able to follow the warnings through the media in cases such as hurricanes or flood risks. Again, they may not be frequently in areas where such warnings are made in squares and places with heavy traffic and may not have access to the right to be warned. The high number of women who have lost their lives may be due to the fact that warning information is given by men to men in public places. In many hurricanes and floods in the past, many women were not aware of the announcement and could not leave their homes without their male relatives and lost their lives while waiting for them to go to a safe place (McIntyre-Mills et al. 2018; Abedin et al. 2013). On August 11, 2021, a flood disaster occurred in Kastamonu Bozkurt-Sinop-Bartın and many other provinces. The social destruction of the disaster in question was quite great. The disaster in question, which can be considered a clear example of global climate change, has become a disaster due to human errors. Due to the flood, 97 people (62 Kastamonu, 15 Sinop, 1 Bartın) lost their lives and 228 people were injured. In addition, this disaster caused considerable material destruction. Most of the dead and injured were women and girls (Doğaner 2022).

In Turkiye, the effects of dense urban populations and uncontrolled urban developments lead to global warming, global warming leads to climate change, climate change leads to changes in temperature and precipitation regimes, and changing precipitation regimes lead to floods and inundations, which in turn cause loss of life and property (McIntyre-Mills et al. 2018). Women's low swimming and climbing skills increase physical vulnerability during disasters. Traditional and cultural clothing may restrict women's movements during disasters and affect their evacuation/escape situations (Gündüz 2022). Similar reasons may be valid among the frequent causes of death among women in flood disasters in our country.

It is clear that Turkiye will be adversely affected by global climate change. In the latitudes where our country is located, it is estimated that there will be increases in temperatures, changes in precipitation regime, rise in sea water level and significant decreases in soil water content. Turkiye is a semi-arid country. Moreover, drought is recognized as the most dangerous natural disaster in the world due to its socio-economic impacts, persistence and difficulty in finding solutions. Considering that water is scarce, the amount and distribution of precipitation is irregular except in some regions, water is limited in large cities and agricultural production, the quality of drinking, using and irrigation water is decreasing day by day as a result of increasing industrial and other environmental pollution and global warming, it is clear that our country will feel the severity of drought much more than today in a very near future (Kadıoğlu and Sanu 2007). Decreased production due to drought will negatively affect the consumption of women and girls. Even in the current production, since boys are seen as the ones who will ensure the continuation of the lineage, they have a higher share of production than girls, eat healthier, and have priority in continuing their education (Ünür 2021). Increasing global warming will cause droughts, floods and storms, leading to a further reduction in resources, especially in Turkiye, which is an agricultural country, and will cause greater problems for the whole society, especially for families whose income sources are limited to agriculture in rural areas, and for women and girls in these families.

16.3 Climate Change Adaptation and Disaster Risk Reduction Studies

Cities are the places where the effects of global climate change are felt the most. After the industrial revolution, the technological revolutions of city life took people away from their natural life. The ecosystem is unable to renew itself due to the destruction caused by the vehicles used by people. In fact, in the report announced by the IPCC in 2007, it was stated that reducing greenhouse gas emissions, focusing on the processes expected to take place and carrying out adaptation studies would be less than the cost of climate change. Sustainability efforts should be prioritized with a holistic approach at local and global scales. In order to reduce vulnerability and carry out adaptation studies, cities that are resistant to disasters and crisis situations due to climate change should be built (Erdoğan and Cantürk 2022). As examples of cities that carry out city studies compatible with global climate change; we can show cities such as Hamburg from Germany, Arizona from the USA, Linz from Austria,

Kopenhang from Denmark, Eco-Viikki from Finland, Nantes from France, Malmö from Sweden, Stockholm from Switzerland. These cities make plans such as using renewable energy sources, improving waste management, and reducing greenhouse gas emissions (Koçan and Alp 2021).

Building Cities Resilient to Climate Change

As seen in the examples above, the aim of adaptation studies is to mitigate the damages associated with disasters and crises caused by climate change. In line with these objectives, it is to mitigate the damage that may occur through policy formulation, finding implementation solutions and projects and to utilize beneficial opportunities. It is the development of systems in which climatic variability is included in the whole process. Adaptation is a process that can be realized at any time and can be influenced by institutions, people and natural resources (Klein et al. 2006).

The Intergovernmental Panel on Climate Change (IPCC) is of the opinion that extreme weather events and disasters due to climate change reduce the rate of economic development. It is predicted that the economies of developing countries will not be able to cope with these situations and new poverty areas will be formed in their cities. It is very important to determine appropriate greenhouse gas mitigation and climate change adaptation (CCA) policies to combat climate change in the context of continued development and sustainability, and to address them together with disaster risk reduction (DRR) actions to ensure resilience to climate change (Sharma and Ravindranath 2019).

16.4 Measures to Ensure a Sustainable Lifestyle

In order to mitigate climate change and reduce the crises caused by climate change, it is necessary to plan sustainable lifestyles and conservation plans that protect the environment and can be created for future generations. In this context, the 1992-Rio Conference, 2005-Kyoto Protocol, Paris Agreement (2015), Habitat III (2016) and COP26 organized in November 2021 are important agreements where the world came together and established sustainability policies. There are also UN conventions, the Paris Climate Agreement, the Sendai Framework for Disaster Risk Reduction, the UN 2030 Agenda SDGs and the UN Habitat New Urban Agenda, which indicate that adaptation to climate change and disaster risk reduction strategies should be carried out in a comprehensive and integrated manner in the studies carried out for sustainable development and to develop policies and establish relations in this context (Hamaber et al. 2017; Tuğaç 2021).

Harmonization efforts in line with these agreements and declared plans are increasing day by day. Local and national governments create action plans and include sustainability strategies in their bureaucratic processes. Processes such as greenhouse gas measurements and calculation of carbon footprint have been structured. These studies should be carried out mostly in urban settlements. Cities, which are most affected by climate change processes and where population density is high, are very important in this regard. City managers should have sufficient knowledge of the scope of climate change and sustainability and take the necessary measures (Koç et al. 2021). Since they are intertwined with the living people, they can formulate the strategies of their local studies by taking into account their social and economic structures. Although this issue is of global importance, the work to be done will create differences at local scales. Therefore, policy makers will be in an advantageous position for the implementation of climate change and adaptation studies (UN-Habitat 2018). In addition, all these studies are to ensure that all segments of the society can build resilience, effective use of resources, and effective cooperation and coordination among stakeholders. And the target group of all these studies is to include vulnerable groups in a sustainable and resilient life through climate change adaptation and disaster risk reduction studies (Tuğaç 2021).

International Measures for Vulnerable Countries in Climate Change Related Disasters

Especially vulnerable segments of the society are highly affected in disasters related to climate change. On a global scale, it is seen that the regions of the world where the least developed countries are located are most affected (Davies et al. 2009). Sendai Framework is the creation of resilient cities against existing and potential risks in the process of combating disasters. This framework requires countries to understand the disaster risks in their local and scope areas, to strengthen disaster and crisis governance, to increase their ability to rehabilitate after disasters and crises and to respond adequately and to aim to be better than before. It will be possible to gradually reduce the loss of life in disaster and crisis situations, and to sustain and sustain the economic, social and physical existence of individuals, public and private sectors, communities.

The integrated work of all public and private units in climate change adaptation and disaster prevention efforts will realize effective progress by identifying vulnerabilities (Tuğaç 2021). The focus of the Sendai declaration plan is the experiences gained from the Hyogo framework action plan. With this protocol, it seems possible to reduce the undesirable effects of disasters with various technological and experience-based information. Loss of life, injuries, economic and social losses due to disasters have increased with the expansion of urban life style. With the developments in technology and science, it has become obligatory to use scientific methods to reduce the undesirable effects of disasters. The Sendai declaration plan makes it obligatory to utilize information and communication technologies (internet, GSM, etc.), new management approaches (Integrated Disaster Management) in order to easily apply these scientific methods. In the earthquakes centered in Kahramanmaraş in our country, the use of technology was very inadequate due to the lack of infrastructure and insufficient number of technological tools. Planning models such as enterprise resource planning, which are included in logistics, should now be used in disaster management models. Such models will ensure the sustainability of the Sendai declaration plan as an integral part of the integrated disaster management process (Macit 2019).

Local governments are establishing ties at the international level in order to strengthen their fight against climate change, reduce the risk of disasters and improve adaptation efforts. Some of these connections can be listed as follows (ICLEI 2020; UCLG-MEWA 2021; Cities Climate Leadership Group 2021; Mengi and İşçioğlu 2018):

I. ICLEI (International Network of Local Governments for Sustainability): A global network of more than 2500 local and regional governments committed to sustainable urban development. It works with partners to create new ways to support sustainable development at the urban scale by building strategic alliances with international organizations, national governments, academic and financial institutions, civil society and the private sector (ICLEI 2020).
II. Covenant of Mayors for Climate and Energy: planning to reduce greenhouse gases at the local level, increasing resilience to global warming and transparently tracking progress made. The Covenant of Mayors is the largest urban climate and energy initiative in the world, covering 6,800 cities and regions from 57 countries. Cities and regions that have signed the Compact share a long-term vision of creating decarbonized and resilient regions that provide sustainable and affordable energy for all (UCLG-MEWA 2021).
III. C40 (Cities' Climate Leadership Group): This group of metropolitan cities come together to plan climate change and adaptation policies. Its aim is to make sustainable policies transparent, cooperative and meaningful and measurable through information sharing. C40, which has brought together over 80 cities in the last 10 years, has the ability to represent a population of 600 million thanks to these cities. From Turkiye, only the city of Istanbul is a member of this network (C40 Cities 2021).

The lack of sufficient financial and human resources of small and local-scale regions will make it difficult to provide sufficient contribution to adaptation and disaster prevention efforts. Because even if small regions have low greenhouse gas emissions and emission rates, they will be more affected by disasters and crisis situations caused by climate change due to the high number of vulnerable groups. The studies that can be carried out can be listed as follows (Esin and Partigöç 2022; Tebaldi et al. 2021; Zuo and Zhao 2014).

- Considering the rise in temperature in urban planning processes, urban plans should be created with wind corridors for natural air conditioning, protecting and increasing green areas in urban planning. Housing should be created by determining the conditions of vulnerable groups and these housing should be built within the scope of green building and sustainability.
- The decrease in water resources with rising temperatures increases the risk of drought. Therefore, water recycling should be well planned. Irrigation systems used in agriculture should be modernized and built in a way that prevents leakage. Urban landscaping and designs should be done in a way that does not need irrigation, and rainwater should be stored and used.

- Melting glaciers due to climate change raise sea levels. Rising salt water threatens fresh water resources. For this reason, an early warning system should be established against the rise of salt water, and the capacity of evacuation and crisis methods should be increased so that it does not affect groundwater.
- Infrastructures should be strengthened and made ready for irregular rainfall regimes. Waste water, municipal water and storm water networks should be separated. Increase crisis response capacity against floods and disasters.
- In agriculture, urban agriculture should be supported, green building and local market practices should be planned. In fisheries, overfishing should be prevented, and food warehouses should have more precautions against different weather conditions such as temperature increases and floods.

16.5 The Importance of Women and Girls in Climate Change Studies

It is thought that the natural events experienced due to climate change are a situation that should be within the world's climate system, but accelerated due to factors such as the increase in global temperature. It is stated that changes have started in the atmosphere, lithosphere and hydrosphere that make up the climate systems. The natural balance is disrupted and climate changes are occurring rapidly due to these reasons. Scientific studies show that these changes caused by global warming may cause serious problems in our future. The climate event that causes these changes is the change in the radiation balance of the earth. Factors affecting radiation are sometimes due to natural processes and sometimes due to human activities (Arı and Aydın 2019).

16.5.1 International Dimension

The fact that the consequences of climate change are a global problem requires that the studies to be carried out should also be global. In order for the policies announced in the local and national context to be successful, it is very important that citizens' awareness of climate change and social awareness are high. In 2050, 68% of the population is expected to live in cities (Allam et al. 2020). Policies and actions should be planned in the light of the knowledge that individuals living in cities have about the impact of climate change. It is very important to show the full participation of the society in the implemented action plans (Ağıralan and Sadioğlu 2021). It is impossible to conduct a healthy and fair study when participation in action plans is based on gender roles. As a result of comprehensive studies on women conducted by the United Nations Environment Program (UNEP) in 1989, it was strikingly revealed that even if they have a low economic level, they have good relations with the environment and care for their family without the thought of harming natural resources (Hamaber et al. 2017). They are able to make friends with their environment and feel close to it.

16.5.2 Impact of Gender Roles

Women are different and closer to natural resources due to their gender roles. In international conferences and declarations such as the Rio Declaration and the Beijing Political Declaration, women have a very important role in the development of sustainable consumption and production patterns in terms of the use of resources (Alıca 2017). Definitions of gender roles have been reshaped by the change in social and economic foundations after the 1960s, when women and men transformed their spheres of activity, redefining the characteristics of people and their relationships with each other. The characteristics of gender roles have increasingly become the focal point of the political agenda due to climate change. Since it is not possible to evaluate human beings apart from the geography they live in, the relationship between human beings and their environment has become one of the basic parameters of gender (Erbaydar et al. 2021). Women try to organize like men for their work and to protect their necessary resources. However, these organizational schemes are different from those of men due to gender roles. For these reasons, the feminist tradition has organized to improve and often expand women's space, to facilitate and transform women's work (Mackenzie 2002).

16.5.3 Ecofeminism Approach

Women who were engaged in feminism studies, who also cared about environmental factors and ecological crises, started to put forward the "Ecofeminism" approach with a pluralist ideology in the 1970s. It emerged as a reaction to men's attitudes towards nature and women. They state that women's relationship with nature is different from men's and that they are in direct contact with it. Environmental problems cannot be solved due to the fact that women are not valued in the public sphere when policy is being formulated and that women's values are ignored. According to ecofeminism, it is seen that capitalism is designed to dominate and exploit women and nature under the shadow of patriarchy and that injustices do not bring solutions. Climate change caused by anthropogenic activities, or climate crisis as it is now called, is rooted in the injustice and inequality of the economic and political system. The scale and impact of the climate crisis has increased due to irresponsible use of natural resources, unequal sharing of the resources obtained and the exclusion of the vast majority of the world in decision-making mechanisms. In addition, new injustices are being added to the existing injustices with the problematic of climate change that cannot be avoided. The urban poor, women, the elderly and children are the most affected. It is necessary to have a structure that is shaped around solutions that are pluralistic, locally oriented and dominated by participatory practice approaches, as opposed to centralized and unilateral explanation of policies (Uncu 2019).

The impact of climate change on human relations is a topic that is rarely considered. This is because technological developments are primarily discussed when climate change is discussed. However, there are also objections due to problems such as the simple division of human society into men and women and the exclusion of differences in the definitions made. By focusing on relationships, it is seen that women are subordinated and neglected because they are shaped by cultural concepts between men and women. Due to such problems, women and men are exposed to climate risks and opportunities at different levels, making climate change definitions controversial (Okur 2023).

Anthropologist Ortner (1974) states that the subordinate status of women in society is universal, based on the assumption that women are closer to nature than men and that men are on the higher ground of culture. In most societies, culture is valued more than nature and therefore men are valued more than women, which he sees as the basis of gender inequality (Ağören 2020).

16.6 The Role of Women in Sustainability Efforts Against Disasters Caused by Climate Change

The importance of the role of women in climate crisis and sustainability studies can actually assume an important role that can transfer knowledge to future generations as they are the breeders, caregivers and educators of family members. While women's education is important for the plans to be created for sustainable development goals, women can be determined as the target audience. Because more than half of the people in the world are women. If the power of women is not used, it will be very difficult to achieve prosperity through sustainable development (Alıca 2017).

Studies conducted in developed countries have shown that women are more aware of climate change than men. The solutions that women and men bring to climate change also differ significantly. While men think that they can solve this problem with technical issues, women state that the culture and style of life should change (Talu 2016; Turhan et al. 2016).

Gender roles in developing and other countries make women and girls even more vulnerable. The Organization for Economic Cooperation and Development (OECD) states that it is often women and girls who struggle to survive and continue during periods of floods or droughts due to the increasing frequency of heavy rains (Dereli 2019). Women are closer to natural resources due to their roles. Seeing the connection with the environment, production and distribution as structured and shaped by a given gender, class/caste/race organization will help us understand women's relationship with climate change. Women are adversely affected by climate change because of their diverse reproductive and productive roles at the household and community level in areas such as land use, water management, forest use, energy provision and use, urban development. Other reasons why women feel the impacts of climate change more are that they are in lower income groups, have less resources needed to combat climate change, and are more dependent on natural resources due to gender role segregation (Abedin et al. 2013). Floods, droughts, cold and heat waves, hurricanes have a major impact on people's lives and livelihoods, especially on issues for which women are responsible. In order to ensure gender equality and overcome the problems, the integration of ignorance or underrepresentation, gender concerns and women's issues into mainstream policies, projects and institutional structures will bring people closer to a solution (Okur 2023).

If rural women benefit from the traditional knowledge of previous generations and make their own livelihoods from the use of natural products, a new sustainable economic model emerges (Bolin 2020). Again, in Turkiye, in Anatolia, women have a vast body of knowledge passed down from previous generations. It is seen that they continue their old cultural habits from seeding, planting, harvesting and cooking. They create a sustainable ecosystem by continuing to produce local products in a traditional way. Since they have put this knowledge into an economic system, they can also transfer it to the next generations. Thanks to the woman's domestic care role, her children have also transferred the necessary information to their memories and behaviors during this period when they learned about their relationship with nature from their mothers (Akgül 2022).

This situation is valid not only in Anatolia but also in all geographies. It is stated that entrepreneurship activities of local women from rural areas are increasing in the world. Women have the traditional and cultural knowledge to process natural resources in a sustainable way. They show examples of an activity that can be a solution to global climate change and resource scarcity, as well as contributing to biodiversity. An economic model in which carbon emission is not excessive, the mutual benefit of human and nature is taken into consideration, and an understanding of sufficiency prevails can be targeted thanks to women (Swiderska 2021).

16.7 Strategies to Protect Women and Girls from Disasters

In the twenty-first century, disasters continue to increase and have a profound impact on societies. However, in the face of recent disasters, people have moved from helplessness to developing strategies to cope with disasters. Recently, with advancing technology, understanding of societies and new strategies for dealing with societies have been developed to respond to these realities in order to be resilient to disasters and these events. One of the fundamental principles in understanding communities in the face of disasters includes a comprehensive gender perspective. Communities and society are heterogeneous structures. They are fundamentally composed of men and women (Dankelman and Fansen 2012). Therefore, women should be involved at every stage of disaster management. Men and women should work together to create disaster-resilient societies (Atalay and Muratoğlu Pehlivan 2023). Addressing gender discrimination against women and girls and empowering women in disaster management is of utmost importance. It is stated that there are inequalities in the fair access of women and girls to resources and that women are less able to act in a natural event due to cultural norms that privilege men over women and boys over girls (Sırma 2023).

The problems faced by women in disasters in Turkiye should be identified and effective solutions should be developed. Despite the importance given by the literature to the study of disasters and women, research is scarce. Gündüz (2024) examined the disaster experiences of women who faced various types of disasters (earthquake, flood, landslide, fire, epidemic disease, drought and terrorist attack) in Turkiye and found that women were unprepared for disasters but took early warnings into consideration. It was also found that women had difficulty in accessing basic needs and security problems after disasters, had misinformation about disasters and were willing to participate in disaster trainings, faced social problems more frequently than men after disasters and were willing to participate in decision-making mechanisms in combating disasters. The Sendai Framework for Disaster Risk Reduction (2015–2030) was adopted in March 2015. In the Sendai declaration plan, four priorities have been identified in general. These priorities are as follows:

i. Understanding disaster risks,
ii. Strengthening governance to deal with disaster risks,
iii. Investments in disaster risk reduction for rapid recovery from disasters,
iv. Improving disaster preparedness for effective response, reconstruction, recovery and recovery activities.

Analyzing these priorities according to their topics will provide a better understanding of the issue. The priorities briefly summarize the expected and targeted objectives as well as what needs to be done (UNISDR 2015).

Priority 1: Disaster risk management should be based on an understanding of the weak and vulnerable characteristics of disaster risks, the capacity, the potential damage to people and assets, disaster characteristics and all environmental dimensions. Such information can be used for risk assessment, prevention, mitigation, preparedness and response.

Priority 2: Disaster risk management at national, regional and global scales is crucial given the four core principles of disaster management: prevention, mitigation, preparedness, response, recovery and restoration. It supports cooperation and partnership.

Priority 3: Public and private sector investments through structural and non-structural measures for disaster risk prevention and reduction are essential to enhance economic, social, health and cultural resilience for people, communities, countries and assets, as well as the environment, and to combat disasters.

Priority 4: With the growing risk of disasters, there is a need to ensure capacities for disaster preparedness, anticipated events, and effective response and recovery at all levels. The recovery, restoration and reconstruction phase represents an important opportunity to better develop disaster response activities, including integrating disaster risk reduction into development plans.

In line with these priorities, there are certain stages in the Sendai Framework for Disaster Risk Reduction (Macit 2019). These are;

The main objectives of the preparation phase include measures to be taken against undesirable situations that may arise as a result of the hazard to be encountered, methods and activities that can be used effectively. The measures and situations mentioned here are determining the risks rather than reducing any risk. Identification of disaster risks and creation of effective action plans are usually done at this stage. It is important for women to receive various psychoeducation in areas such as psychological, cognitive, social skills and crisis management at every moment of the preparation phase.

In the response phase, depending on the magnitude of the disaster, it generally consists of activities starting immediately after the disaster and covering a maximum period of two months. The main objective of the activities to be carried out at this stage is to reach the disaster victims as soon as possible in a way to minimize the loss of life, to provide medical aid, shelter and life-sustaining materials to the injured. Encouraging women who have difficulty in being in social life to receive basic health services and disaster services and ensuring that they know their rights should be ensured by organizing public services and voluntary services.

In the recovery phase, the main objectives are to provide the disaster victims' needs such as communication, transportation, electricity, water, education, etc. for a certain period of time after the disaster and to ensure the continuity of social and economic activities. One of the most important activities of this stage is to find effective solutions in laws and regulations for reducing disaster damages in reconstruction plans. It is of utmost importance that women and girls, who are vulnerable to disasters, are rapidly rehabilitated, placed in safe shelters and empowered economically.

Gender inequality is an unfinished agenda for women's rights. It is still widespread around the world. Deep-rooted gender policies also reproduce gender inequalities within countries. This affects disaster management legislation. According to the Committee on the Elimination of Disasters Convention on the Elimination of Discrimination against Women (CEDAW) General Recommendation No. 37, States Parties should ensure that all disaster-related policies and legislation are gender-sensitive and grounded in human rights, risk reduction and climate change (CEDAW/C/GC/37 2018). Absorb, respond to or recover from the impacts of a hazardous event in a timely and effective manner.

Designing legislation for disaster risk reduction is the first step in increasing safety in society and mainstreaming disaster risk reduction. Legislation provides a framework for disaster risk reduction strategies. Risk reduction and reconstruction activities can be prepared. The law can be used to provide penalties and incentives through Reconstruction should be enabled by defining the rights to enforce standards in construction, land use, tenants and other areas and the rights of people during relief and support (Pelling and Holloway 2006). Besides budgets, policy mandates can be determined through legislative acts. The legislative process should be a constructive period to produce the following. International reports underline the following outcomes. Inequalities between women and men and how existing capacity is wasted and society's long-term development prospects are hindered (Lau et al. 2021; Tyagi et al. 2020). The main internationally recognized strategy to end gender inequality is to promote women's empowerment (Whittington 2021). Sendai's Framework recognizes that "women and their participation are critical to the effective management of disaster risk and the design, resourcing and implementation of gender-sensitive Disaster Preparedness Plans. Policies, plans and programs need to be developed rapidly, with an emphasis on their implementation. Adequate capacity building measures should be taken to empower women to be prepared for disasters as well as to improve their capacity for alternative livelihoods in post-disaster situations" (Sharafat and Nasreen 2022). The Sendai Framework lists "strengthening disaster risk management to manage disaster risk" as one of the four priorities. When we look at these processes in terms of Turkiye, there are extensive and long-term studies in the literature. "Integration of Gender Perspective into Disaster Policies" is a part of the "Turkiye Disaster and Emergency Management Program (UDAP)" supported by the National Earthquake Research Center and carried out by (AFAD) in Turkiye between 2017 and 2020. This study consisted of two stages. In the first phase, a content analysis of Turkiye's disaster report was conducted. Then, Disaster Legislation was analyzed in the context of gender sensitivity. In the second stage of the study, stakeholders' opinions were taken and reported in order to develop recommendations in the context of gender sensitivity. AFAD's 87 legislation on disaster was analyzed. As a result; in the current legislation, women are categorized as risk groups together with children and persons with disabilities in the National Earthquake Report of the Strategy and Action Plan 2023 (NESAP-2023). In the same report, "shelter" is used the most with 258 times, the concept of "woman" is used ten times and the concept of "gender" is used once. In the report, the sensitivity of gender equality in the legislation and the lack of regulations to support women at various stages of the disaster were identified. At the end of the same study, it was suggested that at least one vice president of AFAD, one manager in the municipality and emergency response units should be a woman.

In order to eliminate gender inequality in disaster preparedness in Turkiye and to prepare more robustly for disasters (Çamaş and Turan 2023; Gerdan 2019; Kolukırık et al. 2022; Macit 2019; Yalçın et al. 2022);

- Gender equality should be taken into consideration in every platform related to disaster preparedness.
- Training on gender equality should be provided periodically to the managers and employees of AFAD, civil society organizations (CSOs) and NGOs.
- Gender balance should be taken into consideration in civil defense.
- The content of the training of every institution and organization providing training on the subject should be gender appropriate.
- A working group should be established to develop strategies for integrating gender perspective into disasters and emergencies and emergency policies.
- Missing individuals after a disaster should be accessed by gender and age (women, men, girls and boys, etc.).
- Reports prepared before or after disasters should be gender sensitive. Regulations, including safeguards, should be put in place.

- Providing social assistance and loans for positive discrimination to women especially after the disaster can be suggested.
- For the inclusion of women in the society, empowerment of women and women in working life should be ensured by providing economic freedom through various programs and facilities.
- Women's property rights should be ensured according to whether women and their husbands/partners live together or separately, gender and so on.
- In particular, toilets, breastfeeding rooms and other needs should be provided to women in shelter areas. In addition to basic food and supplies for disasters and emergencies, it will be important for women to have special needs for menstrual periods (sanitary pads, daily pads, etc.) and underwear to prevent pelvic infections.
- In addition, in the chaos that will occur after disasters and emergencies, especially women and their children who have grown up with gender discrimination may not be able to express their medical needs. Therefore, written numbers of health institutions, medical identification cards or dog tags, medication lists, planned meeting place should be established (Elfert 2015).

Disasters and emergencies caused by climate change create extra challenging and adverse conditions for vulnerable groups. They cannot prepare an environment that is as free as in their previous lives due to the restriction of their mobility, forced choices or limited choices. In regions with high gender inequalities, women and girls are highly exposed to these constraints. Women face more difficulties in crisis situations due to their disadvantaged position. Since they do not receive adequate education, they do not know how to cope with natural disasters or crisis situations. Economic hardships due to climate change affect women and men in different ways. Considering that women account for 70% of the world's poverty, it is obvious that they will be more affected by changing processes and will have more difficulty in adapting (Watkins 2007). Therefore, their activities can be very limited in the post-disaster process. Due to their limited activities, women who have to take care of their homes and make a living lose their livelihoods. The fact that such problems occur in already poor regions further deepens the poverty of women.

Due to gender roles, women have different defense methods in disaster and crisis situations. They cannot create a safe space for themselves in times of disaster due to their desire to protect their dependent children and family. They expect to be in a position where men are rescuers and women are rescued. Neumayer and Plümper stated in their research that women face more deaths in disaster and crisis situations in 141 countries. Women are unable to protect themselves due to their economic and social status. Women and girls have the idea that they will be rescued by men and therefore they expect to be rescued in disaster situations. In the 1991 Bangladesh cyclone, one of the reasons why many women and children lost their lives was this thought (Neumayer and Plümper 2007).

Another disaster situation caused by the climate crisis is actually the Covid-19 outbreak and pandemic (Şahin and Erensü 2020). It is stated that violence against women and girls intensified during the pandemic period. As a result of conflicts due to people living in the house, 1 out of 4 women stated that they felt less safe. In addition, 6 out of 10 women reported experiencing sexual harassment and 7 out of 10 women reported increased physical or verbal abuse by their partners (Unwomen 2022). The Malala Fund reports that in 2021, 4 million girls from low- and middle-income families in countries with lagging economies dropped out of school due to the pandemic and climate change. As long as the crises continue, 12.5 million girls will drop out of school in 2025 (Roos 2022).

It is very important to have representation in the solution processes planned and produced for disasters and crisis situations caused by global climate change. Ensuring women's participation in decision-making mechanisms will ensure that women are in a position to easily share the processes they are affected by global climate change and what they need in disaster situations. Since the efforts to make the world livable and sustainable are carried out not for a part of the society but for the whole, they also concern women (Koç et al. 2021).When women make their needs more visible in the public sphere, they will be able to take a more active role against the problems brought by global climate change (Ateş et al. 2018). Ensuring the participation of women in order for the studies to be more qualified will enable the use of social capital and the creation of high-sensitivity projects. In the decisions taken in the Sendai Framework for Disaster Risk Reduction (2015–2030), which Turkiye is among the countries, it is aimed for the society to suffer less damage in disaster and crisis situations and for women to be present at every stage (Baş and Partigöç 2022; Ilgın and Karagül 2022).

References

Abedin MA, Habiba U, Shaw R (2013) Gender and climate change: Impacts and coping mechanisms of women and special vulnerable groups. In: Shaw R, Pulhin J, Jacqueline J (eds) Climate change adaptation actions in Bangladesh. Springer, pp 165–184

Adhikari U, Nejadhashemi AP, Woznicki SA (2015) Climate change and eastern Africa: a review of impact on major crops. Food and Energy Security 4(2):110–132. https://doi.org/10.1002/fes3.61

Ağıralan E, Sadioğlu U (2021) İklim değişikliği farkındalığı ve toplum bilinci: İstanbul örneği. Anadolu Üniversitesi Sosyal Bilimler Dergisi, 21(2), 627–654. https://doi.org/10.18037/ausbd.959287

Ağören GC (2020) Depresyon ve Modern Düşüncenin Eril Dikotomileri. Fe Dergi 12(2):1–14

Akgül Ç (2022) Sürdürülebilir Ekolojik Denge İçin Modern ve Geleneksel Üretim Yöntemlerinin Entegrasyonunda Kırsal Kadının ve Kadın Kooperatiflerinin Önemi: Afşar Balam Kadın Kooperatifi Örneği. Ankara Üniversitesi Çevrebilimleri Dergisi, 9(1), 57–71. Erişim adresi: https://dergipark.org.tr/en/pub/aucevrebilim/issue/68400/1048652. Access date:11.December 2023

Akbaş N, Taner B (2021) Yaşam döngüsünde yönetim: Varoluştaki algının ve değişimin kronolojisi. Ankara Üniversitesi Sosyal Bilimler Dergisi 12(1):132–152

Alıca SS (2017) Kadin Ve Çevre. Türkiye Barolar Birliği Dergisi 2017:151–172

Allam Z, Jones D, Thondoo M (2020) Cities and climate change: climate policy, economic resilience and urban sustainability. 1–32 Springer Nature

Atalay GE, Muratoğlu Pehlivan B (2023) Aestheticizing the pain: a critical analysis of media representation of earthquake victim children in Turkey. Advance online publication, Sociology Lens. https://doi.org/10.1111/johs.12437

Ateş H, Yavuz Ö, Dağli Z (2018) Women's participation in social life and local governments: an evaluation for Turkey. Local Service Local Politics in the Globalization Process. A.Ü. Faculty of Political Sciences Publications, 137–164 Access address: https://www.researchgate.net/profile/Zulkif-Dagli/publication/334398597. Access date: December 2023

Arı İ, Aydın L (2019) Recommendations for the preparation and effective implementation of local climate change action plans in Turkey. Igdir Univ J Soc Sci 395–414. Access from: https://www.researchgate.net/publication/338793252. Access date: 14 December 2023

Aydın G, Genç FN, Sürdürülebilir Kalkınma Bağlamında Afet Ve İklim Değişikliği Politikaları. Çankırı Karatekin Üniversitesi İktisadi ve İdari Bilimler Fakültesi Dergisi 13(2):563–590

Baş E, Partigöç NS (2022) İklim değişikliğine uyum sürecinde kent planlamanın rolü. Resilience 6(1):127–143. https://doi.org/10.32569/resilience.1026712

Basu R, Malig B, Ostro B (2010) High ambient temperature and the risk of preterm delivery. Am J Epidemiol 172(10):1108–1117

Bilgili M (2017) Toplumsal cinsiyet, bedenin mekansallığı ve cinsiyetçi mekanlar. Felsefe-Yazın Dergisi, Ekim-Kasım 23:19–28. Access from: https://www.researchgate.net/publication/330170792. Access date: Aralık 2023

Bolin A (ed) (2020) Women's empowerment through collective action: How forest and farm producer organisations can make a difference. Rome, FAO and London, IIED. https://doi.org/10.4060/ca8713en

Bozkurt V (2023) Depremin toplumsal boyutu. Avrasya Dosyası 14(1):77–99. Access from: https://dergipark.org.tr/en/pub/avrasyadosyasi/issue/78803/1303147. Access date: 10 December 2023

Castañeda JV, Bronfman NC, Cisternas PC, Repetto PB (2020) Understanding the culture of natural disaster preparedness: exploring the effect of experience and sociodemographic predictors. Nat Hazards 103:1881–1904

Cities Climate Leadership Group (2021) C40 Cities Climate Leadership Group. Cities Climate Leadership Group from: https://www.c40.org/ Cities Climate Leadership Group date: 11 December 2023

Çamaş T, Turan M (2023) A new model in disaster management: Turkey national risk shield model. J Environ, City and Climate 2(4):238–261

Davies M, Guenther B, Leavy J, Mitchell T, Tanner T (2009) Climate change adaptation, disaster risk reduction and social protection: Complementary roles in agriculture and rural growth?. IDS Working Papers 2009(320):01–37. https://doi.org/10.1111/j.2040-0209.2009.00320_2.x

Dankelman I, Fansen W (2012) Gender, environment and climate change: understanding the linkages. In: Gender and climate change: An introduction. Routledge, pp 21–54

Dereli M (2019) The effects of global climate change on tourism sector with green economy approach (Master's thesis. Trakya University, Institute of Social Sciences, Edirne, Turkey

Doğaner A (2022) Sürdürülebilir Ekonomik Kalkınma İçin Türkiye'de Dirençli Şehirler. Çevre Şehir Ve İklim Dergisi 1(2):59–80

Ebi KL, Capon A, Berry P, Broderick C, de Dear R, Havenith G, Honda Y, Kovats RS, Ma W, Malik A, Morris NB, Jay O (2021) Hot weather and heat extremes: health risks. The Lancet 398(10301):698–708

Elfert M (2015) UNESCO, the Faure report, the Delors report, and the political utopia of lifelong learning. Eur J Educ 50(1):88–100

Erbaydar NP, Onal EI, Kaya E (2021) Analysis of gender sensitivity of Turkey's disaster legislation to create a disaster-resilient society. Tohoku J Exp Med 254(4):261–266. https://doi.org/10.1620/tjem.254.261

Erdoğan Z, Cantürk S (2022) Understanding the climate-conflict-migration nexus: immigration from climate-conflict zones to Turkey. SİYASAL: J Polit Sci 31(1):137–155. https://doi.org/10.26650/siyasal.2022.31.994670

Esin BAŞ, Partigöç NS (2022) İklim Değişikliğine Uyum Sürecinde Kent Planlamanın Rolü. Resilience 6(1):127–143. https://doi.org/10.32569/resilience.1026712

Fatema SR, Islam MS, East L, Usher K (2019) Women's health-related vulnerabilities in natural disasters: a systematic review protocol. BMJ Open 9(12):1–5. https://doi.org/10.1136/bmjopen-2019-032079

Gagnon-Lebrun F, Agrawala S (2006) Progress on adaptation to climate change in developed countries. Organisation for economic co-operation and development. http://www.oecd.org/env/cc/37178873.pdf
Gerdan S (2019) Bir Sosyal Sorumluluk Alanı Olarak Afet Eğitimleri. Int J Manage Admin 3(5):101–110
Glasser R (2019) Preparing for the era of disasters. Australian Strategic Policy Institute, Canberra, Australia, pp 3–24
Gündüz F (2022) Women in disasters and lessons to be learned with a gender perspective (The Case of Haiti and Japan Earthquake). IBAD J Soc Sci 12: 440–460. https://doi.org/10.21733/ibad.1039215
Güneş EB (2023) Assessing the climate vulnerabilities of Syrian immigrants living in Istanbul from a climate justice perspective (Master's thesis, Middle East Technical University)
Gürçam S (2022) Analysing the global fight against climate change and the Turkish context. Lectio Socialis 6(1):39–54
Gündüz F (2024) A phenomenological study on women's disaster experience in Türkiye. Int J Disaster Risk Reduction 100. 104196
Hamaber J, Grafakos S, Tollin N (2017) Sustainable urbanization in the Paris agreement'–a comparative review of Nationally Determined Contributions for Urban Content
Habitat UN (2018) Tracking progress towards inclusive, safe, resilient and sustainable cities and human settlements. SDG 11 synthesis report-high level political forum 2018. Access adress https://discovery.ucl.ac.uk/id/eprint/10087018/1/SDG%2011%20Synthesis%20Report_web2.pdf access date 12.12.2023
Ilgın HÖ, Karagül D (2022) Afet Süreçlerinde Kadınlara Yönelik Toplumsal Cinsiyet Eşitsizliğinde Sivil Toplum Kuruluşu Çalışanlarının Deneyimleri: Çanakkale İli Örneği. JOEEP: J Emerging Econ Policy 7(2):85–103
Kadioğlu M, Sonu BH (2007) Küresel İklim Değişimi ve Türkiye. Güncel Yayıncılık aş 110:98–108
Karahasan BC, Pinar M (2023) Climate change and spatial agricultural development in Turkey. Rev Dev Econ 27:1699–1720. https://doi.org/10.1111/rode.12986
Kaygusuz K, Toklu E (2012) Energy issues and sustainable development in Turkey. J Eng Res Appl Sci 1(1):1–25
Klein RJT, Alam M, Burton I, Dougherty WW, Ebi KL, Fernandes M, Huberlee A, Rahman AA, Swartz C (2006) Application of environmentally sound technologies for adaptation to climate change. Technical Paper FCCC. TP/2006/2, United Nations Framework Convention on Climate Change Secretariat, pp 157–168. https://doi.org/10.1007/978-94-007-1770-1_9
Kipay SS (2023) Deprem gerçeği ve kadın sağlığı üzerine etkileri. İzmir Kâtip Çelebi Üniversitesi Sağlık Bilimleri Fakültesi Dergisi 8(2):855–860. Access adress: https://dergipark.org.tr/tr/pub/ikcusbfd/issue/78150/1267045. Access Date: 07 December 2023
Koç G, Natho S, Thieken AH (2021) Estimating direct economic impacts of severe flood events in Turkey (2015–2020). Int J Disaster Risk Reduction 58:102222
Koçan N, Alp F (2021) Ekokent statüsündeki kentler ve özellikleri. Düzce Üniversitesi Orman Fakültesi Ormancılık Dergisi 17(1):1–23. Access freom: https://dergipark.org.tr/tr/pub/duzceod/issue/63286/894692. Access date: 11 December 2023
Kolukırık S, Arslan DA, Yımaz GG (2022) Orman yangınlarının toplumsal etkileri ve görünümü: Medya paylaşımlarında 2021 büyük Antalya-Manavgat yangını. Afet ve Risk Dergisi 5(2):560–580. https://doi.org/10.35341/afet.1086418
Lau JD, Kleiber D, Lawless S, Cohen PJ (2021) Gender equality in climate policy and practice hindered by assumptions. Nat Clim Chang 11(3):186–192
Lee YS, Saito NT, Todres J (2020) The fallacy of contract in sexual slavery: a response to ramseyer's" contracting for sex in the pacific war". Mich J Int'l l 42:291
Local Governments for Sustainability (ICLEI) (2020) Access from: https://www.iclei.org/en/About_ICLEI_2.html. Access date: December 2023
Mackenzie S (2002) Kentte kadınlar. B. Duru & A. Alkan (Ed. & Çev.), 20:249–283

Macit İ (2019) Bütünleşik afet yönetiminde Sendai çerçeve eylem planının beklenen etkisi. Doğal Afetler Ve Çevre Dergisi 5(1):175–186. https://doi.org/10.21324/dacd.421564

Mengi A, İşçioğlu D (2018) Küreselleşme Sürecinde Yerel Hizmet Yerel Siyaset. In: Güler M, Karanuh Z (eds) D. Kentsel Yaşam Kalitesinin Belirleyicisi Olarak Belediye Hizmetleri ve Stratejik Planlar, pp 285–310

McIntyre-Mills NRA, Romm RRA, Nantes CY (eds) (2018) Balancing individualism and collectivism: collected papers from special integration group for international systems sciences plus 16 contributors. Contemporary Systems Series. Springer, pp 199–209. 978-3-31958014-2

Neumayer E, Plümper T (2007) The gendered nature of natural disasters: the impact of catastrophic events on the gender gap in life expectancy, 1981–2002. Ann Assoc Am Geogr 97(3):551–566. https://doi.org/10.1111/j.1467-8306.2007.00563.x

Okur M (2023) Sosyal, beşeri ve idari bilimler alanında gelişmeler (H. A. Kutlu, Ed.). Platanus Yayıncılık, 1.Baskı, Ankara, 57–77

Özşahin E (2013) Türkiye'de yaşanmiş (1970–2012) doğal afetler üzerine bir değerlendirme. 2. Türkiye Deprem Mühendisliği ve Sismoloji Konferansı, pp 25–27

Pelling M, Holloway AJ (2006)x Legislation for mainstreaming disaster risk reduction. Tearfund. Access address: https://www.researchgate.net/publication/305992622_Legislation_for_mainstreaming_disaster_risk_reduction. Access date: 08 December 2023

Pink RM, Pink RM (2018) Europe: UK, Italy, Greece. The climate change crisis: solutions and adaption for a planet in Peril, Routledge, pp 217–255

Roos EJ (2022) Call for leadership: climate change and the future health of girls. J Pediatr Adolesc Gynecol 35(2):107–108

Rossati A (2017) Global warming and its health impact. The Int J Occupational Environ Med 8(1):7. https://doi.org/10.15171/ijoem.2017.963

Sırma ÇS (2023) Being a woman while the climate is changing: vulnerability due to gender roles in climate change. In: Academic research and evaluations in social sciences – II, 47–62. Özgür Publication. https://doi.org/10.58830/ozgur.pub309.c1415

Sharafat S, Nasreen S (2022) Performance assessment regarding UNESCO sustainable development goals 4 & 5 in pakistan: targets, current positions & challenges. J Soc Sci Adv 3(4):199–211

Sharma J, Ravindranath NH (2019) Applying IPCC 2014 framework for hazard-specific vulnerability assessment under climate change. Environ Res Commun 1(5):051004. https://doi.org/10.1088/2515-7620/ab24ed

Sohrabizadeh S, Tourani S, Khankeh HR (2016) Women and health consequences of natural disasters: challenge or opportunity? Womens Health 56:977–993. https://doi.org/10.1080/03630242.2016.1176101

Swiderska K (2021) Here's why indigenous economics is the key to saving nature. International Institute for Environment and Development. Access date: August, 20, 2023. Access address: https://www.iied.org/heres-why-indigenous-economics-key-saving-nature

Şahin Ü, Erensü S (2020) Covid-19 Pandemisini ve İklim Krizini Birlikte Okumak. Mercator Politika Notu. Access adress: https://ipc.sabanciuniv.edu/Content/Images/CKeditorImages/20201220-23123921.pdf. Access date:10 December 2023

Tebaldi C, Ranasinghe R, Vousdoukas M, Rasmussen DJ, Vega-Westhoff B, Kirezci E, Kopp RE, Sriver R, Mentaschi L (2021) Extreme sea levels at different global warming levels. Nat Climate Change 11(9):746–751

Turhan E, Cerit Mazlum S, Şahin Ü, Şorman AH, Cem Gündoğan A (2016) Beyond special circumstances: climate change policy in Turkey 1992–2015. Wiley Interdisc Rev: Climate Change 7(3):448–460

Talu N (2016) İklim Değişikliği ve Toplumsal Cinsiyet Politika Belirleme Süreçleri. Yasama Dergisi 33:68–87. Access from: https://dergipark.org.tr/tr/pub/yasamadergisi/issue/54466/741353#article_cite. Access date: 08 December 2023

Tamers SL, Streit J, Pana-Cryan R, Ray T, Syron L, Flynn MA, Castillo D, Roth G, Geraci C, Guerin R, Schulte P, Howard J (2020) Envisioning the future of work to safeguard the safety,

health, and well-being of the workforce: a perspective from the CDC's national institute for occupational safety and health. Am J Industrial Medicine 63(12):1065–1084
Tham S, Thompson R, Landeg O, Murray KA, Waite T (2020) Indoor temperature and health: a global systematic review. Public Health 179:9–17. https://doi.org/10.1016/j.puhe.2019.09.005
Thurston AM, Stöckl H, Ranganathan M (2021) Natural hazards, disasters and violence against women and girls: a global mixed-methods systematic review. BMJ Glob Health 6(4):e004377
Topgül S (2012) Kadın ve Doğa İlişkisi: Ekofeminizm. Sosyoloji Dergisi 27:71–83. Acces from: https://dergipark.org.tr/tr/pub/sosder/issue/40999/495422#article_cite. Acces date 08 December 2023
Tuğaç Ç (2021) Bütünleşik afet riski azaltımı ve iklim değişikliğine uyum yaklaşımı üzerine bir değerlendirme. Afet Ve Risk Dergisi 4(1):1–19. https://doi.org/10.35341/afet.891478
Tyagi R, Vishwakarma S, Rishi M, Rajiah S (2020) Reducing inequalities through education and skill development courses. Reduced Inequalities 1–13
Uncu BA (2019) İklim için kentler yerel yönetimlerde iklim eylem planı. Dijital Düşler Basım San. ve Tic. AŞ, 93
United Cities and Local Government Middle East and West Asia Section (MEWA) (2021) Access from http://uclg-mewa.org/tr/. Access date: 12 December 2023
UNISDR (2015) Sedai Framework for Disaster Risk Reduction 2015–2030. United Nations Office for Disaster Risk Redcution, Geneva, Switzerland, Access address: https://www.unisdr.org/files/43291_sendaiframeworkfordrren.pdf. Access date: December 2023
Unwomen (2022) Tackling violence against women and girls in the context of climate change. Access address: https://www.unwomen.org/sites/default/files/2022-06/Tackling-violence-against-women-and-girls-in-the-context-of-climate-change-tr.pdf. Access date: December 2023
Ünür E (2021) Afet Yönetiminde Toplumsal Cinsiyet Eşitsizliği: Doğal Afetlerde Cinsiyete Dayalı Zarar Görebilirlik Farkı. Istanbul Aydin University Journal of Social Sciences/istanbul Aydin Üniversitesi Sosyal Bilimler Dergisi 13(2):351–381
Watkins K (2007) Human development report 2007/8. Fighting climate change: Human solidarity in a divided world. UNDP-HDRO Human Development Report. Access from https://hdr.undp.org/en/2007-2008. Access date: 08 December 2023
Whittington V (2021) Gender and human rights within unesco's international heritage discourse: an analysis of the world heritage and intangible heritage conventions. Heritage & Society 14(2–3):242–266
Yalçın E, Höbek Akarsu R, Ural A, İlkin E (2022) İklim Değişikliklerinin Kadın Üreme Sağlığına Etkileri. Gazi II. Uluslararası Sağlık Bilimleri Kongresi, p 109
Zuo J, Zhao ZY (2014) Green building research–current status and future agenda: a review. Renew Sustain Energy Rev 30:271–281

Part IV
Capacity Building

Chapter 17
Capacity Building in Serbia for Disaster and Climate Risk Education

Vladimir M. Cvetković and Vanja Šišović

Abstract The territory of the Republic of Serbia is exposed to various natural and technical–technological hazards that increasingly endanger people and their material goods. An integrated disaster risk management system (preparedness, mitigation, response and recovery) has not yet taken root in Serbia to a sufficient extent, although there are certain institutional and legal prerequisites for its smooth functioning. Abandoning the reactive way of preparing for disasters and increasingly implementing proactive measures, societies are desperately trying to improve their resilience in resource-scarce situations and mitigate future consequences of disasters. In Serbia, there is insufficient resistance to disasters, reflected in the insufficiently improved ability of Serbian society exposed to dangers to respond to them and recover from the consequences of disasters promptly and efficiently. Despite this, many efforts are in vain, considering that the population's awareness of various aspects of disaster management remains very low. The aim of this chapter refers to the comprehensive analysis and description of capacity building in Serbia for Disaster and Climate Risk Education. The undoubted importance of education in the field of disasters, which is increasingly receiving priority positions on various agendas around the world, emphasizes the necessity of continuous research on the perspectives of formal and informal education on disasters. One of the most effective tools used in all areas of integrated disaster risk management is education. Admittedly, it should be borne in mind that education is a complex and interactive psychological process in which the transfer of knowledge from different fields is carried out in a planned and clearly grounded manner in the form of a kind of transfer.

Keywords Disasters · Climate risk · Capacity building · Education · Serbia

V. M. Cvetković (✉)
Faculty of Security Studies, University of Belgrade, Gospodara Vučića 50, 11040 Belgrade, Serbia
e-mail: vmc@fb.bg.ac.rs

V. M. Cvetković · V. Šišović
Scientific-Professional Society for Disaster Risk Management, Dimitrija Tucovića 121, 11040 Belgrade, Serbia

V. M. Cvetković
International Institute for Disaster Research, Dmitrija Tucovića 121, 11040 Belgrade, Serbia

A. Yildiz and R. Shaw (eds.), *Disaster and Climate Risk Education*, Disaster Risk Reduction, https://doi.org/10.1007/978-981-97-5987-3_17

17.1 Introduction

In today's period, more than ever before, developed and underdeveloped countries face the consequences of various natural and man-made disasters (Cvetković et al. 2023; El-Mougher et al. 2023; Iftikhar and Iqbalx 2023; Rajani et al. 2023; Sergey and Gennadiy 2022; Shibru et al. 2022). Abandoning the reactive way of preparing for disasters and increasingly implementing proactive measures, societies are desperately trying to find ways to improve their level of resilience in resource-scarce situations and mitigate future consequences of disasters (Aktar et al. 2021; Cruz and Ormilla 2022; Dukiya and Banji Adeleye 2022; El-Mougher 2022; Jaiye and Benjamine 2021; Janković 2021; Kabir et al. 2022; Mohammed and Maysaa 2022; Öcal 2021; Odero and Mahiri 2022).

Despite this, many efforts are in vain, considering that the population's awareness of various aspects of disaster management remains at a very low level (Cvetković and Dragićevic 2014; Cvetković et al. 2019b; Cvetković et al. 2020a, b; Cvetković and Stanišić 2015; Ocal et al. 2020). The implications of neglecting to find innovations in the field of education and training of the population leave behind tragic and very expensive consequences (Carla 2019; Cvetković 2019a, b; Goyal 2019; Kumiko and Shaw 2019; Mano and Rapaport 2019; Öcal 2019). In one of the studies, several types of deficiencies were identified that currently hinder the promotion of a culture of preparedness in the face of a particular disaster. These are (1) decision-making skills, (2) teamwork skills, (3) understanding of regulations, (4) compliance with regulations, and (5) radiation anxiety (Salleh and Yusof 2013). Consequently, it can be said that these factors represent a small part of the core competencies (knowledge, skills and attitude) in building a culture of preparedness (Cvetković et al. 2022a, b, c; Cvetković and Martinović 2021; Cvetković and Nikolić 2021; Cvetković et al. 2022a, b, c; Cvetković et al. 2021a, b, c, d, e; Cvetković et al. 2021a, b, c, d, e; Cvetković and Filipović 2020; Cvetković and Martinović 2020; Cvetković et al. 2021a, b, c, d, e; Cvetković et al. 2021a, b, c, d, e; Cvetković and Todorović 2020; Cvetković and Janković 2021; Janković and Cvetković 2020; Krnjić and Cvetković 2021a, b).

Man has always found ways to protect himself and his neighbours from destructive phenomena that came from nature and his built environment (Cvetković and Janković 2020; Cvetković and Jovanović 2020; Cvetković et al. 2020a, b; Cvetković et al. 2018a, b, c, d; Cvetković and Bošković 2021; Cvetković et al. 2017a, b, c). Of course, it is difficult to say what, in that evolutionary sense, was destructive to man and his existence. Were these phenomena from nature that surprised him and very often subdued him, or did everything take on a different dimension when accelerated industrialization occurred? Not so rarely do people notice that we have become a society of risk, that absolute security has become such a hard-to-achieve phenomenon and that serious efforts are needed to adequately and timely prevent and, in the worst case, mitigate the risks of disasters. Elaborations of the concept of disaster risk reduction have indeed reached enormous proportions and serious discussions are taking place in different scientific disciplines, geographies and dialects (Kulatunga 2010).

A disaster is an event in a certain time and space, in which society or one part of society suffers damage and social disruption so that all or some basic social values are violated. Such consequences arise because the event exceeds the usual protection measures. The concentration of disasters in a certain area defines a series of geographical areas that are identified as a series of concentric zones that show the impact and response to disasters. In practice, this zone of influence is more difficult to determine than it seems. For example, earthquake damage can cause power loss in areas where there is no physical damage. Therefore, defining the boundaries of the impact zone is a significant problem for disaster response researchers, as well as a problem for those trying to estimate where the effects of an earthquake actually occurred, let alone those trying to predict where an earthquake will actually occur (Lindell 2013).

The prevailing and widely acknowledged perspectives of experts, juxtaposed with the opinions of non-expert political leaders, regarding the necessity or dispensability of training and educating the population, are becoming increasingly central in the arena of scholarly and professional discourse. These viewpoints take a prominent position in the ongoing clash of opinions on the scientific and professional stage. Experts typically advocate for increased investments in the education sector, while political leaders, often faced with pressures on other budgetary needs, may have different priorities. This dynamic creates discussions about the proper allocation of resources and challenges in achieving long-term education goals. More and more voices are being heard calling for the community to focus on designing and implementing non-structural measures aimed at improving citizens' behaviour and response to disasters. States can invest huge financial resources in various structural measures in order to mitigate certain aspects of disasters, but all this is in vain if, there are no trained individuals who will know how to protect themselves and their loved ones in the right way and in a timely manner. Precisely for this reason, it can be said that the secret of effective disaster risk management lies in the fact of comprehensive, proactive action on people by improving their mental predispositions for proper disaster response.

The undoubted importance of education in the field of disasters, which is increasingly receiving priority positions on various agendas around the world, emphasises the necessity of continuous research on the perspectives of formal and informal education on disasters (Cvetković et al. 2020a, b; Cvetković and Filipović 2020; Cvetkovic and Martinović 2020; Cvetković and Svrdlin 2020). The focus of this chapter entails a theoretical examination of the function of schools in mitigating the risk of disasters. Its primary objective is to provide a scholarly analysis of the current scenario, challenges, and potential avenues for future development in this realm. Employing a scientific description, the chapter aims to elucidate the multifaceted aspects associated with the role of schools. Furthermore, it expands its scope to encompass various dimensions, wherein diverse educational approaches in the field of disaster studies across global nations are expounded upon. In pursuit of this scientific objective, the study aims to provide a comprehensive framework for understanding disaster education, emphasizing the pivotal role that schools play as fundamental components of formal education systems. Beyond mere organization, the research endeavours to delve into the nuanced aspects of disaster education's significance, shedding light on its multifaceted dimensions within the educational landscape. Furthermore, the analysis acknowledges the dynamic nature of non-formal education, recognizing its diverse channels and methodologies. The exploration extends beyond the traditional classroom setting, encompassing community-based initiatives and awareness programs. By scrutinizing both formal and non-formal education, the study aims to offer a holistic perspective on their collective impact in effectively reducing the risks associated with disasters.

17.2 Disaster Risk Reduction Through Formal and Informal Education

In contemporary societies, the operation of social entities is inconceivable without the adoption of suitable structural and non-structural measures for reducing the risk of disasters (Cvetković and Filipović 2017; Cvetković et al. 2017a, b, c). Certainly, one of the most important measures of non-structural risk reduction from increasingly serious and frequent consequences of disasters is education. Education has become one of the most powerful weapons applied in all spheres of integrated disaster risk management. Admittedly, it should be borne in mind that education is a complex and interactive psychological process in which the transfer of knowledge from different fields is carried out in a planned and clearly grounded manner in the form of a kind of transfer.

Starting from different forms of education, such as formal, non-formal and informal education, each of these types of education has its clear advantages and disadvantages that should be taken into account when creating educational programs (Jakovljević et al. 2015). A great deal of academic thinking is directed towards finding a genuine connection between the success and quality of education in the field of disaster studies and integrated disaster risk reduction. In all these studies (Cvetković et al. 2017a, b, c; Cvetković and Filipović 2018a, b, c; Ivanov and Cvetković 2014; Stanišić and Maksić 2014), the general conclusion is that disaster education is one of the basic prerequisites for improving the resilience of society at all levels. However, the importance of education should not be taken literally without including other important elements in the disaster risk reduction process. Namely, all structural measures to reduce the risk of disasters can be implemented, and if citizens are not sufficiently trained to use such measures, there may be inconsistencies in the functioning of the protection systems provided for in this way.

There is a growing awareness of the impact of disasters on society. The establishment of the UN International Strategy for Disaster Reduction, "Enabling all societies to be resilient to the effects of natural hazards and related technological and environmental disasters, to reduce human, economic and social losses", has proven to be an effective platform for worldwide discussion. The publication Living with Risk was a turning point. Various extraterritorial entities have already released an extensive array of reports and publications about the handling of disasters.

The response during a disaster can be broken down into an alarm phase and also into a recovery or reconstruction phase. During the critical alarm phase, authorities vested with competence must navigate a complex series of decisions, process a diverse array of information, and subsequently assess the reliability of this information. They are tasked with determining the pertinence of the advice provided, carefully weighing its relevance to their specific circumstances. Extensive research consistently underscores that the pivotal juncture for formulating an effective response lies in the recognition of the threat as tangible and imminent. Within this context, citizens play a crucial role as they actively engage in evaluating the credibility of authorities or other entities disseminating warnings. The availability of disaster information is of great importance. There are three well-known myths about citizen response that continue to be presented as the basis for government disaster preparedness. The first myth tells about the panic of citizens in disasters. In fact, what is realistic is that the initial fear that strikes citizens in emergency situations quickly stops because they start looking for ways to ensure their own and others' safety. Contrary to another prevailing myth suggesting citizens' helplessness and dependence on external assistance, this research contradicts such notions. The study reveals that citizens are far from powerless; in fact, they often take the initiative in initiating search and rescue operations. Addressing a third noteworthy myth, it pertains to the occurrence of robberies during and after disasters, despite the fact that the governments of nearly all countries are actively developing legislation aimed at safeguarding the interests of law-abiding citizens (Helsloot and Ruitenberg 2004). Recognizing and dispelling such misconceptions is crucial for enhancing disaster and climate risk education, fostering a more informed and resilient global community.

17.2.1 The Importance of Educational Policies for Disaster Risk Reduction

Policies serve all spheres of social existence to systematically guide people's activities to achieve some projected goals. Viewed from that perspective, disaster risk reduction education policies occupy a special place in non-structural disaster risk reduction measures (Chicatto et al. 2015). Unfortunately, there is no educational policy in Serbia that would comprehensively and systematically design and implement certain measures aimed at reducing the risk of disasters (Cvetković 2023). Regarding climate risk education policies in Serbia, we could say that Serbia recognizes the importance of climate risk education and has implemented policies to raise awareness about climate-related challenges. Also, the Serbian government is committed to integrating climate risk education into the national curriculum to ensure that citizens are well-informed about the environmental threats the country faces. Initiatives in Serbia aim to foster a culture of environmental responsibility through comprehensive climate risk education policies. Besides that, educational institutions in Serbia are incorporating climate change topics into various subjects, providing students with a holistic understanding of the risks and impacts associated with climate change. On the other side, Serbia's climate risk education policies seek to empower communities by equipping them with the knowledge and skills needed to adapt to changing environmental conditions. Public awareness campaigns in Serbia play a crucial role in communicating climate-related information, encouraging sustainable practices, and fostering a sense of urgency regarding environmental issues. Climate risk education in Serbia emphasizes the interconnectedness of global climate issues and the role individuals can play in mitigating climate change.

The International Institute for Educational Planning at UNESCO has developed a guide for designing disaster education, which contains five framework chapters (Gwee et al. 2011): the first chapter provides a general overview (introduction, prevention and preparation for disasters, challenges in disasters and reconstructions, capacity building, disaster education for all); the second chapter refers to the supplement and content (rural population, gender, ethnicity/religion/children with developmental disabilities, facilities, open and distance learning, non-formal education); the third chapter concerns teachers and students (identification, selection, educational workers, teaching and learning methods, social support for learning); the fourth refers to curriculum and learning (curriculum, health and hygiene education) and the fifth chapter refers to capacity management (assessment of needs and resources, planning process, project management, legal work frameworks, budget and financial management).

Educational policies should include teaching plans and programs, ways or modalities of performing theoretical and practical exercises, as well as proposals for workshops depending on the age of the students (Cvetković 2023). One step further, it is necessary to clearly indicate in them the thematic units that should be based on all the identified hazards that were recognized in the National Disaster Risk Assessment in the Republic of Serbia. In addition, it is important to point out that the educational policy should include ways of support from the competent state authorities in terms of financing and sustainable development of this type of education (Cvetković et al. 2018a; b).

Starting from local specificities and regionalization of the risks themselves, educational policies should also be harmonized with the concrete risks of the given area. This actually means that it is unnecessary to burden the students of a certain area with the characteristics of dangers that are not recognized at the national or regional-local level. In order to develop educational policies, it is first necessary to carry out an analysis of the existing situation in terms: (a) potential dangers characteristic of the given area; (b) existing educational plans and programs; (c) potential modalities of student education in the field of emergency situations; (d) available resources for financing such a system.

In the second step, it is necessary to carry out the analysis, selection and prioritization of the thematic units and then of the specific teaching units with the experts of the appropriate profiles. In doing so, it is necessary to take into account the content of the teaching unit itself, which should include more social than natural aspects of protection from such events. More concretely, students should not be burdened with content from the etiology of various natural and technical–technological hazards but should immediately focus on the genesis of the manifestation of harmful effects and ways of protection against them. In relation to the spatial dimension, educational policies to reduce the risk of disasters could be adopted at the national, regional or local level. Certainly, it is recommended that educational policies be coordinated with each other along horizontal and vertical lines.

17.2.2 Strategic Aspects of the Importance of Disaster Education

The general impression and view are that disasters, extraordinary events and all dangers affect all parts of the world, and all peoples do not know borders in terms of religion, nation, or culture. The existence and respect of universal measures and mechanisms for joint action to combat and prevent the occurrence of disaster risks is necessary. It is crucial to harmonize policies at the national level when it comes to disasters. Investing in sustainable development is no less important and important, and this applies not only to environmental management but also to political and economic management, from the family to the international level (Trim 2004).

Also of primary importance is the cooperation of competent institutions at the state level. It is also necessary to eradicate corruption at all levels in order to prevent risks. Moreover, individuals engaged in disaster management should pay careful attention to the media. The influence wielded by the media, when wielded irresponsibly, can give rise to numerous challenges. These challenges may include the dissemination of panic, the unjust discrediting of those providing accurate guidance, and the hindrance of the implementation of essential measures during emergency situations. It is imperative to recognize and address the potential ramifications of media dynamics in disaster scenarios to ensure effective communication and response strategies. If individuals do not have a developed understanding of the requirements for adopting preventive measures, infrastructure costs in all sectors of human existence may not be adequate. Even if the building has state-of-the-art fire protection equipment, poorly informed and informed people will not be able to use it in an emergency. It is necessary to nurture community knowledge in addition to individual awareness of dangers and ways to mitigate them, starting with the understanding that local self-government plays an irreplaceable role in this. Citizens must draw attention to the following issues: (a) the nature and characteristics of various natural and technological dangers that are present around them; (b) the importance of taking actions to prepare and mitigate the effects of disasters; (c) protective measures implemented before, during and after disasters; and (g) potential for risk mitigation using various risk transfer mechanisms (insurance). Public education, group seminars and training sessions, as well as the creation and execution of numerous campaigns, can help increase awareness (Cvetković 2020).

The goals of training for appropriate and correct response to disasters can be multiple (Alexander 2002, p. 2): (a) testing and evaluation of the allocation and coordination of local and regional resources; (b) demonstration of the possibility of alerting the public in certain periods; (v) testing alternative communication systems to be activated and functional within 90 min from the occurrence of the disaster; (g) demonstration of the ability of the personnel in the headquarters to function at the operational level and make valid decisions; (d) testing the possibility of enabling the initial assessment of disasters within four hours; (đ) demonstrating the ability to provide food, communications and administrative, logistical support to the headquarters staff; (e) demonstrating the ability to verify the existence of certain events or dangers and (h) assessing the possibility of the participation of various intervention and rescue services (Alexander 2002).

17.3 Schools and Disaster Risk Reduction

17.3.1 The Role of Primary Schools in Disaster Risk Reduction and Climate Risk Education

Elementary schools represent a kind of pillar of education in every country (Suarmika et al. 2022). The basic characteristic of primary education is reflected in the continuous transfer of knowledge from different subjects, that is, scientifically based disciplines, in terms of their awareness of certain thematic entities. Students of a certain age attend classes in them, and all teaching content is adapted to their age level, that is, appropriate didactic and methodical methods are applied. Leaving aside the basic teaching content, unfortunately, in Serbia, there is no subject within which students in primary schools would be comprehensively educated on the issues of etiology of endangerment and protection of people in emergencies (Cvetković et al. 2016; Cvetković 2016a, b). Nevertheless, based on the all-purpose support of the Ministry of Internal Affairs of the RS, and the Department for Emergency Situations, certain lectures are organized on various topics such as floods, fires, landslides, earthquakes, etc.

Judging by the insufficiently systematized education of students in the field of emergencies in the RS, one can point out the insufficient responsibility of society to protect the most vulnerable categories, such as children up to the age of 14 and young people up to the age of 18. In developed countries, such as Germany and Japan, there are special plans and programs following which the education of students of that age is carried out, and the education methods themselves are very original and include various demonstration exercises such as demonstrations of how to extinguish fires, reacting in earthquake situations etc.

Climate risk education refers to the incorporation of climate-related topics into educational curricula, aiming to equip individuals, particularly students, with knowledge and understanding of the risks associated with climate change. This form of education covers a range of subjects, including the science of climate change, its impacts on ecosystems and communities, and strategies for mitigation and adaptation. In many cases, climate risk education is increasingly recognized as essential for building a resilient and environmentally conscious society. Primary schools play a crucial role in laying the foundation for environmental awareness and responsibility. While the extent to which climate risk education is integrated into primary school curricula can vary globally, there is a growing trend towards incorporating these topics at early educational levels.

17.3.2 The Role of Secondary and Higher Schools in Disaster Risk Reduction and Climate Risk Education

In contrast to primary education, secondary and higher education present the opportunity to employ a broader range of methods and educational resources for enhancing awareness and knowledge in the realm of emergencies. The more advanced educational levels allow for the implementation of diverse approaches, facilitating a richer learning experience through varied instructional materials and methodologies. Starting from the very characteristics of students to whom knowledge is transferred, it is possible to use more available terminology, which, unlike students in the primary school system, is simply impossible to use due to their understanding. Of course, the modalities of education in the field of disasters can differ according to the types of secondary schools, i.e. whether they are vocational or general education schools (Cvetković et al. 2019a; Cvetković et al. 2018a, b, c, d; Cvetković, Noji, et al. 2018a, b, c, d).

Consequently, in secondary technical schools, students will receive certain prior knowledge in the field of technical–technological hazards, while in secondary schools of the general education system, it will be much more difficult (Cvetković and Filipović 2020; Cvetkovic and Martinović 2020; Cvetković and Svrdlin 2020). Despite organized attempts by academic organizations to introduce a subject that could be called "Safety culture", one gets the impression that society is still not sufficiently interested in this type of education. The justification for such bad decisions is based on the argument that such teaching content can be implemented within the framework of other subjects, and indirectly, one gets the impression that the decision-makers themselves are not aware of the importance of such a subject (Cvetković et al. 2020a, b; Cvetković et al. 2018a, b, c, d; Cvetković 2017; Cvetković 2019a, b).

Leaving aside the numerous imperfections of the functioning of society in Serbia, the question arises as to how all this is reflected in the level of the society's resilience to disasters. Practice shows that by far, the largest number of people were saved in the first few hours after the manifestation of the disaster by neighbours or other people who happened to be nearby. Therefore, the members of the intervention and rescue services will not be able to provide help to all endangered people in a short period, and therefore it is necessary to make people active participants in the protection and rescue system and not passive victims who will wait for help from the competent intervention and rescue services. To achieve this, it is necessary that there are organized attempts by society to raise awareness and knowledge to a higher level and that the necessary strategic orientations be carried out. On the contrary, societies that are more developed and in which knowledge has taken an appropriate position better recognize the relevance of such activities, while on the other hand, everything in Serbia is exposed to devaluation and undermining of society's resilience. In 2007, an international conference on school safety was held, which recognized the importance of ensuring the education of every child, as well as living in a safe environment to achieve zero per cent mortality of children in schools due to natural disasters. For these reasons, the following priorities are foreseen: disaster education will be implemented in schools; school facilities and other infrastructure must be disaster-resistant; strengthen the education of local community members to make them more resilient to disasters; make schools safer (Cvetković and Filipović 2018a, b, c).

The significance of secondary and higher educational institutions in advancing climate risk education in Serbia cannot be overstated, as they play a pivotal role in promoting environmental awareness and fostering sustainable practices. The evolution of ecological education in the country has traversed distinctive phases, progressing from marked anthropocentrism to the present era of dominant ecocentrism. While primary schools establish the groundwork for environmental consciousness, secondary and higher schools assume a critical function in augmenting students' comprehension and encouraging a more comprehensive engagement with issues related to climate (Stanišić 2019). Within the realm of the educational process, noteworthy advancements have been made in secondary and higher education by seamlessly integrating ecological and sustainable development content into existing subjects. Moreover, the inclusion of elective courses dedicated to ecological education significantly contributes to a more holistic curriculum. Beyond the confines of traditional classrooms, extracurricular activities have undergone enrichment with ecological content, aligning with overarching educational policies that delineate planned initiatives in this domain. Nevertheless, persistent challenges are discernible. A conspicuous gap exists between the acquired ecological knowledge and the individual actions taken in daily life to preserve the environment. Additionally, interdisciplinary approaches and project-based learning, validated by numerous studies for their efficacy, are not yet universally implemented in pedagogical practices.

To effectively address these challenges and chart a course for future developments, recommendations advocate for a concentrated effort to create opportunities that facilitate authentic, receptive, and reflective student engagement with nature. This necessitates the implementation of successful pedagogical approaches, including interdisciplinary methods and project-based learning. The augmentation of extracurricular activities, encompassing the enhancement of existing initiatives such as field trips and nature classes, alongside the introduction of novel programs like ecological camps, holds promise in bridging the divide between theoretical knowledge and direct experiential learning in nature (Stanišić 2019). Notably, financial considerations are underscored, accentuating the imperative for increased investments in ecological education. Substantial funding would empower educational institutions to bolster their capacities, enrich educational resources, and provide comprehensive training for teaching staff. In summation, the recommendations underscore the pivotal role of a multifaceted approach encompassing pedagogical strategies, extracurricular activities, and financial support to fortify the role of secondary and higher educational institutions in advancing climate risk education in Serbia.

17.3.3 Disaster Risk Reduction and Environmental Education Programs

In the scholarly discourse, considerable emphasis is placed on the pivotal role that schools play in equipping children to cope with disasters, as underscored by various researchers (Suarmika et al. 2022; Yusuf et al. 2021). The significance of this role is magnified by specific contextual factors. Generally, the vulnerability of youth to disasters diminishes significantly when they are equipped with pertinent information, and essential resources, and are actively engaged in disaster preparedness and response activities. Moreover, access to personal and community support further fortifies their resilience in the face of potential disasters. It is worth noting that the active participation of young individuals in processes related to disaster risk reduction extends beyond individual benefits, extending its impact to national and local levels. Involving youth in such initiatives, not only enhances their capacity to navigate and withstand disasters, but it also yields broader advantages for the community and the nation as a whole. This proactive engagement establishes a foundation for fostering a culture of preparedness, community cohesion, and an informed citizenry capable of responding effectively to unforeseen challenges. Consequently, schools serve as vital institutions for imparting not only academic knowledge but also practical skills and a sense of collective responsibility, contributing to the overall resilience of communities in the face of potential disasters.

For example, in southern Thailand, during the disaster caused by the tsunami in 2004, young people who actively participated in the management processes during the mentioned event had a significant role and comprehensive support in the processes of direct or indirect assistance to their own and other people's families, as well as to local communities. By transferring knowledge and experience among young people, it is possible to reduce the risk of disasters through processes of mutual support and guidance. A study conducted in Serbia (Krnjić and Cvetković 2021a, b) revealed that a significant majority of participants rely on the Internet as their primary source of information regarding disasters, with almost half of the respondents (185 individuals, or 49.2%) expressing this preference. Additionally, 105 respondents (27.9%) reported obtaining information through public channels, while 40 participants (11.2%) indicated that family conversations were a source of awareness. Interestingly, a mere 28 students (7.4%) mentioned learning about disasters through formal education at school. The smallest proportion, comprising 16 respondents (4.3%), attributed their awareness of disasters to social and video games. These findings highlight the diverse sources from which individuals gather information about disasters, underscoring the need for comprehensive and varied educational approaches.

Environmental education in Serbia plays a crucial role in shaping students' awareness of environmental protection. The initial recognition of its importance occurred at the First Congress of Ecologists in Yugoslavia in 1973, as noted by Radović (1996). Since this recognition became official, there has been significant progress in integrating the content and activities of environmental education into the regular curricula of primary schools throughout the country. Although environmental topics have been present in school programs for the past two decades, it has been observed that students are not sufficiently motivated to translate their knowledge into practical action. This challenge underscores the need for stronger integration and coordination among different subjects to stimulate student engagement. Enhancing environmental education requires not only teaching strategies that encourage active student participation but also promoting awareness of the importance of preserving the environment. Through the collaborative efforts of teachers, curriculum development, and motivation, a deeper understanding of environmental challenges can be achieved, fostering long-term changes in attitudes toward the environment.

17.3.4 Comparative Practice of School Education in the Field of Disaster Risk Reduction

According to research, children can learn how to take disaster precautions and the best place for them to learn these skills is at school (Coates 2021; Johnson et al. 2014; Luna 2012; Nakano et al. 2020). In 2004, the English Government sent to all households a corresponding brochure with advice on how to react in disaster conditions, as well as advice on how to turn off gas, electricity, etc. Therefore, in that brochure, citizens are advised on how to best protect themselves and help others. However, in the period after that and everything that followed, it can be said that there was no significant effect (Page et al. 2008). Namely, after one year, on July 7, 2005, a bomb attack took place in London, on that occasion, 52 passengers were killed, and many people were directly or indirectly affected by this event. In the survey after the mentioned event, respondents were asked whether they had read the brochure. However, slightly more than half of the respondents refused to receive the brochure, even though the government sent a copy to every household. Only 300 people (52.3%) confirmed receiving the brochure, while 239 (41.6%) people said they had read the same brochure. It can be said that the receipt of the brochure influenced the collection of supplies but not the development of plans for adequate disaster response (Page et al. 2008).

The literature has recommended certain components that are necessary for the development of disaster education programs, including (1) adopting a disaster management perspective that supports teaching protective behaviours before and during disasters, (2) using a gradual sequence of learning activities throughout the school years, and (3) incorporating activities that facilitate interaction between children and parents (Ronan and Johnston 2001).

In Germany, there are two master's degree programs directly related to emergency and disaster management, which are implemented within the framework of two schools and several universities. Namely, continuing education in the field of disaster prevention and disaster management is offered by the University of Bonn in cooperation with the Federal Office for Civil Protection and Disaster Relief (BBK). It is interdisciplinary and teaches actions and key competencies that are useful and necessary for sustainable disaster risk reduction and, therefore, sustainable disaster management. The course is intended for university graduates who have at least three years of relevant professional experience and wish to obtain further academic qualifications (https://www.uni-bonn.de/).

In the Czech Republic, certain aspects of disaster reduction are incorporated into the curriculum within physics and chemistry subjects from primary to secondary school, catering to students aged 10 to 15. Secondary schools specifically integrate components of disaster prevention into their programs. However, educating adults poses a greater challenge. Although the mass media occasionally covers disaster risk reduction, it lacks a systematic approach. Notably, public interest in disaster risk reduction tends to surge during actual disasters (Education 2007). This indicates a need for more structured and consistent adult education programs and media engagement in promoting disaster awareness and preparedness. When it comes to China, it should be emphasized that Article 30 of the "Law on Emergency Response of the People's Republic of China" stipulates that schools of all levels and types incorporate education about emergencies into their curriculum, educate students about disaster knowledge and foster students' awareness of safety and self-rescue and the ability to rescue each other. To improve safety awareness and the ability to respond to disasters in the whole society, it is emphasized that it is necessary to educate students about knowledge about disasters in the stage of school education. Colleges and universities should educate students on disaster prevention and emergency rescue knowledge according to the characteristics of disciplines and majors (http://www.jyb.cn/rmtzcg).

In China, May 12 is designated as the National "Disaster Prevention and Reduction Day", to conduct disaster prevention education for the general public and students and promote disaster prevention awareness, disaster prevention knowledge and disaster prevention skills. Education on disaster prevention in schools only started with the adoption of the five-year plan for comprehensive national prevention and disaster risk reduction. Education on disaster prevention is incorporated into the national education system and education on disaster prevention and knowledge on disaster mitigation. On the other hand, compared to elementary and middle schools and kindergartens, colleges and universities are relatively slow in disaster prevention education. In the implementation of disaster prevention education, there is a common phenomenon of emphasizing theory over practice and emphasizing form over real effect, the construction of a curriculum system for disaster education is lacking (http://www.jyb.cn).

An extensive seven-hour program has been meticulously crafted for students across secondary, primary, and preschool education levels in France. Within secondary education, each school designates a "correspondent" for the program, a role that can be assumed by either a teacher or a parent. The primary responsibility of these assistants is to support teachers in addressing challenges and to enhance awareness of the principles outlined in the Prevention-First Gestures program among their peers. Additionally, these assistants are tasked with organizing local risk exhibitions. At the elementary school level, the lead teacher imparts training to students on appropriate reactions and behaviours during emergencies. Meanwhile, in preschool settings, the designated correspondent guides students to observe adults and practice fundamental "first gestures". Between 1996 and December 2002, a substantial number of individuals, exceeding 6,200, in the Grasse area actively participated in diverse training programs. Since 1999, the program has engaged over 3,300 students, encompassing 663 prevention assistants, 153 lead teachers, and 166 school teachers. This concerted effort has involved the participation of more than 5,000 students and 300 adults in comprehensive training programs, underscoring the commitment to promoting disaster preparedness and awareness (Education 2007).

17.4 Curricula and Programs in the Field of Disaster Risk Reduction

17.4.1 Basic Characteristics of Educational Programs in the Field of Disasters

Globally, a plethora of educational programs in the field of disaster studies have been established, as evidenced by scholarly works such as those by Boon and Pagliano (2014), Faupel et al. (1992), and Johnson et al. (2016). Within this domain, Johnson et al. (2014) conducted a significant study that methodically delved into the landscape of evaluating the effectiveness of disaster education programs for children. The crux of their research lay in conducting a meticulous review of the methodological literature to ascertain how scholars and practitioners presently measure and assess the impact of disaster education programs through evaluations. A systematic exploration of published literature unveiled 35 relevant studies, which were then analysed to formulate a comprehensive categorization of operational components within the existing body of research.

This categorization encompassed various facets, including the types and sources of evaluations, research methods and designs, research participants, outcome indicators, analysis approaches, and inherent research limitations. By methodically examining these elements, the study aimed to contribute to a nuanced understanding of the current state of evaluating disaster education programs for children, shedding light on both the strengths and limitations within the existing research landscape. As asserted by Johnson et al. (2014), a substantial insight emerges from their analysis:

the majority of insights into the efficacy of disaster education programs for children derive from quantitative studies primarily centred on assessing children's awareness of disaster risks, protective measures, and their self-reported preparedness actions. The effectiveness of programs in most descriptive and quasi-experimental studies is often gauged by the percentage of positive shifts in children's accurate responses in surveys. Likewise, correlational studies frequently establish a link between positive outcomes, such as enhanced household preparedness, and children's engagement in disaster education programs. Despite these findings, the authors underscore the noteworthy prevalence of methodological limitations across many studies. Notably, they highlight a critical gap in empirical evidence concerning how disaster education programs influence children's contributions to household preparedness, their self-protective capabilities, and their inclination to prepare for disasters in adulthood, despite indications of valuable knowledge changes. In addition to emphasizing the imperative to identify and refine program theory and meaningful outcome indicators, the authors propose various avenues for future research. This includes exploring the multifaceted ways in which disaster education programs shape children's roles in household preparedness, bolster their capacity for self-protection, and influence their long-term preparedness behaviours as they transition into adulthood.

The multidimensionality of resilience (Cvetković et al. 2021a, b, c, d, e; Cvetković and Filipović 2018a, b, c; Cvetković and Filipović 2017), which is based on the national, local and individual level, is primarily based on having an appropriate level of knowledge and awareness in the field of disasters, which indicates the necessity of having developed and high-quality educational programs in the field of disaster studies. Nevertheless, an examination of assessments of disaster education initiatives aimed at children has unearthed substantial voids in the evidential foundation regarding the effectiveness of these programs (Back et al. 2009). Both the US Federal Emergency Management Agency (FEMA) and UNICEF have meticulously catalogued a diverse array of educational initiatives for children in the context of disasters across the globe. These initiatives encompass formal and informal programs within communities, schools, and extracurricular activities, often receiving support from either governmental or private entities (FEMA 2010). The growing emphasis on the development and investment in disaster education programs for children is indicative of the international consensus that these educational endeavours yield discernible benefits for individual and community disaster resilience. However, numerous scholars assert that a critical dearth of formal evaluations hampers the ability to assess the efficacy of these programs in achieving desired learning and behavioural outcomes (Ronan and Towers 2014).

In line with this, Johnson et al. (2014) contend that, to fulfil the ambitious objectives of reshaping the culture of safety and resilience, children's disaster education programs must exhibit both efficacy and scalability. A majority of existing studies gauge the outcomes of ad hoc disaster education initiatives implemented for a limited number of children. Given the pivotal aim of the Hyogo Framework for Action 2005–2015 to integrate disaster education into school curricula, there is a compelling need for research into national curriculum integration processes. Such research would not only aid in identifying scalable models but would also facilitate the incorporation of disaster education within broader academic subjects, such as science, geography, and social studies (Johnson et al. 2014). This strategic approach holds the potential to pave the way for widespread adoption and replication of successful models on an international scale.

As outlined by Cvetković (2020), a pivotal aspect of curricula lies in the elevation of citizens' awareness on several fronts: (a) comprehending the nature and attributes of diverse natural and technical–technological hazards that encompass their surroundings; (b) recognizing the significance of adopting measures to alleviate and prepare for disasters; (c) understanding the protective measures to be implemented in different phases—pre-disaster, during the event, and in the aftermath; and (d) grasping the potential for risk mitigation through diverse mechanisms, including risk transfer methods like insurance. The cultivation of heightened awareness is a multifaceted endeavour that can be actualized through public information dissemination, the orchestration of collective training sessions and workshops, and the conception and execution of various campaigns explicitly tailored to augment the awareness levels of both individual citizens and society at large (Cvetković 2020).

17.4.2 Representation and Diversity of Disaster and Climate Risk Topics in Curricula

The significance of education in disaster risk reduction has received notable attention across various international platforms, encompassing agendas, frameworks, conferences, and United Nations programs. Spanning from the International Decade for Natural Disaster Reduction to the International Strategy for Disaster Reduction, and progressing through the Hyogo Framework for Action 2005–2015 to the Sendai Framework for Disaster Risk Reduction 2015–2030, the global consensus underscores the pivotal role education plays in mitigating the impact of disasters. In Japan, formal education takes a dual-pronged approach to addressing natural disasters, involving both subject-specific learning and extracurricular activities. Within formal education, a focus on particular subject areas, particularly the social sciences and scientific disciplines related to the natural landscape, the causes of natural disasters, and their repercussions on human populations, is evident. However, despite recognizing the necessity for an interdisciplinary approach, the treatment of natural disasters often remains compartmentalized within specific subject domains. Natural education and social studies, for instance, tend to address the natural and social environments separately, lacking a comprehensive integration of the two.

Concurrently, extracurricular educational activities occurring beyond regular teaching hours place a heightened emphasis on practical engagement rather than delving deeply into the scientific underpinnings of natural disasters. This practical approach manifests in activities like disaster drills and the development of disaster prevention manuals. Notably, the emphasis in these activities is often placed on pragmatic aspects, such as life-saving techniques and evacuation training, focusing on actions during and after a natural disaster (Fujioka 2016). In terms of curriculum development for disaster risk reduction, a technical guide divides the process into four distinct stages. This systematic breakdown facilitates a clear and easily understandable depiction of the development process, presenting it in a format that can be seamlessly adapted into a checklist for various stakeholders involved in the educational landscape (UNICEF 2014):

(a) Initial planning and preparation of the terrain: determining the need for curriculum development and building a broad-based consensus around the need; uniting interested parties around a common need; conducting "state-of-the-art" research into the existing curriculum, its functioning and implementation through a review of the curriculum, a basic study or a needs assessment; consensus building around specific needs revealed through "state-of-the-art" research; determining the focus (curriculum and class location) for curriculum development; establishing a schedule, with milestones and deadlines, for the curriculum development process; forming a team for curriculum development, determining the roles and responsibilities of the team members and establishing a modus operandi for cooperation, teamwork and meetings.
(b) Crafting the curriculum involves several key steps: delineating the desired learning outcomes, encompassing knowledge, skills, attitudes, and behaviours; meticulously selecting and arranging curriculum content to effectively achieve the specified outcomes; transforming chosen content into age-appropriate learning materials; creating engaging learning activities supported by relevant materials to bring about decisive outcomes; scrutinizing and assessing existing teaching materials and activities to ascertain their potential inclusion in the curriculum, thus avoiding redundancy; actively seeking feedback from stakeholders, including panels of seasoned educators, on curriculum materials and making revisions as needed.
(c) Executing the curriculum involves a phased approach: pinpointing schools and educators for the initial pilot implementation of the new curriculum; providing comprehensive training to pilot teachers to adeptly instruct in alignment with the new curriculum; overseeing, monitoring, and evaluating the pilot implementation; refining teaching materials and training activities based on the evaluation outcomes; conducting subsequent rounds of teacher training and pilot testing, gradually expanding to involve a larger pool of learners and educators; executing widespread teacher training initiatives following their participation in 'training of trainers' events; incorporating new materials and activities into pre-service teacher training; ensuring the formal endorsement of the curriculum by national, regional, or local educational authorities.

Oversight, assessment, and enhancement of the curriculum: Formulating strategies for systematic data collection to periodically evaluate the impact and calibre of the updated curriculum. This involves assessing its efficacy in attaining anticipated learning outcomes and discerning any unanticipated effects, whether positive or negative. Compiling comprehensive reports on the monitoring and evaluation processes is essential. Additionally, instituting mechanisms for routine revisions to the curriculum based on the insights garnered from the evaluation process is crucial. This iterative approach ensures a dynamic and responsive curriculum that aligns with evolving educational needs and goals.

The incorporation of environmental content into primary school curricula is prevalent, with ecological themes seamlessly integrated across various subjects. Unlike some educational systems, there is not a dedicated subject specifically focusing on Ecology or Environmental Education at the primary school level. Instead, the emphasis lies on embedding ecological content within the frameworks of different subjects, primarily those falling under the umbrella of natural sciences (Stanišić and Maksić 2014). Despite early proposals advocating for Ecology as a distinct and compulsory subject, efforts faced challenges. The First Congress of Ecologists in Yugoslavia initially proposed the idea of making ecology a mandatory subject at all educational levels. However, subsequent international gatherings showcased diverse opinions on this matter. Presently, the educational approach favours the integration of ecological topics within existing subjects rather than having a standalone subject. This current perspective emphasizes that ecological principles should be interwoven into various disciplines, offering students a multidimensional understanding of environmental issues. Critics argue that designating a specific subject to Ecology might not necessarily lead to a more profound understanding if treated in isolation. Instead, the integrated approach allows students to explore ecological concepts within the context of different subjects, fostering a more interconnected understanding of the environment. While there were early initiatives to establish Ecology as a standalone subject in our primary schools, the prevailing trend leans towards integrating ecological content across various subjects. This approach reflects a holistic perspective, aiming to provide students with a comprehensive and interconnected understanding of ecological principles within the broader context of their education.

17.4.3 Education of Teaching Staff for the Needs of Education in the Field of Disasters

To actualize education focused on disasters, it becomes imperative for educational strategies not only to envision the methods of instruction but also to intricately consider the educational background of the instructors themselves. This sparks a pivotal discourse revolving around the question of whether educators delivering such content should be drawn from those currently teaching established subjects like geography or chemistry, or if the role is better suited for individuals with a university background in the field of security. This deliberation unfolds as a critical consideration, prompting an exploration of the merits and drawbacks inherent in each proposed alternative.

Teachers who are already teaching within the framework of the existing curricula could, within their thematic units, implement teaching in the field of emergency situations as well. In addition, resources would be saved because there would be no need for additional employment. On the other hand, if teachers were hired for these needs who would be specialized in this type of teaching in the mentioned field, additional perfection would be achieved in terms of the teaching units themselves. It is advisable for individuals to have graduated from the Faculty of Security and successfully completed the study program focusing on the management of emergency situations and environmental security. In this sense, it is necessary to define an educational policy as soon as possible, within which the prerequisites that should be met by teachers who would perform such classes would be defined in advance. In other developed countries, different practices are applied, according to which there are different modalities mentioned above. Certainly, it is recommended to carry out all necessary additional analyses within certain project activities and to clearly determine all advantages and disadvantages.

17.5 Conclusion

The role of schools in reducing the risk of disasters is an unavoidable topic in all world agendas where innovative solutions for mitigating the consequences of disasters are discussed. Today, it has become unthinkable to discuss structural disaster risk reduction measures without referring to the importance of education in that process. Unfortunately, in Serbia, little attention is paid to educating citizens about disaster risk reduction, which could be the result of mistaken beliefs that it is one of the marginal topics in the disaster risk management process itself. Although calls for the introduction of the subject safety culture in schools could be heard in several places, the decision-makers explicitly refuse to include such a subject in primary and secondary schools in Serbia. The reasons for such decisions can be found in insufficient knowledge of the area of risk reduction and not giving adequate importance to such non-structural measures of risk reduction. Unfortunately, there is still no sufficient consensus in Serbia regarding the necessity and importance of introducing the subject of security culture. One gets the impression that all other content is prioritized more than security content.

The reason for such a situation can also be found in the fact that there is still a socialist view of security, according to which citizens should not take measures, but the competent state authorities, such as the police, fire-rescue units, emergency medical services, etc. However, practice shows that the largest number of people were saved by neighbours or other friends who happened to be nearby at the time when the harmful effects of disasters, such as shock and heat waves, radioactive radiation, biological contamination, etc., were manifested. Also, one of the famous examples of the power of knowledge in the field of disaster studies is the girl Tilly Smith, who saved her peers on the shores of the Indian Ocean when she warned them that the retreat of water towards the centre of the ocean meant that a tsunami would occur. She then warned the teacher who started the evacuation of the children who survived, thanks to that. Regardless of all that, it is necessary to continue improving theoretical knowledge and implementing practical training in Serbia. Certainly, in future planning documents, it is very important to highlight the importance of education in the field of disaster studies and to further develop all possible ways of its realization. In addition, it is important to improve curricula and programs in the field of emergencies. To achieve this, it is recommended to introduce the subject safety culture in primary and secondary schools.

The significance of this chapter lies in shedding light on the critical role of schools in disaster risk reduction, a topic of undeniable importance in global discussions on innovative strategies to mitigate the aftermath of disasters. The text emphasizes that contemporary deliberations on structural measures for disaster risk reduction are incomplete without acknowledging the pivotal role of education in this process. The paper underscores a concerning trend in Serbia, where minimal attention is directed towards educating citizens about disaster risk reduction. This negligence may stem from misconceptions that view it as a peripheral aspect within the broader disaster risk management framework. Despite calls for the inclusion of a safety culture subject in schools, decision-makers in Serbia resist its incorporation into primary and secondary education, citing insufficient knowledge of risk reduction and a lack of recognition for non-structural risk reduction measures. The text identifies a potential reason for this situation in the lingering socialist perspective that advocates citizens should not take individual measures but rely on competent state authorities. However, practical examples reveal that, in reality, community members, such as neighbours and friends, play a crucial role in saving lives during disasters.

Highlighting the power of knowledge in disaster studies, the text references the inspiring case of Tilly Smith, who, armed with her understanding, saved her peers from a tsunami. Despite such examples, the need for continuous improvement in theoretical knowledge and practical training in Serbia is emphasized. The paper advocates for future planning documents to underscore the importance of education in disaster studies, urging the development of various avenues for its implementation. Furthermore, it stresses the need to enhance curricula and programs related to emergencies, recommending the introduction of a safety culture subject in primary and secondary schools. In essence, the paper calls for a paradigm shift in Serbia towards recognizing the central role of education in fostering resilience and preparedness in the face of disasters.

The integration of disaster and climate risk education into educational curricula is crucial. It is recommended to incorporate specific modules on disasters and climate risks into relevant academic programs across all levels of education. This integration could span subjects such as geography, biology, chemistry, and other natural and social sciences. Empowering educators is essential for effective implementation. Regular training sessions and seminars should be organized to enhance teachers' understanding and delivery of content related to disasters and climate risks. Additionally, specific courses or certification programs focusing on this area should be developed for teachers. Active community involvement is key. Schools and educational institutions should actively engage local communities in disaster education programs. Workshops, lectures, or simulations could be organized to raise awareness about natural risks and disaster preparedness among the local population. Leveraging modern technologies is imperative. The introduction of technological resources, such as interactive guides, simulations, and online courses, can enhance disaster and climate risk education. Developing applications or platforms that facilitate access to information on climate risks and provide guidance on appropriate actions is recommended. Establishing partnerships with governmental bodies, non-governmental organizations, and international collaborators will facilitate the sharing of best practices and resources.

Building a network that supports ongoing education on disasters and climate risks is critical. Enhancing capacity in disaster and climate risk education in Serbia is vital for constructing a resilient society. The holistic approach involves integrating this field into the educational system, empowering educators, actively engaging local communities, and leveraging modern technologies. These initiatives not only prepare citizens at all levels for the challenges associated with disasters and climate change but also contribute to overall societal security and sustainability. Collaboration with relevant partners and continuous monitoring and adaptation of these initiatives will further strengthen national capacities in this critical area.

References

Aktar MA, Shohani K, Hasan MN, Hasan MK (2021) Flood vulnerability assessment by flood vulnerability index (FVI) method: a study on Sirajganj Sadar Upazila. Int J Disaster Risk Manage 3(1):1–14

Alexander DE (2002) Principles of emergency planning and management. Oxford University Press on Demand

Back E, Cameron C, Tanner T (2009) Children and disaster risk reduction: taking stock and moving forward

Boon HJ, Pagliano PJ (2014) Disaster education in Australian schools. Aust J Environ Educ 30(2):187–197

Carla SRG (2019) School-community collaboration: disaster preparedness towards building resilient communities. Int J Disaster Risk Manage 1(2):45–59

Chicatto JA, Vieira R, Bohn N (2015) Disaster risk management and non-structural civil defense's actions in the municipality of blumenau/SC. Pesquisa Em Educação Ambiental 10(2):124–143

Coates R (2021) Educational hazards? The politics of disaster risk education in Rio de Janeiro. Disasters 45(1):86–106

Cruz RDD, Ormilla RCG (2022) Disaster risk reduction management implementation in the public elementary schools of the department of education, Philippines. Int J Disaster Risk Manage 4(2):1–15

Cvetković V (2019a) Risk perception of building fires in Belgrade. Int J Disaster Risk Manage 1(1):81–91

Cvetković V (2020) Upravljanje rizicima u vanrednim situacijama [Disaster risk management]. Beograd: Naučno-stručno društvo za upravljanje rizicima u vanrednim situacijama

Cvetković VPS, Janković B (2021) Private security preparedness for disasters caused by fires. J Criminalistic and Law, NBP 26(1)

Cvetković V, Bošković N, Ocal A (2021) Individual citizens' resilience to disasters caused by floods: a case study of Belgrade. PREPRINT (Version 2) available at Research Square. https://doi.org/10.21203/rs.3.rs-923368/v2

Cvetković V, Dragašević A, Protić D, Janković B, Nikolić N, Milošević P (2022a) Fire safety behavior model for residential buildings: implications for disaster risk reduction. Int J Disaster Risk Reduction 75:102981

Cvetković V, Dragićevic S (2014) Spatial and temporal distribution of natural disasters. J Geogr Institute/Geogr Inst "Jovan Cvijić" 64(3):293–309
Cvetković V, Filipović M (2017) Research into students' knowledge of the proper way of acting during natural disasters. Bezbjednost, Policija i Građani 13(2):105–121
Cvetković V, Filipović M (2018) Koncept otpornosti na katastrofe [Concept of disaster resilience]. Ecologica 25(89)
Cvetković V, Filipović M, Popović D, Ostojić G (2017a) Činioci uticaja na znanje o prirodnim katastrofama [Factors influencing knowledge about natural disasters]. Ecologica 24(85):121–126
Cvetković V, Jakovljević V, Gačić J, Filipović M (2017b) Obuka građana za reagovanje u vanrednim situacijama [Citizen training for emergency response]. Ecologica 24(88):856–882
Cvetković V, Janković B (2020) Private security preparedness for disasters caused by natural and anthropogenic hazards. Int J Disaster Risk Manage 2(1)
Cvetković V, Jovanović M (2020) Examination of the factors that influence public perception of mythically-based human behavior in disaster conditions. Glasnik Srpskog Geografskog Drustva 100(2):161–179
Cvetković V, Kevin R, Shaw R, Filipović M, Mano R, Gačić J, Jakovljević V (2019a) Household earthquake preparedness in Serbia—a study from selected municipalities. Acta Geographica 59(2):27–43
Cvetković V, Martinović J (2021) Upravljanje u nuklearnim katastrofama [Nuclear Disaster Management]. Naučno-stručno društvo za upravljanje rizicima u vanrednim situacijama, Beograd
Cvetković V, Milašinović S, Lazić Ž (2018a) Examination of citizens' attitudes towards providign support to vulnerable people and voluntereeing during disasters. J Social Sci, TEME 42(1):35–56
Cvetković V, Nikolić M (2021) The role of social networks in disaster risk reduction: a case study Belgrade. Bezbednost 61(3):25–42
Cvetković V, Nikolić N, Nenadić RU, Ocal A, Zečević M (2020a) Preparedness and preventive behaviors for a pandemic disaster caused by COVID-19 in Serbia. Int J Environ Res Public Health 17(11):4124
Cvetković V, Nikolić N, Ocal A, Martinović J, Dragašević A (2022b) A predictive model of pandemic disaster fear caused by coronavirus (COVID-19): implications for decision-makers. Int J Environ Res Public Health 19:652
Cvetković V, Noji E, Filipović M, Marija MP, Želimir K, Nenad R (2018b) Public risk perspectives regarding the threat of terrorism in belgrade: implications for risk management decision-making for individuals, communities and public authorities. J Criminal Invest Criminology/revija Za Kriminalistiko in Kriminologijo 69(4):279–298
Cvetković V, Öcal A, Ivanov A (2019b) Young adults' fear of disasters: a case study of residents from Turkey, Serbia and Macedonia. Int J Disaster Risk Reduction 35:101095
Cvetković V, Öcal A, Lyamzina Y, Noji E, Milošević G, Jovanović M (2020) Public risk perception of nuclear power in Serbia: fear of exposure to radiation versus social benefits. Advance. Preprint, advance-13082240
Cvetković V, Öcal A, Lyamzina Y, Noji E, Nikolić N, Milošević G (2021b) Nuclear power risk perception in Serbia: fear of exposure to radiation versus social benefits. Energies 14:2464
Cvetković V, Pavlović S, Janković BD (2021c) Factors of influence on preparedness of the private security members for fire emergencies. NBP-J Criminalistics and Law 26(1):35–59
Cvetković V, Ristanović E, Gačić J (2018c) Citizens attitudes about the emergency situations caused by epidemics in Serbia. Iran J Public Health 47(8):1213–1214
Cvetković V, Roder GOA, Tarolli P, Dragićević S (2018d) The role of gender in preparedness and response behaviors towards flood risk in Serbia. Int J Environ Res Public Health 15(12):2761
Cvetković V, Stanišić J (2015) Relationship between demographic and environmental factors with knowledge of secondary school students on natural disasters, SASA. J Geogr Inst Jovan Cvijic 65(3):323–340

Cvetković VM (2016a) Fear and floods in Serbia: citizens preparedness for responding to natural disaster. Zbornik Matice Srpske Za Drustvene Nauke 155–156:303–324
Cvetković VM (2016b) The impact of the level of motivation on the preparedness of citizens of the Republic of Serbia to respond to natural disaster caused by flood. Vojno Delo 68(3):141–171
Cvetković VM (2017) Crises: readiness of state, local community and citizens. Vojno Delo 69(7):122–136
Cvetković VM (2019) First aid disaster kit for a family: A case study of Serbia. Archibald Reiss Days 9(1)
Cvetković VM, Bošković N (2021) Conceptual bases and dimensions of resilience to disasters. Zbornik radova Naučno-stručnog društva za upravljanje rizicima u vanrednim situacijama i Međunarodnog instituta za istraživanje katastrofa, pp 175–191
Cvetković VM, Filipović B (2020) The survey of citizen attitudes toward preparedness for disasters caused by wildfires: case study: Prijepolje. Žurnal Za Bezbjednost i Kriminalistiku 2(2):11–24
Cvetković VM, Filipović M (2018) The role of the family in reducing risks of natural disasters. Nauka, bezbednost, policija (NBP) 23(1):71–85
Cvetkovic VM, Martinović J (2020) Innovative solutions for flood risk management. Int J Disaster Risk Manage 2(2):71–100
Cvetković VM, Radovanović MP, Milašinović SM (2021d) Disaster risk communication: attitudes of Serbian citizens. Sociološki Pregled 55(4):1610–1647
Cvetković VM, Romanić S, Beriša H (2023) Religion influence on disaster risk reduction: a case study of Serbia. Int J Disaster Risk Manage 5(1):66–81
Cvetković VM, Svrdlin M (2020) The vulnerability of women from the consequences of a natural disaster: a case study of Svilajnac. Bezbednost, Beograd 62(3):43–61
Cvetković VM, Tanasić J, Ocal A, Kešetović Ž, Nikolić N, Dragašević A (2021e) Capacity development of local self-governments for disaster risk management. Int J Environ Res Public Health 18(19):10406
Cvetković VM, Todorović S (2020) Comparative analysis of disaster risk management policies in the region of South-East Europe, in press
Cvetković V (2023) Otpornost na katastrofe - Vodič za prevenciju, reagovanje i oporavak (Disaster resilience - A Guide to Prevention, Response, and Recovery). Beograd: Naučno-stručno društvo za upravljanje rizicima u vanrednim situacijama
Cvetković V, Tarolli P, Roder G, Ivanov A, Ronan K, Ocam A, Kutub R (2017) Citizens education about floods: a Serbian case study. Paper presented at the VII International scientific conference Archibald Reiss days
Cvetković V, Milašinović S, Gostimirović L (2016) The impact of age on flood preparedness in Serbia. Int J Soc Sci Humanities, USA 6(10):10–23
Cvetković V, Rikanović S, Knežević S (2022) Resilience of society in disasters caused by nuclear accidents. In: IAI Academic Conference Proceedings, Budapest Conference, 5 May 2022
Cvetković VM, Filipović M (2018c) Ispitivanje uloge porodice u edukaciji dece o prirodnim katastrofama. Nauka, Bezbednost, Policija 23(1):71–85
Dukiya JJ, Banji Adeleye AB (2022) Remote sensing and GIS assessment of domestic fuel energy supply: a threat to global DRR crusade in south-western Nigeria. Int J Disaster Risk Manage 4(2):45–59
Education D (2007) Building Research Institute (BRI) and National Graduate Institute for Policy Studies (GRIPS). Tokyo, Japan
El-Mougher MM (2022) Level of coordination between the humanitarian and governmental organizations in Gaza Strip and its impact on the humanitarian interventions to the Internally Displaced People (IDPs) following May escalation 2021. Int J Disaster Risk Manage 4(2):15–45
El-Mougher MM, Abu Sharekh DSAM, Abu Ali MRF, Zuhud DEAAM (2023) Risk management of gas stations that urban expansion crept into in the Gaza strip. Int J Disaster Risk Manage 5(1):13–27
Faupel CE, Kelley SP, Petee T (1992) The impact of disaster education on household preparedness for Hurricane Hugo. Int J Mass Emerg Disasters 10(1):5–24

Federal Emergency Management Agency (2010) Youth emergency preparedness education paper. Philadelphia, PA: Regional Advisory Council, FEMA Region III

Fujioka T (2016) Disaster prevention education in the Japanese school curricula in recent years: current status and future challenges. In: Disaster resilience of education systems. Springer, pp 39–52

Goyal N (2019) Disaster governance and community resilience: the law and the role of SDMAs. Int J Disaster Risk Manage 1(2):61–75

Gwee QR, Shaw R, Takeuchi Y (2011) Disaster education policy: current and future. Emerald Group Publishing Limited, In Disaster education

Helsloot I, Ruitenberg A (2004) Citizen response to disasters: a survey of literature and some practical implications. J Contingencies Crisis Manage 12(3):98–111

Iftikhar A, Iqbal J (2023) The factors responsible for urban flooding in Karachi (a case study of DHA). Int J Disaster Risk Manage 5(1):81–103

Ivanov A, Cvetković V (2014) The role of education in natural disaster risk reduction. Horizons—Int Sci J 10(16):115–131

Jaiye DJ, Benjamine O (2021) Building resilience through local and international partnerships, Nigeria experiences. Int J Disaster Risk Manage 3(2):11–24

Jakovljević V, Cvetković V, Gačić J (2015) Prirodne katastrofe i obrazovanje (Natural disaster and education). Univerzitet u Beogradu, Fakultet bezbednosti, Beograd

Janković B (2021) The role of the police in disasters caused by pandemic infectious diseases. Int J Disaster Risk Manage 3(1):41–50

Janković B, Cvetković V (2020) Public perception of police behaviors in the disaster COVID-19—the Case of Serbia. Policing An Int J Police Strategies Manage

Johnson VA, Ronan KR, Johnston DM, Peace R (2014) Evaluations of disaster education programs for children: a methodological review. Int J Disaster Risk Reduction 9:107–123

Johnson VA, Ronan KR, Johnston DM, Peace R (2016) Improving the impact and implementation of disaster education: programs for children through theory-based evaluation. Risk Anal 36(11):2120–2135

Kabir MH, Hossain T, Haque MW (2022) Resilience to natural disasters: a case study on southwestern region of coastal Bangladesh. Int J Disaster Risk Manage 4(2):91–105

Krnjić I, Cvetković V (2021a) Investigating student's attitudes and preferences towards disaster learning multimedia to enhance preparedness. Bull Serbian Geogr Soc 101(2):79–96

Krnjić I, Cvetković V (2021) The role of multimedia content in educating young people about disasters. Pravni i bezbednosni aspekti upravljanja rizicima od prirodnih i antropogenih katastrofa—Legal and security aspects of natural and man-made disaster risk management (Vol. 1): Scientific-Professional Society for Disaster Risk Management, Belgrade

Kulatunga U (2010) Impact of culture towards disaster risk reduction. Int J Strateg Prop Manag 14(4):304–313

Kumiko F, Shaw R (2019) Preparing international joint project: use of Japanese flood hazard map in Bangladesh. Int J Disaster Risk Manage 1(1):62–80

Lindell MK (2013) Disaster studies. Curr Sociol 61(5–6):797–825

Luna EM (2012) Education and disaster. In: The Routledge handbook of hazards and disaster risk reduction. Routledge, pp 750–760

Mano R, Kirshcenbaum A, Rapaport C (2019) Earthquake preparedness: a social media fit perspective to accessing and disseminating earthquake information. Int J Disaster Risk Managem 1(2):19–31

Mohammed E-M, Maysaa J (2022) International experiences in sheltering the Syrian refugees in Germany and Turkey. Int J Disaster Risk Manage 4(1):1–15

Nakano G, Suwa S, Gautam A, Yamori K (2020) Long-term evaluation of proactive attitudes toward disaster education in Nepal. Int J Disaster Risk Reduction 50:101866

Öcal A (2019) Natural disasters in Turkey: social and economic perspective. Int J Disaster Risk Manage 1(1):51–61

Öcal A (2021) Disaster management in Turkey: a spatial approach. Int J Disaster Risk Manage 3(1):15–22

Ocal A, Cvetković V, Baytiyeh H, Tedim F, Zečević M (2020) Public reactions to the disaster COVID-19: a comparative study in Italy, Lebanon, Portugal, and Serbia. Geomat Nat Haz Risk 11(1):1864–1885

Odero NA, Mahiri I (2022) The complacency of flood victims, socio economic factors, and effects and vulnerabilities of floods in Lower Kano Plains, Kisumu County, Kenya. Int J Disaster Risk Manage 4(2):59–77

Page L, Rubin J, Amlôt R, Simpson J, Wessely S (2008) Are Londoners prepared for an emergency? A longitudinal study following the London bombings. Biosecur Bioterror 6(4):309–319

Radović I (1996) Ekološko obrazovanje i zaštita životne sredine [Ecological education for environmetal protection]. Vaspitanje i Obrazovanje, 5–6, 3–20; Alexander DE (2002) Principles of emergency planning and management. Oxford University Press on Demand

Rajani A, Tuhin R, Rina A (2023) The challenges of women in post-disaster health management: a study in Khulna District. Int J Disaster Risk Manage 5(1):51–66

Ronan KR, Johnston DM (2001) Correlates of hazard education programs for youth. Risk Anal 21(6):1055–1064

Ronan KR, Towers B (2014) Systems education for a sustainable planet: preparing children for natural disasters. Systems 2(1):1–23

Salleh SH, Yusof NAM (2013) A preliminary study on prepared culture amongst healthcare practitioners for nuclear and radiological emergency in Malaysia. Procedia Soc Behav Sci 81:416–421

Sergey K, Gennadiy N (2022) Methodology for the risk monitoring of geological hazards for buildings and structures. Int Jo Disaster Risk Manage 4(1):41–49

Shibru M, Operea A, Omondi P, Gichaba M (2022) Impact of 2016–2017 drought on household livestock assets and food security: the case of pastoralists and agro-pastoralists in Borana zone, southern Ethiopia. Int J Disaster Risk Manage 4(1):49–69

Stanišić J, Maksić S (2014) Environmental education in serbian primary schools: challenges and changes in curriculum, pedagogy, and teacher training. J Environ Educ 45(2):118–131

Stanišić J (2019) Ekološko obrazovanje u Srbiji: trenutno stanje i pravci razvoja. Obrazovanje u funkciji modernizacije društva, pp 30–33.

Suarmika PE, Arnyana IBP, Suastra IW, Margunayasa IG (2022) Reconstruction of disaster education: the role of indigenous disaster mitigation for learning in Indonesian elementary schools. Int J Disaster Risk Reduction 72:102874

UNICEF (2014) Towards a learning culture of safety and resilience: Technical guidance for integrating disaster risk reduction in the school curriculum. UNESCO

Yusuf R, Yunus M, Maimun M, Fajri I (2022) Environmental education: a correlational study among environmental literacy, disaster knowledge, environmental sensitivity, and clean-living behavior of post tsunami disaster in aceh communities, Indonesia. Polish J Environ Stud 31(1); Aktar MA, Shohani K, Hasan MN, Hasan MK (2021) Flood vulnerability assessment by flood vulnerability index (FVI) method: a study on Sirajganj Sadar Upazila. Int J Disaster Risk Manage 3(1):1–14

Chapter 18
Experiences of First Responders in Enhancing Disaster Education Capacity

Sefa Mızrak and Ramazan Aslan

Abstract First responders contribute to the community to be more resilient in disaster situations by using their disaster response skills. However, they encounter environmental, social, institutional, organizational and psychological problems while responding to disasters. The experiences and observations of those involved in disaster response are critical for the creation of more effective strategies for disaster education and training. This section revealed the experiences of sixteen people who participated in the response activities after the earthquakes that affected more than ten cities in Türkiye on February 6, 2023. The study was designed using a phenomenological model. Data were collected with semi-structured interview forms in September 2023 and analyzed with the Maxqda 2020 qualitative data analysis program. Consequently, four key themes emerged: benefits of disaster education, challenges and solution proposals about disaster education, training experiences in the disaster processes and suggestions for training topics. Participants stated that disaster education increased institutional capacity and provided social benefits. They also expressed many negative opinions about the content of the disaster training they received and its implementation during earthquakes and offered suggestions for future disaster training based on their experiences. The findings of this study will enhance the disaster education capacity in Türkiye and other countries experiencing similar challenges.

Keywords Earthquakes · Disaster education · Disaster experience · First responders · Türkiye

---

S. Mızrak (✉)
Faculty of Health Sciences, Emergency Aid and Disaster Management Department, Gümüshane University, Gümüşhane, Türkiye
e-mail: sefamizrak1@gmail.com

R. Aslan
Faculty of Health Sciences, Emergency Aid and Disaster Management Department, Ardahan University, Ardahan, Türkiye
e-mail: ramazan.att.aslan@gmail.com

A. Yildiz and R. Shaw (eds.), *Disaster and Climate Risk Education*, Disaster Risk Reduction, https://doi.org/10.1007/978-981-97-5987-3_18

18.1 Introduction

Disasters pose a major threat to the sustainability of communities in the short and long term. Disasters reduce per capita income in developing countries and hinder economic growth (Klomp and Valckx 2014). Natural disasters cause a decline in agricultural production in poor countries (Coulibaly et al. 2020). Due to disasters, injuries, deaths and disabilities occur in a significant part of the population (Ahmed et al. 2021). Depression, anxiety and trauma symptoms negatively affected the normal living conditions of earthquake victims and reduced their quality of life (Park and Bae 2022). Infected food and water sources, poor living conditions and weather conditions after disasters cause the emergence of infectious diseases in the disaster area (Suk et al. 2020). To reduce human and financial loss, disaster training that enables people to protect themselves in the event of a disaster is necessary (Torani et al. 2019). Disaster training increases people's knowledge about disasters and their disaster response skills (Hung et al. 2021). Disaster education helps children better understand the risks and dangers around them and pay attention to disaster preparedness (Yildiz et al. 2023).

Scientists have offered many suggestions to increase the scope and quality of disaster education. For example, topics related to disasters and emergencies should be included in education and training course contents, and central government, local government and non-governmental organizations should work in coordination for disaster education (Sudan Aran and Erol 2022). Disaster education topics should be included in the higher education curriculum (Hasan et al. 2022). Expert and experienced people should provide disaster education (Mızrak 2018; Sudan Aran and Erol 2022). Teachers, who have an important role in the development of society, should take courses on natural disasters during their university education and they should participate in activities related to protection from natural disasters (Gezer and Şahin 2022).

Regular training should be given to families to prepare them for disasters, and materials such as brochures, posters and public service announcements should be prepared to increase the disaster preparedness level of people with low education levels (Tercan 2022). Institutions responsible for disaster management should inform the public through their websites and social media accounts (Özdemir and Şahinöz 2021). To ensure sustainable education, boards containing disaster education should be made on streets, schools, public buses and bus stops to inform the public (Bekler et al. 2022). Disaster training should be organized regularly so that workplaces are ready for disasters and staff are informed about disaster plans (Konakçı et al. 2023). Nurses' professional knowledge and skills regarding disasters should be increased, and nurses should be more involved in disaster planning (Şentürk et al. 2020). Hospital staff should receive regular training on disasters, and an examination should be administered after the training (Canaslan et al. 2022).

Victims and survivors need human resources, first aid, emergency care and shelter services after disasters (Basu et al. 2017). Therefore, people in the disaster area, as well as officials and people coming to the disaster area from other regions, participate in disaster response efforts (Bekircan et al. 2023; Köçer and Aslan 2023). After the tornado, volunteer citizens and aid organizations supported debris removal and cleaning efforts (Mason et al. 2017). In the 2018 Kerala flood disaster in India, fishermen participated in rescue operations due to their maritime experience and ability to use sea vessels effectively (Joseph et al. 2020). Lifeguards from England and Australia were involved in flood mitigation (sandbagging, pre-emptive evacuations) and response efforts (sand bagging, pre-emptive evacuations) (Peden et al. 2022). Official teams responding to disasters have expectations from people in the disaster area, such as sharing information, complying with recommendations, being prepared and educated and having organizational ability (Elkady et al. 2022). Strengthening local responders and increasing their disaster knowledge provides better disaster risk and crisis management (Harris et al. 2018).

Understanding the experiences of first responders is important for establishing plans and procedures for coordination and communication in future emergency response efforts (Mason et al. 2017). The experiences and recommendations of first responders will provide insight into preparing for future disasters (Altuntaş et al. 2023). Studies on first responders working in humanitarian crises offer solutions to reduce their mental problems (Winders et al. 2021). Investigating the experiences of first responders reveals their deficiencies in disaster education (Tunç 2023). The study, conducted by interviewing volunteers working in forest fires, revealed that volunteers encountered great difficulties in responding to forest fires and those volunteers and victims did not have sufficient knowledge about forest fires (Köçer and Aslan 2023).

The purpose of this study is to explore the experiences of first responders and experts in different fields regarding disaster training after the 6 February 2023 Kahramanmaraş earthquakes. In the study, data will be collected by interviewing the participants through a semi-structured interview form. With the phenomenological design, which is a type of qualitative research, participants' expressions will be presented with themes, categories and codes. The results of the study will help develop disaster education policies applied to first responders working in different areas for disaster risk reduction.

18.2 Effects of the Earthquakes that Occurred in Türkiye on February 6, 2023

Türkiye is a country prone to many natural disasters such as earthquakes, fires, avalanches, floods and landslides (Fig. 18.1). Earthquakes caused the death of 1.119 people in Türkiye between 2000 and 2022, and the earthquakes that occurred in February 2023 caused the death of more than 50 thousand people (Emergency Events Database, 2023). These earthquakes, which caused destruction in more than ten provinces, were named the Kahramanmaraş earthquakes. The damage to the highways, excessive snowfall and the vehicles of people coming to reach their relatives

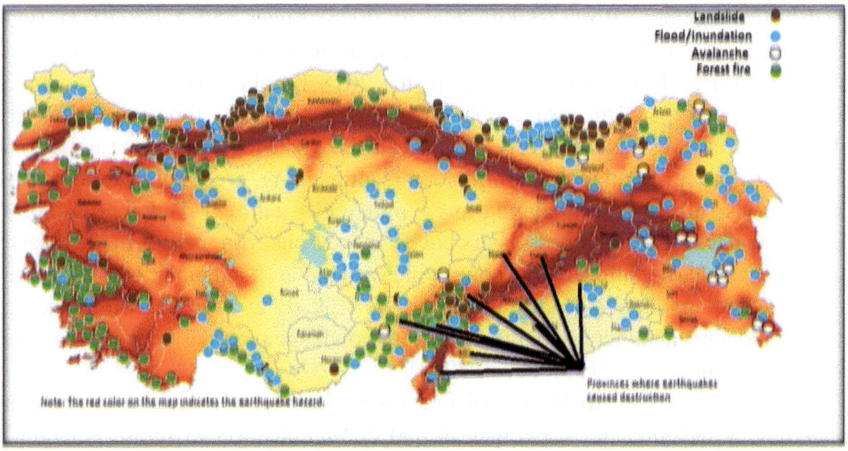

Fig. 18.1 Türkiye disaster risk map (Ministry of Interior of the Republic of Türkiye 2022)

in the earthquake zone made it difficult to reach the earthquake area by road (Yilmaz 2023). People from many countries worked in earthquake zones to help earthquake victims (Kahraman 2023). Since earthquakes cause serious environmental, physical and economic damage, earthquake victims and first responders encountered many problems with equipment, coordination, management, shelter, psychological and security issues (Usta et al. 2023). Earthquakes caused damage to hundreds of buildings (Nemutlu et al. 2023). Şen (2023) stated that these earthquakes will deeply affect the entire country's economy.

18.3 Method

18.3.1 Research Design

The identification of human experiences and the explanation of phenomena by bringing together different experiences are best achieved through qualitative research designs, particularly the phenomenological design (Merriam and Tisdell 2016). As this research aims to determine the experiences of personnel involved in disaster response regarding on-the-job training (OJT), a phenomenological design has been adopted.

18.3.2 Study Group

The criteria sampling and maximum variation sampling techniques were used in determining the study group. Participation in emergency response activities as part of a professional team during the Kahramanmaraş earthquakes was identified as the criterion. Maximum diversity was sought in the identified core emergency aid tasks (search and rescue, emergency medical assistance, shelter, nutrition and distribution of in-kind aid). Data saturation was assessed after contacting and interviewing each participant. Due to the common practice of conducting qualitative studies with 10–15 participants and the inability to obtain new information in subsequent interviews, data collection was halted after the 16th participant (Johnson and Christensen 2014: 383). A total of 13 h and 57 min of interviews were conducted, ranging from a minimum of 30 min to a maximum of 99 min. The average interview duration was determined to be 52 min.

To ensure participant confidentiality, the demographic data of participants were presented in ranges and approximate values. The majority of participants indicated having a university education related to disaster management. Professional experience is generally six years or less. Other demographic data are presented in Table 18.1.

18.3.3 Data Collection Technique and Data Analysis

The primary data collection technique used in gathering data was a semi-structured interview form, which is a key data collection method in qualitative research (Peräkylä and Ruusuvuori 2018). The interview form was developed based on a literature review and researcher experiences, and necessary improvements were made after each analysis following each interview. Data were collected through online video interviews with each participant on agreed-upon dates. The conducted interviews were recorded with participant consent.

The recorded interviews were analyzed using Maxqda 2020 Qualitative Analysis software. In the analysis process, all expressions that could be meaningfully abstracted were initially coded. Subsequently, similar codes obtained in each interview were merged and identified. After the final interview, codes were grouped under categories and themes. The visualization of the obtained themes was done using the Hierarchical Code-Subcode Model.

18.3.4 Validity of the Study and Ethical Dimension

Throughout the entire research process, maximum effort has been exerted to ensure participant confidentiality. During in-depth interviews, consensus on potential codes

Table 18.1 Demographic characteristics of participants

Code	Professional experience	Field duty duration	Field duty	Education	Duty location (Province)
P1	1–3 years	Over 30 days	Shelter	Bachelor's	Hatay
P2	Over 10 years	15 days	Emergency health	Bachelor's	Adıyaman
P3	4–6 years	Over 30 days	Nutrition	Bachelor's	Hatay
P4	Over 10 years	5 days	Emergency health	High School	Adıyaman
P5	1–3 years	10 days	Emergency health	High School	Hatay
P6	1–3 years	15 days	Shelter	Bachelor's	Kahramanmaraş
P7	1–3 years	15 days	Shelter	Bachelor's	Hatay
P8	1–3 years	Over 30 days	Shelter	Bachelor's	Adıyaman
P9	4–6 years	5 days	Search and rescue	Bachelor's	Adıyaman
P10	4–6 years	10 days	Search and rescue	Bachelor's	Kahramanmaraş
P11	1–3 years	Over 30 days	Nutrition	Bachelor's	Gaziantep
P12	Over 10 years	Over 30 days	Search and rescue	Associate's	Adıyaman
P13	1–3 years	Over 30 days	Shelter	Bachelor's	Adıyaman
P14	Over 10 years	15 days	Search and rescue	Bachelor's	Diyarbakır
P15	4–6 years	Over 30 days	Emergency health	Bachelor's	Adıyaman
P16	4–6 years	5 days	Emergency health	Bachelor's	Hatay

was sought with participants to ensure a correct understanding of expressions, and participant confirmation was obtained accordingly. The evaluation of interviews with a co-researcher and the direct inclusion of participant expressions in presenting findings contributed to the validity of the study.

Ethical approval for the research was obtained from the Gümüşhane University Presidency Scientific Research and Publication Ethics Board on 25/10/2023 (Approval No: E-95674917–108.99–215,193).

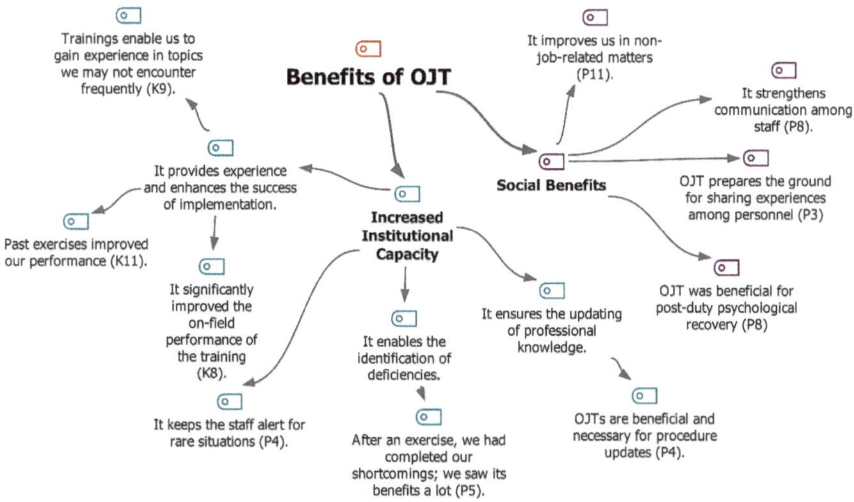

Fig. 18.2 Hierarchical code-subcode model illustrating the benefits of OJTs

18.4 Results

As a result of the conducted analysis, codes were consolidated under four separate themes through the consensus of two researchers. These themes are "Benefits of OJTs," "OJT Challenges and Solution Proposals," "Training Experiences in the Disaster Processes," and "Suggestions for Training Topics."

18.4.1 Benefits of OJTs

Under this theme, the benefits perceived by participants based on their experience with OJTs were compiled. All participants expressed that well-planned OJTs are beneficial. The benefits expressed by participants have been categorized into two groups: social benefits and increased institutional capacity. The codes and participant statements related to the theme are shown in Fig. 18.2.

18.4.2 OJT Challenges and Solution Suggestions

Coding resulted in the grouping of statements related to the problems frequently encountered in the current OJTs and proposed solutions to these problems under the theme "Challenges in OJTs and Solution Suggestions." Codes were categorized under three main categories: Institutional, participants, and training format and content.

In the Institutional category, problems such as perceiving OJTs as a formality, low quality of trainers, and inadequacy in the number and frequency of OJTs were identified.

In the participants category, issues were raised about the preference for new personnel in participant selection, neglect for older personnel, excessive class size and the lack of importance given to training by participants.

In the training content and format category, it was found that participants had negative experiences with online training; they believed that the intensity of practical applications in OJTs should be increased, and they thought that the training scope needed better organization.

The codes and participant statements related to the theme are shown in Fig. 18.3.

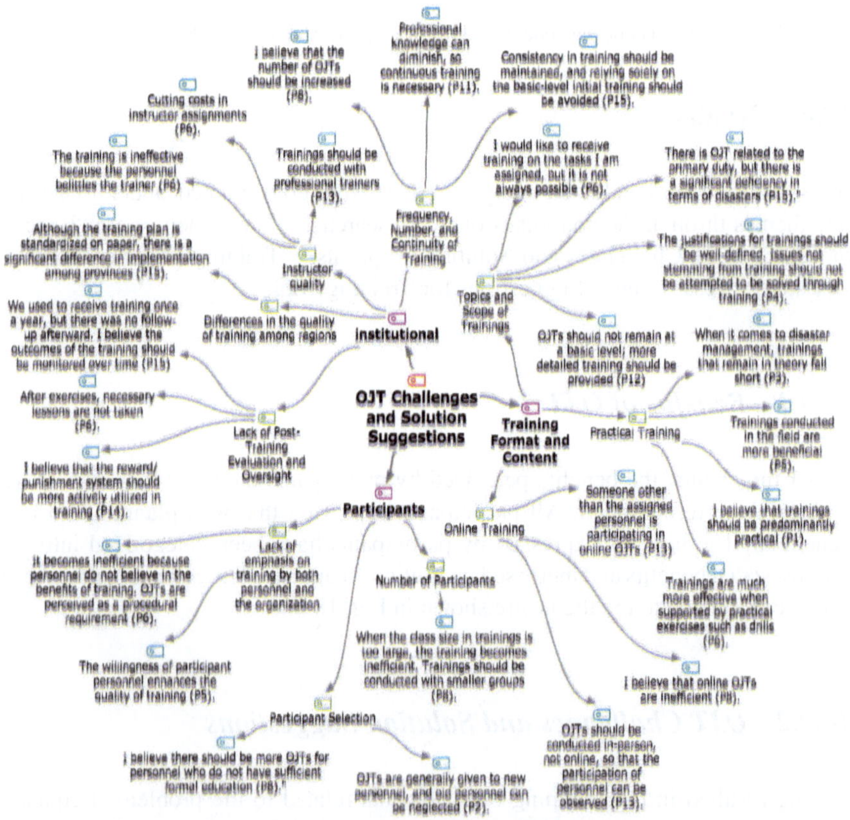

Fig. 18.3 Hierarchical code-subcode model illustrating the OJT challenges and solution proposals

18.4.3 Training Experiences in the Disaster Processes

Participants' experiences with OJTs during the disaster process have been grouped under the theme "Training Experiences in the Disaster Process." The obtained codes were categorized into four groups: Experiences related to field duties, strategies to overcome training gaps, post-disaster measures and the impact of the disaster on the training process.

In the category of experiences related to field duties, it was found that personnel mostly had to work in areas where they lacked sufficient knowledge and training. Personnel with inadequate training struggled, and despite having sufficient theoretical knowledge, they made mistakes due to insufficient practical experience.

In the category of strategies to overcome training gaps, the most common strategy was found to be consulting experienced personnel or acquaintances to overcome training deficiencies. It was also observed that when this training gap is not addressed, personnel are reluctant to go back to duty.

Despite some institutions trying to address identified deficiencies by organizing OJTs after disasters, this situation was expressed by a limited number of participants. The majority of participants stated that lessons were not learned from the disaster.

When examining the impact of the disaster on the training process, it was found that many planned OJTs could not be conducted due to the disaster. Consequently, newly appointed personnel, especially, faced significant challenges in performing their duties.

The codes and participant statements related to the theme are shown in Fig. 18.4.

18.4.4 Suggestions for Training Topics

The areas where participants indicated a partial or significant lack of knowledge or skills in themselves and their team members, based on their field experiences, have been grouped under this theme. These expressions, considered as training suggestions, have been categorized under the themes of Technical Training, Individual Preparedness and Sustaining Life, Psycho-Social Skills, Working in the Disaster Zone, and Management and Coordination.

The first category, **Technical Training**, compiles the observed deficiencies in personnel's primary professional practices. All participants in the study emphasized that all personnel involved in disaster response should receive at least basic search and rescue training. It was identified as a finding that search and rescue personnel need training in medical intervention for disaster victims, and existing NMRT teams need more training in dealing with trauma patients. Additionally, personnel actively involved in search and rescue reported observing problems in the use of search and rescue equipment in the field. They suggested that personnel should be trained in the use of a variety of types and models of equipment. Personnel working in shelter and aid distribution areas expressed a serious lack of training and experience

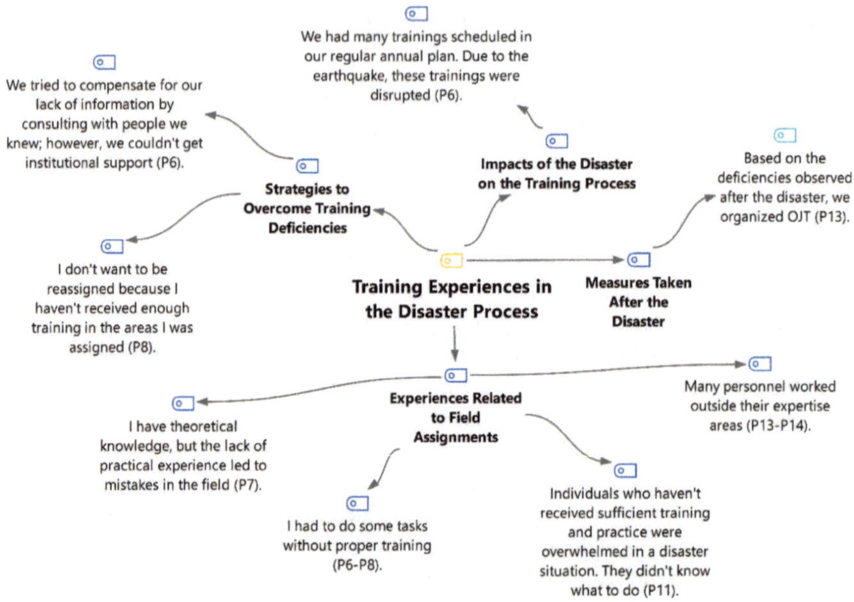

Fig. 18.4 Hierarchical code-subcode model illustrating the training experiences in the disaster processes

in logistics management. They also mentioned that personnel with low IT skills encounter problems in this field.

In the category of **Individual Preparedness and Sustaining Life**, the focus is on personnel preparation before going on duty and the living conditions in the disaster zone. The majority of participants emphasized a significant lack of training in individual preparation for disasters. It was noted that, especially in the initial stages of deployment, personnel did not bring enough equipment with them, resulting in significant problems in nutrition, shelter, heating and hygiene in the field. A participant who felt well-prepared individually expressed that they did not face difficulties in the field thanks to their preparation. Some participants highlighted the necessity of providing training on living in the disaster zone in addition to individual preparation.

Under the category of **Psychosocial Skills**, the codes for psychology, ethics, communication, teamwork and stress management are included. Psychology, highlighted as the first category, is an area that the majority of participants emphasized. Participants who stated that they lack the skills to provide psychological first aid to disaster victims have emphasized the serious training deficiency in this regard. Numerous experiences have been observed regarding the difficulties personnel face in managing their psychology. In this context, it is believed that training programs allowing personnel to assess their own and their team members' psychological states and take necessary precautions would be beneficial. In terms of **ethics**, some participants reported witnessing discriminatory practices in earthquake-affected areas, particularly related to refugees, involving issues of race, gender, hometown and language. Integrating disaster ethics training into OJT programs is seen as crucial in addressing such issues. Regarding **communication**, personnel who are especially required to work with unfamiliar individuals have serious communication issues within the team. It has been observed that personnel who previously knew each other work more harmoniously. Many personnel expressed feeling inadequate in communicating with the public, especially with angry crowds. Another communication problem identified is the lack of knowledge of sign language and local languages. In the area of **teamwork** and professionalism, personnel who have worked with unfamiliar individuals in the field have pointed out significant deficiencies in teamwork and professionalism among individuals. They mentioned issues such as wanting to choose their work colleagues, an increase in personal/emotional relationships within the team, the use of a harsh communication style, and the formation of cliques as variables that lead to significant problems in teamwork. Regarding **stress management**, participants expressed feeling highly stressed during their assignments, particularly when dealing with angry crowds and reported experiencing burnout over time. Many acknowledged their inadequacy in stress management and emphasized the need for more practical training in this area.

The category of **working in a disaster zone** compiles expressions from personnel regarding their field duties and procedures. Experiences are categorized into two codes: site hazards/personal protection and working order in a disaster zone. In terms of personal protection, the majority of participants expressed that the use of personal protective equipment is widespread. However, some personnel mentioned not having enough knowledge about the hazards that disaster areas may pose. Particularly, negative experiences related to epidemic diseases in shelter centers were highlighted. Participants emphasized the importance of including specific hazards in OJT programs. Regarding the working order in disaster areas, participants stressed the need to inform personnel about shift systems and working hours before the assignment. There were requests to establish a system to teach personnel about the working order in advance.

The final category in this theme is **management and coordination**, consisting of codes related to coordination and responsibilities of organizations. Under the coordination title, the lack of a common language between institutions and the personnel's insufficient knowledge about coordination structures like the Disaster Coordination Center for Health Services has been highlighted.

The codes and participant statements related to the theme are shown in Fig. 18.5.

The majority of participants have expressed that most disaster personnel are not familiar with the institutional duties and responsibilities outlined in Turkey National Response Plan. Due to this lack of knowledge, personnel have mentioned encountering significant issues in inter-agency communication. They emphasized that all disaster response personnel should be aware of their duties and responsibilities first, followed by those of the potential institutions with which they will communicate.

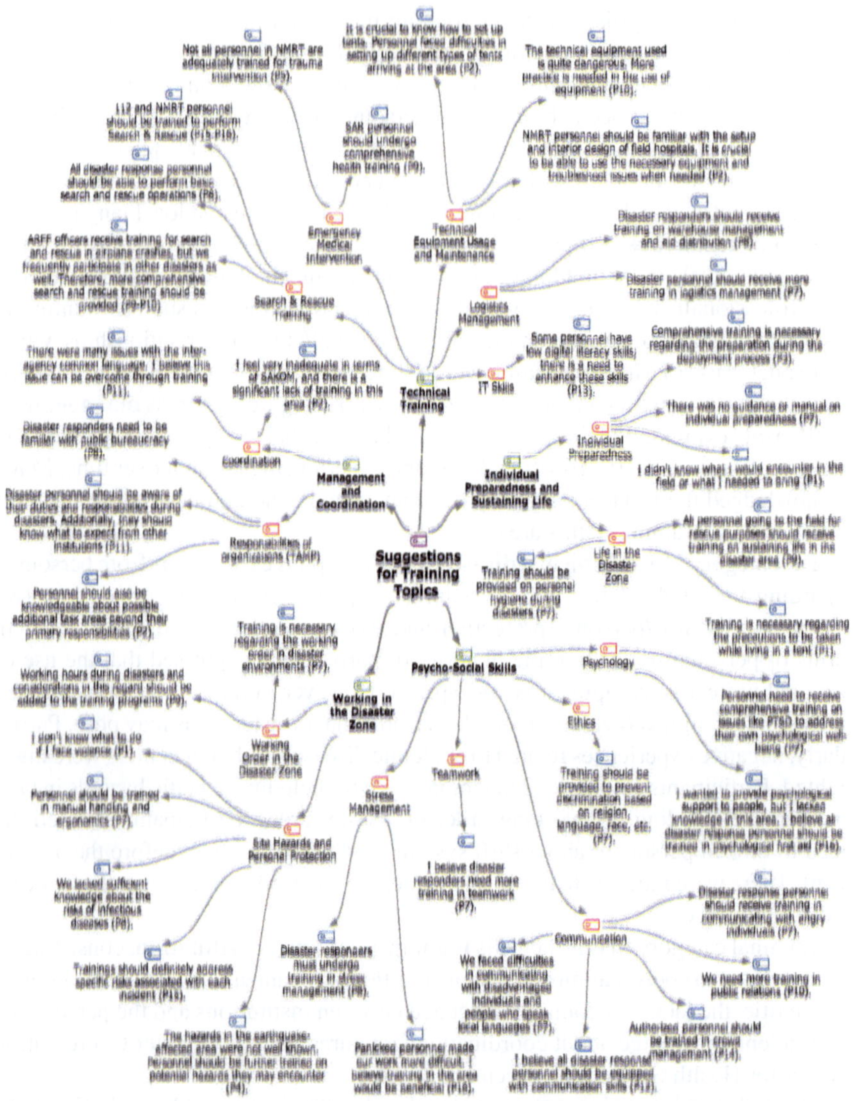

Fig. 18.5 Hierarchical code-subcode model illustrating the suggestions for training topics

18.5 Discussion

This study revealed that first responders, who have different roles after the earthquake, have deficiencies in disaster education and need more comprehensive and advanced disaster education in the future. First, the first responders in this study realized the benefit of disaster education. The results of the qualitative study conducted with first responders after the 2011 Joplin-Duquesne Tornado emphasized the importance of disaster preparedness (Mason et al. 2017). The study conducted after the storm and lava flow revealed that it was a learning opportunity for the community to take part and help each other in case of disaster (Ching et al. 2020). The self-satisfaction and confidence level of first responders who participated in flood rescue efforts increased (Joseph et al. 2020). Current disaster education needs should be determined in more detail, especially through comprehensive studies conducted after disasters.

The participants of this study stated that they encountered many problems related to disaster management while working together in earthquake regions. Similarly, after two earthquakes in China, nurses working in the disaster area stated that they had poor coordination, inadequate planning and self-efficacy problems related to disasters (Wenji et al. 2015). Three government disaster response teams that took part in the response efforts after the earthquake that occurred in Türkiye in 1999 stated that they encountered problems such as coordination, organization, communication, health, security and logistics (Altıntas and Delooz 2004). In the 2015 Nepal earthquakes, coordination and cooperation were the most important problems among teams (Cook et al. 2018). A systematic review study revealed that medical rescue teams mainly experience organizational, individual, environmental and health, logistics, communication and information problems during disasters (Kuday et al. 2023). Since the severity of disasters can vary depending on factors such as individual, society, time and environment, different effects may occur after each disaster. For this reason, disaster training should be organized appropriate to the problems to be encountered at local and national levels after disasters.

First responders in this study suggested that disaster training strategies should be developed in the areas of technical knowledge, management, coordination, individual preparedness and survival. Nurses stated that they needed training on mental health preparedness, self-protection and disaster knowledge (Wenji et al. 2015). Nurses working in the region affected by the Kahramanmaraş earthquakes stated that the personnel working in the earthquake region should have knowledge about crisis, risk, stress, emergency management, coordination and psychological coping (Altuntaş et al. 2023). Disaster management experts who had responded to different disasters observed that national and foreign health teams had deficiencies in disaster training and education (Djalali et al. 2014). Since personnel work in many areas such as security, nutrition, transportation, health, infrastructure, search, rescue and assistance during disasters, disaster education should be given specifically to the profession of these people. Current disaster education needs should be determined in more detail, especially through comprehensive studies conducted with the community, educators and first responders after disasters.

Participants emphasized that the training was beneficial for their psychological recovery and that more training was needed on this subject. Similarly, post-traumatic stress disorder occurred in many healthcare workers who responded to the Kahramanmaraş earthquakes (Satilmis et al. 2024). Some of Hurricane Katrina's first responders, such as police, fire, emergency medical services and city workers, had symptoms of trauma, depression, alcohol use, and conflict with partners six and nine months after the hurricane (Osofsky et al. 2011). Joyce et al. (2019) found that first responders with lower psychological resilience had more symptoms of trauma and depression. Scientific studies stated that psychological support should be provided to disaster response personnel (Aitken et al. 2009; Skryabina et al. 2021). Therefore, the psychological resilience of first responders before and after a disaster should be constantly monitored and strengthened.

18.5.1 Limitations

This study has some limitations. The study was conducted not in one province or region where earthquakes occurred, but in all eleven provinces. First responders may have encountered problems specific to provinces or regions. Additionally, this study did not focus on a specific professional group of first responders. Future studies can investigate the experiences of only one professional group of first responders in a region affected by earthquakes.

18.5.2 Recommendation

The practical disaster training capacity of first responders should be developed to increase the efficiency and quality of post-disaster response services. Standard disaster training contents should be prepared according to the duty types of first responders and these contents should be updated according to changing conditions. Training should be organized for first responders to work in coordination and cooperation with each other and other teams. The psychological and mental health of first responders should be constantly monitored and necessary support training programs should be organized to help them cope with stress. In order to self-evaluate and identify features that need to be improved, the difficulties and problems encountered by first responders after disaster response should be identified. Scientists and disaster managers should plan appropriate disaster training to provide solutions to the problems encountered by first responders.

18.6 Conclusion

This study revealed the thoughts, attitudes and suggestions of first responders who worked in different positions in Türkiye's deadliest earthquake disaster on February 2023 regarding disaster training. Participants in this study worked for days in shelter, health, nutrition, search and rescue services after the earthquakes. The results obtained from the data analyzed with the qualitative method showed how important disaster education is for effective disaster risk reduction implementations. Participants emphasized that disaster training increased the work efficiency of first responders during disasters and improved them socially. Participants suggested that the number of disaster training should be increased and disseminated, the content of the training should be improved, and the training should be made continuous. They encountered problems such as lack of knowledge, inexperience and working in a field in which they did not receive any training. This study revealed that training planning is necessary to improve the coordination, cohesion and social relations of first responders in the disaster area. This study is important for policies to be implemented to increase the knowledge and skills of first responders regarding disaster response.

References

Ahmed S, Hasan MZ, Pongsiri MJ et al (2021) Effect of extreme weather events on injury, disability, and death in Bangladesh. Clim Dev 13:306–317. https://doi.org/10.1080/17565529.2020.1772705

Aitken P, Leggat P, Robertson A et al (2009) Health and safety aspects of deployment of Australian disaster medical assistance team members: results of a national survey. Travel Med Infect Dis 7:284–290. https://doi.org/10.1016/j.tmaid.2009.03.005

Altıntas KH, Delooz H (2004) The problems faced by three government disaster response teams of Ankara City during the Marmara earthquake–1999 response. Eur J Emerg Med 11:95–101. https://doi.org/10.1097/01.mej.0000124508.23774.8e

Altuntaş S, Çiçek Korkmaz A, Efe N et al (2023) Problems experienced by nurses working in the region affected by Kahramanmaraş earthquakes in Turkey and their recommendations for risk reduction: a descriptive and cross-sectional study. Int J Disaster Risk Reduct 99:104115. https://doi.org/10.1016/j.ijdrr.2023.104115

Basu M, Ghosh S, Jana A et al (2017) Resource mapping during a natural disaster: a case study on the 2015 Nepal earthquake. Int J Disaster Risk Reduct 24:24–31. https://doi.org/10.1016/j.ijdrr.2017.05.020

Bekircan E, Usta G, Torpuş K (2023) The effect of psychological first aid intervention on stress and psychological resilience in volunteers participating in 2023 earthquakes centered in Kahramanmaraş, Turkey. Curr Psychol. https://doi.org/10.1007/s12144-023-05239-z

Bekler T, Cifci S, Bekler FN, Demirci A (2022) Çanakkale settlement risk reduction studies and evaluation of disaster awareness. Turkish J Earthq Res 4:73–97. https://doi.org/10.46464/tdad.1041575

Canaslan H, Erdoğan Ö, Taşlıdere B (2022) The Evaluation of hospital disaster and emergency plans practitioner trainings of the staff in public, private and university hospitals. J Disaster Risk 5:229–238. https://doi.org/10.35341/afet.1032175

Ching A, Morrison L, Kelley M (2020) Living with natural hazards: tropical storms, lava flows and the resilience of island residents. Int J Disaster Risk Reduct 47:101546. https://doi.org/10.1016/j.ijdrr.2020.101546

Cook ADB, Shrestha M, Htet ZB (2018) An assessment of international emergency disaster response to the 2015 Nepal earthquakes. Int J Disaster Risk Reduct 31:535–547. https://doi.org/10.1016/j.ijdrr.2018.05.014

Coulibaly T, Islam M, Managi S (2020) The impacts of climate change and natural disasters on agriculture in African Countries. Econ Disasters Clim Chang 4:347–364. https://doi.org/10.1007/s41885-019-00057-9

Djalali A, Ingrassia PL, Della CF et al (2014) Identifying deficiencies in national and foreign medical team responses through expert opinion surveys: implications for education and training. Prehosp Disaster Med 29:364–368. https://doi.org/10.1017/S1049023X14000600

Elkady S, Hernantes J, Muñoz M, Labaka L (2022) What do emergency services and authorities need from society to better handle disasters? Int J Disaster Risk Reduct 72:1–25. https://doi.org/10.1016/j.ijdrr.2022.102864

Emergency Events Database (2023) Data. https://public.emdat.be/data

Gezer M, Şahin İF (2022) Earthquake education: the knowledge levels of social studies teacher candidates on earthquake. Erzincan Univ J Educ Fac 24:97–106

Harris C, McCarthy K, Liu EL et al (2018) Expanding understanding of response roles: an examination of immediate and first responders in the United States. Int J Environ Res Public Health 15. https://doi.org/10.3390/ijerph15030534

Hasan MK, Moriom M, Shuprio SIM et al (2022) Exploring disaster preparedness of students at university in Bangladesh. Nat Hazards 111:817–849. https://doi.org/10.1007/s11069-021-05080-2

Hung MSY, Lam SKK, Chow MCM et al (2021) The effectiveness of disaster education for undergraduate nursing students' knowledge, willingness, and perceived ability: an evaluation study. Int J Environ Res Public Health 18:10545. https://doi.org/10.3390/ijerph181910545

Johnson B, ve Christensen L (2014) Eğitim araştırmaları: Nicel, nitel ve karma yaklaşımlar (S. B. Demir, Çev.) (4. Baskı). Ankara: Eğiten Kitap

Joseph JK, Anand D, Prajeesh P et al (2020) Community resilience mechanism in an unexpected extreme weather event: An analysis of the Kerala floods of 2018, India. Int J Disaster Risk Reduct 49:101741. https://doi.org/10.1016/j.ijdrr.2020.101741

Joyce S, Tan L, Shand F et al (2019) Can resilience be measured and used to predict mental health symptomology among first responders exposed to repeated trauma? J Occup Environ Med 61:285–292. https://doi.org/10.1097/JOM.0000000000001526

Kahraman S (2023) A comprehensive health economic analysis of the 2023 Turkey earthquake. Disaster Med Public Health Prep 17:3–6. https://doi.org/10.1017/dmp.2023.160

Klomp J, Valckx K (2014) Natural disasters and economic growth: a meta-analysis. Glob Environ Chang 26:183–195. https://doi.org/10.1016/j.gloenvcha.2014.02.006

Konakçı G, Cömert MB, Özgörsoy Uran BN, Nemli A (2023) Investigation of disaster preparedness of workers working in different ındustry. Anatol J Emerg Med 6:53–60

Köçer MS, Aslan R (2023) Evacuation experiences of volunteer search and rescue teams in forest fires: 2021 Mediterranean Forest Fires. J Disaster Risk 6:829–851. https://doi.org/10.35341/afet.1236021

Kuday AD, Özcan T, Çalışkan C, Kınık K (2023) Challenges faced by medical rescue teams during disaster response: a systematic review study. Disaster Med Public Health Prep 17:e548. https://doi.org/10.1017/dmp.2023.217

Mason AM, Drew S, Weaver D (2017) Managing Crisis-induced uncertainty: first responder experiences from the 2011 Joplin-Duquesne Tornado. Int J Disaster Risk Reduct 23:231–237. https://doi.org/10.1016/j.ijdrr.2017.04.012

Merriam SB, Tisdell EJ (2016) Qualitative research: a guide to design and implementation (fourth edition). The Jossey-Bass higher and adult education series. San Francisco, CA: Jossey-Bass

Mızrak S (2018) Education, disaster education and community disaster resilience. MSKU J Educ 5:56–67. https://doi.org/10.21666/muefd.321970

Ministry of Interior of the Republic of Türkiye (2022) DEMA Draws Türkiye's Disaster Risk Map. https://www.icisleri.gov.tr/afad-turkiyenin-afet-risk-haritasini-cikardi

Nemutlu ÖF, Saru A, Balun B (2023) comparison of actual loss of life and structural damage in 06 February 2023 Kahramanmaraş Earthquakes (Mw 7.7-Mw 7.6) with estimated values. Afyon Kocatepe Univ J Sci Eng 23:1222–1234. https://doi.org/10.35414/akufemubid.1302254

Osofsky HJ, Osofsky JD, Arey J et al (2011) Hurricane Katrina's first responders: the struggle to protect and serve in the aftermath of the disaster. Disaster Med Public Health Prep 5. https://doi.org/10.1001/dmp.2011.53

Özdemir A, Şahinöz T (2021) Tools used to raise disaster awareness in community: a qualitative study. J Disaster Risk 4:121–134. https://doi.org/10.35341/afet.975612

Park JM, Bae SM (2022) Impact of depressive, anxiety, and PTSD symptoms in disaster victims on quality of life: the moderating effect of perceived community resilience. Int J Disaster Risk Reduct 69:1–10. https://doi.org/10.1016/j.ijdrr.2021.102749

Peden AE, Mayhew A, Baker SD (2022) Experiences, beliefs, and attitudes of lifeguards from Australia and the United Kingdom toward lifeguard involvement in flood mitigation and response. Int J Disaster Risk Reduct 76:103013. https://doi.org/10.1016/j.ijdrr.2022.103013

Peräkylä A, ve Ruusuvuori J (2018) Analyzing talk and text. N. K. Denzin ve Y. S. Lincoln (Der.), The SAGE Handbook of Qualitative Research (pp. 1163–1201). SAGE Publications, Inc.

Satilmis D, Yildiz E, Cevik E (2024) Posttraumatic stress disorder in health - care workers after two. Turkish J Emerg Med 24:27–32. https://doi.org/10.4103/tjem.tjem_192_23

Skryabina E, Betts N, Reedy G et al (2021) UK healthcare staff experiences and perceptions of a mass casualty terrorist incident response: a mixed-methods study. Emerg Med J 38:756–764. https://doi.org/10.1136/emermed-2019-208966

Sudan Aran A, Erol I (2022) The effect of education on the awareness of emergency action plans: an application on occupational health and safety. Ankara Üniversitesi SBF Derg 77:679–707. https://doi.org/10.33630/ausbf.1134690

Suk JE, Vaughan EC, Cook RG, Semenza JC (2020) Natural disasters and infectious disease in Europe: a literature review to identify cascading risk pathways. Eur J Public Health 30:928–935. https://doi.org/10.1093/eurpub/ckz111

Şen S (2023) The effect of Kahramanmaraş Earthquakes on the economy. J Dipl Strateg 4:1–55

Şentürk S, Büyükdavraz GG, Yıldırım Keskin A (2020) Knowledge, opinion and approaches of intensive care nurses about disaster management. J Inonu Univ Heal Serv Vocat Sch 8:527–546. https://doi.org/10.33715/inonusaglik.753643

Tercan B (2022) Investigation of ındividuals' preparedness for disasters in disaster resilience: Erzincan province example. J Disaster Risk 5:261–269. https://doi.org/10.35341/afet.1063790

Torani S, Majd P, Maroufi S et al (2019) The importance of education on disasters and emergencies: a review article. J Educ Health Promot 8:85. https://doi.org/10.4103/jehp.jehp_262_18

Tunç H (2023) The nurses' experiences working in the earthquake area: a qualitative research. Turkish J Heal Sci Life 6:122–132. https://doi.org/10.56150/tjhsl.1362173

Usta G, Aslan R, Gündüz F, et al (2023) On-site evaluation of emergency response carried out after earthquakes with epicentral base Kahramanmaraş/Pazarcık-Elbistan in Türkiye: an observational study. Turkish Geogr Rev 83:145–154. https://doi.org/10.17211/tcd.1360428

Wenji Z, Turale S, Stone TE, Petrini MA (2015) Chinese nurses' relief experiences following two earthquakes: implications for disaster education and policy development. Nurse Educ Pract 15:75–81. https://doi.org/10.1016/j.nepr.2014.06.011

Winders WT, Bustamante ND, Garbern SC et al (2021) Establishing the effectiveness of ınterventions provided to first responders to prevent and/or treat mental health effects of response to a disaster: a systematic review. Disaster Med Public Health Prep 15:115–126. https://doi.org/10.1017/dmp.2019.140

Yildiz A, Dickinson J, Priego-Hernández J et al (2023) Effects of disaster education on children's risk perception and preparedness: a quasi-experimental longitudinal study. Geogr J 1–19. https://doi.org/10.1111/geoj.12556

Yilmaz S (2023) Transportation model utilized in the first week following the Kahramanmaraş earthquakes in Turkey - transport health centers. Scand J Trauma Resusc Emerg Med 31:1–3. https://doi.org/10.1186/s13049-023-01108-7

Chapter 19
Teachers' Responsive Actions to Protect Their Typhoon-Affected Schools: An Empirical Investigation of Teachers' Experiences of Typhoon Odette

Saki Arima-Shirai, Edwin F. Marquez, Nana Sugaya, Lyn Alexandria T. Novilla, Mitsuko Shikada-Otsuyama, and Rajib Shaw

Abstract This chapter aims to investigate the factors influencing school personnel's protective actions in relation to schools' response to hazards, particularly focusing on the aftermath of Typhoon Odette (internationally known as Rai) in December 2021 in the Philippines. The research analyzes interview results from three schools affected by Typhoon Odette by employing models for protective action decision-making to identify the variables that played a role in promoting school personnel's responses as part of their school disaster risk reduction and management practices. This study also delves into the relevance between the training experiences of the personnel and the subsequent protective behaviors displayed during the typhoon. The findings reveal that training provided by the Philippines Department of Education in partnership with a Japanese nongovernmental organization did not significantly influence the protective actions taken by school personnel during Typhoon Odette. Instead, the study identifies institutional arrangements within the Department of Education and collaborations with external agencies as crucial factors influencing the school personnel's actions. This chapter contributes to the existing literature by examining the specific factors that drive individual school personnel to undertake protective actions in the context of school-based disaster risk reduction. The focus on the protective behavior of personnel as responsible persons for their schools' safety and the sustainability of education offers valuable insights into the understanding of how future training within the Department of Education can enhance disaster risk reduction and management practices of school personnel.

S. Arima-Shirai (✉) · E. F. Marquez · N. Sugaya · L. A. T. Novilla · M. Shikada-Otsuyama · R. Shaw
SEEDS Asia, 1-7-7-307 Okamoto, Kobe City, Hyogo Prefecture, Japan

R. Shaw
Graduate School of Media and Governance, Keio University, 5322 Endo, Fujisawa-shi, Kanagawa, Japan

A. Yildiz and R. Shaw (eds.), *Disaster and Climate Risk Education*, Disaster Risk Reduction, https://doi.org/10.1007/978-981-97-5987-3_19

Keywords School disaster risk reduction and management · Typhoon Odette (Rai) · Protective actions · Philippines · Teacher training · Education continuity

19.1 Introduction

The central part of the Philippines, the Visayas Region, was devastated by Typhoon Odette (internationally known as Rai) in December 2021. This typhoon left over ten million learners and approximately 30 thousand schools affected (Department of Education [DepEd], 2022) and impacted their ability to resume learning and education after an almost two-year closure of education facilities due to the novel Coronavirus pandemic (COVID-19) (Dela Cruz and Lema 2021). While a large number of schools were affected, the responses observed across the impacted schools' disaster risk reduction and management (DRRM) personnel varied (SEEDS Asia, 15 January 2022).

The existing literature focusing on internal motivation for self-protection, such as evacuation, puts an emphasis on individual protective behavior and not on varied actions by school teachers intended to protect their institution (Dasci Sonmez and Gokmenoglu 2022; Fujimoto et al. 2019; Lindell and Perry 1993, 2003, 2012; Najafi et al. 2017; Vinnell et al. 2021). Given this, this chapter aims to investigate the factors influencing school personnel's actions beyond their personal safety to protect their schools in the context of Typhoon Odette.

19.1.1 Background and Context

This chapter focuses on school personnel's experience of Typhoon Odette in December 2021 in the Philippines. The authors had long-term collaborative relationships with the local Department of Education (DepEd) Region VII Office, and the research was conducted in the context as follows.

19.1.2 Typhoon Odette and Its Impact on Schools

Typhoon Odette made its landfall in the Philippines on December 16, 2021. Its violent winds, accompanied by torrential rain, characterized the typhoon and damaged its path along the Visayas region, including Cebu Island, particularly in the southern part (DepEd 2022). This typhoon was, for many in the south, the first "super typhoon," as they were not directly stricken by Typhoon Yolanda (internationally known as Haiyan), which had devastated northern Cebu in November 2013 (National Disaster Risk Reduction and Management Council 2014).

Odette affected a total of 29,671 schools across 11 regions of the Philippines, impacting 14,882,985 learners (DepEd 2022). This typhoon struck the education sector severely in terms of education continuity. In fact, many schools were preparing for partial face-to-face class resumption after the almost two-year closure of schools in response to Covid-19. The DRRM Coordinator of a Schools Division Office [SDO] in Cebu personally reported to SEEDS Asia, a Japan-based nongovernmental organization, the damage caused by Odette, stating that all schools required repairs in the roofs and clean-up of trees that were uprooted by the winds of over 100 km/hr (International Federation of Red Cross and Red Crescent Societies 2022). Electricity and networks were cut off, and generators were priced as high as triple or sold out in many retail shops in Cebu.

The immediate impacts of Typhoon Odette on schools, as reported by DRRM Coordinators of affected SDO's, were characterized as follows:

1. The physical infrastructure of education facilities was damaged to an extent that face-to-face class resumption was impossible without intensive cleaning up; and
2. Communication was cut off due to large-scale blackouts and network service failures, resulting in delays in reporting to upper offices (SDO per city/province, Regional Office per region, and Central Office at the national level where decisions are made to download emergency funds for recovery), requests for help, and tracking of learners and personnel.

Given the more frequent and severe typhoon risk in the Philippines due to higher sea surface temperature (Holden and Marshall 2018), it is presumed an increasing number of schools are susceptive to weather risks, and lessons learned from Typhoon Odette have implications for climate-responsive education settings.

19.1.3 Emergency Support Project by SEEDS Asia in Response to Typhoon Odette

SEEDS Asia is a nongovernmental organization that has implemented joint projects with the DepEd, particularly Regional Office VII, since 2014 in response to the locally identified long-term needs for disaster risk reduction education. There were two projects from 2014 to 2017 and 2017 to 2020, after which SEEDS Asia suspended its operations in the Philippines due to the pandemic.

When Typhoon Odette took place in December 2021, upon receiving requests for help from project partners (DepEd personnel), SEEDS Asia commenced information gathering, issuing an appeal, and writing a short-term project proposal to Japan Platform (another Japan-based NGO that specializes in humanitarian supports in disaster or conflict affected regions/nations by mobilizing resources sourced from Japanese private and government donors) in December 2021. Initial supports offered by SEEDS Asia to typhoon-affected SDO's included coordination with a Manila-based church for delivering medicines that were scarce in Cebu and procuring and shipping power generators from Cagayan de Oro, Mindanao, to Cebu. In the course of these small-scale projects, coordination and negotiation between SEEDS Asia and Japan Platform continued for the commencement of a seven-month commitment in February 2022. The emergency support project consisted of two components: the provision of power generators to SDO's to assist them with their administration to support typhoon-affected schools in their jurisdictions and renting clean-up kits (chainsaws, safety equipment, etc.) to accelerate the schools' thrusts towards reopening school grounds for their learners. Alongside the clean-up activities, damage to school facilities was investigated and assessed by an architect, and a proposal for typhoon-safer construction of school buildings was compiled.

By the end of the project, 19 schools received and used power generators for the purposes of modular printing, power provision to evacuees who were taking shelter in classrooms, recovering facilities (i.e., welding of water pipes), and groundwater pumping. 105 schools were visited with procured clean-up tools, and recovery operations took place to remove trees and debris from the school grounds before the commencement of face-to-face classes in September. August is the month for the annual Brigada Eskwela (School Maintenance Week) in the Philippines, where community members and other stakeholders volunteer to contribute to making schools conducive to learning (DepEd 2023). In this occasion in preparation of schools for the new semester, clean-up kits were utilized by school personnel, learners and their families, community members, local government units, and other stakeholders.

19.1.4 Findings from Typhoon Odette

Upon visits by SEEDS Asia staff members to deliver generators and clean-up kits, the experiences of DepEd personnel from Typhoon Odette were shared. There were many could-haves, should-haves, and would-haves in their action and inaction: people did not expect such a large typhoon to occur in southern Cebu, unlike the north that had been devastated by Typhoon Yolanda in 2013; schools wished they had requested help from their local governments to prune tree branches prior to the landfall; and some schools felt left behind by their SDO due to their locations. Many school personnel expressed concerns about their challenging situations stemming from intergovernmental coordination (i.e., permits and workforce required from the local government for pruning of trees) and conflictual DepEd policies that caused the dilemma between and confusion of disaster preparedness (pruning of trees) and school beautification (planting of trees).

Despite many disclosures of regrets, there were improvements that were reported as a result of the lessons from Typhoon Yolanda in 2013: overall enhancement of early warning across government agencies, and the decision-making regarding school suspension that allows School Heads' own judgment based on their risk perception; greater collaborations between disaster management and education sectors;

social media insinuating disaster awareness, exposing more people to various disasters taking place worldwide; and learners sharing their learning from Disaster Risk Reduction Education (DRRE) with family members.

Internal discussions in SEEDS Asia revealed that some schools demonstrated a high level of disaster preparedness and response and swiftly shifted towards recovery, while others appeared to have no clue of where to start their response and recovery. When SEEDS Asia visited the latter, many of them had not begun cleaning up till a good few months after the typhoon's occurrence or had not arranged external help for the clean-up drive with SEEDS Asia.

There were a variety of responses to Typhoon Odette from schools. The differences that influenced how "well" or "ill" schools responded and recovered from the typhoon include: some had developed a common understanding about how they would respond to a disaster with external stakeholders (local government, SDO, community, etc.); and/or the leadership and creativity of the School Head and DRRM Coordinators of their SDO. It was further discussed among SEEDS Asia that DepEd's and local governments' focus had been more on response, as the education sector was urged to take countermeasures against COVID-19 infections rather than preparedness; therefore, school personnel were not fully alert to an emerging typhoon risk.

The determinants of the different responses of schools to Typhoon Odette have been unexplained. Some schools were SEEDS Asia's partner schools in either/both projects in the past. This chapter seeks to analyze the link between SEEDS Asia's previous capacity building activities and schools' actual capacity demonstrated in responding to Typhoon Odette.

Teachers play an important role in terms of ensuring education continuity and minimizing the impact of disasters on their students' future. There are policies, rules and agreements put in place to standardize and enhance DRRM implementation within DepEd; however, Typhoon Odette revealed different actions taken by school personnel, resulting in different outcomes for respective schools. This research will focus on the context and triggers of school personnel's action and inaction to protect their school from the typhoon's devastating impact; therefore, it will provide implications for how DepEd can strengthen their disaster preparedness, response and recovery among their schools by incorporating the findings into their capacity building initiatives.

19.2 Literature Review

This chapter focuses on the working model for risk-aversive behavior by Fujimoto et al. (2019). Additionally, existing research on teachers' awareness on the importance of disaster risk reduction will be reviewed as well as a model on effective disaster education leading to disaster risk reduction action.

19.2.1 Model for Risk Aversive Behavior and Protective Action Decision-Making

Tatsuki et al. (2004), in their research on the residents' risk-aversive behavior against earthquakes in Marikina City, Philippines, incorporated the conceptual framework for the prediction of intentions and behaviors by Fishbein and Ajzen (1975) and the notion of schema, posited by Neisser (1978) interpreted as "one phase of an ongoing activity which relates the perceiver to his (*sic*) environment." (ibid., p. 23). Tatsuki et al. (2004) highlighted the crucial role of schemas in forming individuals' recognition of threat (perceived risk consisting of the estimation of probability and consequences) and that of normative beliefs, which provoke individuals to interpret their surrounding phenomena as disasters. Fujimoto et al. (2019) examine a further developed model on protective action decision making in which external stimuli and disaster schemas are deemed to influence internal processing towards protective behavior. This working model further applies the concept of "milling" or "making sense of the situation" by Mileti and Sorensen (1990) that is influenced by and influences individuals' motivation for protective behaviors. In summary, external stimuli or disaster information from various sources and channels will be influenced and influence individual disaster schemas, and these two variables will be further processed internally through threat perceptions, protective action perceptions, and reference group perceptions. These perceptions are internally processed through fear and milling, then lead to protection motivation, resulting in risk aversive behavior, influenced by situational facilitators and impediments. The findings of Fujimoto et al.'s research revealed that the reference groups' threat perceptions influence individuals' act of milling (the surroundings are more influential than the fear or the assessment of protective actions, urging one to reconfirm the situation), and disaster schemas are most effective in leading to risk aversive behavior (2019).

19.2.2 Teachers' Disaster Preparedness Belief

Dasci Sonmez and Gokmenoglu (2022) analyze public school teachers' perceptions in relation to disaster preparedness in Türkiye. Being built upon Inal and Dogan's (2018) improved model on general disaster preparedness belief scale, data were gathered in order to identify the study sample teachers' levels of perceived benefits, perceived susceptibility, perceived barriers, perceived severity, cues to action, and self-efficacy. The research findings effectively demonstrate that the perceived benefits, perceived susceptibility, perceived barriers, and self-efficacy are relatively high among the teachers, while the cues to action "moderate," and perceived severity "low," indicating that these teachers are highly aware of the importance of disaster preparedness but not self-confident with their actual preparedness action (Dasci Sonmez and Gokmenoglu 2022).

19.2.3 KIDA Tree Model

KIDA stands for Knowledge, Interest, Desire, and Action, signifying the importance of encouraging knowledge acquisition to promote interest in taking, and eventually desire to take, action in disaster risk reduction (Shaw et al. 2009). Built upon the AIDMA (Attention, Interest, Desire, Memory, and Action) model by Samuel Roland Hall (1924) which was developed to explain consumer behavior, the KIDA Tree Model refers to the roots and sources of nutrition from the ground as knowledge, interest, and desire underpinning a tree's growth as action. The model highlights the importance of planning disaster education activities, taking into consideration these steps (Shaw et al. 2009) (Fig. 19.1).

While Fujimoto et al.'s model (2019) is useful in identifying the variables that determine individuals' action or inaction to protect themselves from disaster risks and potential efficient approaches to promote such action, there is no evaluation of the internal processes between these variables and behavior that go beyond self-protection and serve to a wider community and for an extended period of time than evacuation, for instance, for school personnel to engage in school disaster management before, during and immediately after a disaster. Dasci Sonmez and Gokmenoglu's (2022) research highlights teachers' awareness on disaster preparedness, however, did not attest the findings in an actual disaster scenario nor did it question specific roles as DRRM practitioners for their schools. The KIDA Tree Model, on the other hand, could be used to explain individuals' disaster schemas posited by Fujimoto et al. (2019), while the interrelevance between knowledge, interest, desire and action is not quantitatively verified. In this regard, this chapter focuses on the

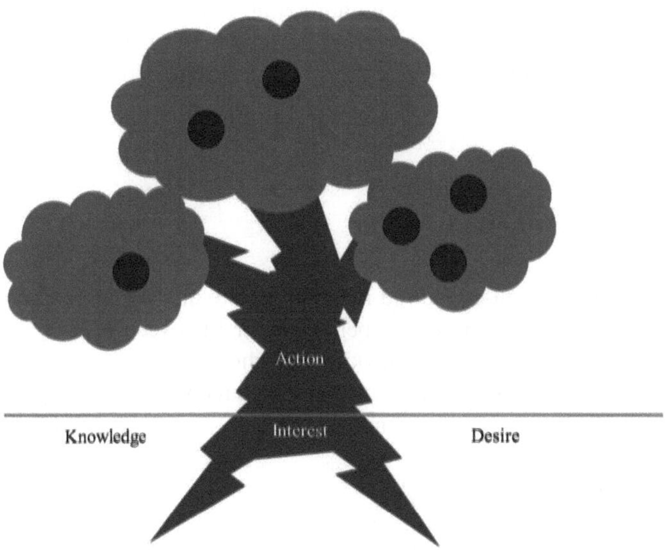

Fig. 19.1 Kida Tree Model adapted from Shaw et al. (2009)

actions of school personnel that were impacted by Typhoon Odette in Cebu, Philippines, and analyze the variables that influenced the action and inaction by them as education management practitioners towards ensuring education continuity in their damaged schools, highlighting their risk aversive behaviors and focusing on their disaster schemas.

19.3 Method

19.3.1 Study Samples and Survey Method

Three semi-structured interviews were held with three DepEd schools in Cebu Island that were affected by Typhoon Odette. The summary of the sample schools is shown in Table 19.1. The three schools were selected by the SDO DRRM Coordinators in consideration of their distance from the central part of Cebu Island, Cebu City, geographical commonalities (inland), and different training experiences with SEEDS Asia: one school being a direct partner of its previous project (School B), one with a SDO Coordinator who worked directly with SEEDS Asia throughout the said project (School C), and one whose SDO Coordinator with partial direct work with SEEDS Asia.

Data were collected from semi-structured interviews conducted by SEEDS Asia. Two schools were visited for face-to-face interviews on August 23, 2023, and one was visited on-site for an online interview with the lead author overseas on October 12, 2023. The semi-structured interviews consisted of a modified set of questions based on the theoretical model by Fujimoto et al. (2019) and a timeline described by the interviewees as to when they received external information that led to protective behaviors as individuals and as education practitioners to protect their schools (Table 19.2).

19.3.2 Analysis of Data

Data collected in the semi-structured interviews were analyzed to explore the relevance between triggers of protective behaviors, and between protective behaviors and training experiences. Decision-making by school DRRM personnel in relation to the school's disaster preparedness, response, and recovery was highlighted on the basis of Fujimoto et al.'s theoretical model (2019).

Table 19.1 Sample schools for semi-structured interviews

Name of the school	A. Camp 4 Elementary School	B. Uling Elementary School	C. Gabi Elementary School
Date of interview	August 23, 2023	August 23, 2023	October 12, 2023
Location	Camp IV, Talisay City	Uling, City of Naga	Gabi, Cordova, Cebu Province
SDO	Talisay City Schools Division Office	City of Naga Schools Division Office	Cebu Province Schools Division Office
Interviewee (s)	• Interviewee 1: School DRRM Coordinator • Interviewee 2: School Guidance Counsellor • Interviewee 3: SDO DRRM Coordinator	• Interviewee 4: School DRRM Coordinator • Interviewee 5: School Alternate DRRM Coordinator • Interviewee 6: SDO DRRM Coordinator	• Interviewee 7: School DRRM Coordinator
Number of learners as of the date of the interview	579	833	600
Participation in SEEDS Asia's previous project on DRRM	No	Yes	No
Visit by SEEDS Asia after Typhoon Odette	No	Yes, for a clean-up activity	No
Reported damage to school from Typhoon Odette (RADaR consolidation, DepEd, 2021)	Totally damaged classrooms: 0 Classrooms with major damage: 0 Classrooms with minor damage: 1	Totally damaged classrooms: 0 Classrooms with major damage: 0 Classrooms with minor damage: 1	Totally damaged classrooms: 1 Classrooms with major damage: 2 Classrooms with minor damage: 7

Table 19.2 Interview questionnaire

• Attitude towards learning • Disaster risk reduction literacy • Previous disaster experience • Hazard awareness • Risk awareness • Perceptions of protective behaviors • Responsibilities as school personnel	• Sources of information • Social capital • Communications with others • Actions before, during, and after the typhoon • Past training undertaken • Knowledge resources utilized during the typhoon

19.3.3 Ethics

The authors of this research ensured to maintain transparency in collecting data from the interview respondents. The purpose of the interviews was explained to all the respondents and consents were gained from the respondents that their positions and offices (schools) would be listed in this research.

19.4 Results

The interviews with the three schools were analyzed in order to explore the relevance between training experiences and school DRRM personnel's protective behaviors for their respective schools (preparedness and responses concerning schools' DRRM) in Typhoon Odette.

19.4.1 Summary of Schools' Disaster Profiles, and School DRRM Coordinator's Decision-Making and Protective Behaviors Before and During Typhoon Odette

Camp 4 Elementary School: This school is located in a mountainous area in Talisay City, exposed to flood, landslide (rockfalls), and potential earthquake risk. The school is a designated evacuation site for the community. Two days prior to Typhoon Odette's landfall, a pre-disaster risk assessment (PDRA) had been spearheaded by Talisay City in the morning, and by DepEd Division DRRM Coordinator in the afternoon. One day before the landfall of the typhoon, the school personnel started preparing their classrooms for evacuees by placing vital documents in elevated locations, printing materials provided by the Department of Social Work and Development for evacuees for their registrations, and surrendering classroom keys to the barangay (Philippine's smallest administrative division) in charge of evacuation center operations. The school DRRM Coordinator went home at 4:30 PM after her duties on the day of the landfall but returned to the school at 5 PM to attend to evacuees whose numbers exceeded the capacity of the classrooms. She also reminded the classroom occupants of the house rules and collected data on the evacuees in order to report to the Division DRRM Coordinator to apply for the Quick Response Fund (QRF). The School DRRM Coordinator had to go home to send the data, as the school did not have internet or phone signals.

Uling Elementary School: The school is the only direct participant of SEEDS Asia's Project on DRRM and is situated on a national road in the City of Naga. The school has not been affected by major disasters as far as the current school DRRM Coordinator is concerned, except for soil erosion near classrooms, in the surrounding roads, and in the houses. The interviewees have stated that they felt ground shaking from the Bohol Earthquake in 2013, and are concerned about landslide risks like the actual incident in another area within the city (in Sitio Sindulan, in 2018). The school's classrooms are not recommended as an evacuation center for the community due to safety concerns, and only its covered court was opened for evacuees.

The barangay handled the evacuees in the midst of Typhoon Odette's onset. Therefore, the school personnel did not come to school for camp management prior to or during the typhoon. Before the landfall, the school personnel of Uling Elementary School prepared for the typhoon's arrival by securing important documents in likely safe locations in a classroom (but the ceiling of that classroom collapsed, and the documents were soaked with rainwater). On the coming morning of the landfall, the School Head and another staff member visited the houses of the personnel one by one and checked if any roof of the school buildings had been damaged. Uling Elementary School was one of the schools that SEEDS Asia visited for clean-up activities post-Odette.

Gabi Elementary School: This school is in the Municipality of Cordova, sitting next to Mactan Island, under the jurisdiction of DepEd Cebu Province SDO. Despite being surrounded by sea channels, the entire Cordova is deemed risk-free from flooding, according to Interviewee 7 and the Mines and Geosciences Bureau Regional Office (Mines and Geosciences Bureau Regional Office, n.d.). The School DRRM Coordinator of Gabi Elementary School stated that Cordova experiences only typhoons and earthquakes. The school is designated as an evacuation area for the community. Once the school was aware that the typhoon was approaching, the School Head spearheaded preparedness activities among the staff, such as pruning trees near the school buildings. On the day before the typhoon's arrival, the School DRRM Coordinator and other staff members engaged in the preparation of the classrooms for evacuees by labeling them, providing information to the nearest houses around the school, and safekeeping school equipment for evacuees' use. After the preparation of the classrooms, the school facilities were handed over to the barangay, and school personnel went home at around 6:30 PM on the day of the landfall, December 16.

19.4.2 External Stimuli and Disaster Information

The interview respondents respectively named various sources of information on Typhoon Odette, such as news on television, advisories from and the website of the Philippine Atmospheric, Geophysical and Astronomical Services Administration (PAGASA), "tsismis" (a Filipino word for gossiping and/or chit-chat), and communication within the DepEd organization. All three schools emphasized their reliance on the DepEd Division and/or District DRRM Coordinators in screening information.

In Talisay City, where School A belongs, the Division DRRM Coordinator (Interviewee 3) is a member of several group chats on Messenger for announcements across Talisay City, for all School DRRM Coordinators in the City, and for the DRRM Coordinators in DepEd Region VII where the PAGASA weather station's updates are directly sent out. Interviewee 3 selects information from outside DepEd Talisay City Division and cascades the selected information to the group chat consisting of all the District DRRM Coordinators in his Division; then, the District DRRM Coordinators share the same information with their School DRRM Coordinators.

Interviewee 2 stated that she no longer needs television for weather updates because they are not localized in Talisay City and more actively refers to the group chat. She also used a group chat to discuss the possible impact of the typhoon with her colleagues. At School B, the group chats consisting of all School DRRM Coordinators in the City of Naga, as well as the school's DRRM Team, were named as the most reliable source of information concerning the typhoon by Interviewee 5. School C's interviewee 7 stated that the group chat among the Cordova Municipal DRRM Office, School DRRM Coordinators, and their alternates was named as the third most common information source following PAGASA and the Municipal DRRM Office. This School DRRM Coordinator (Interviewee 7) also utilized a group chat to communicate with learners' families by encouraging them to be prepared for the typhoon. These remarks by the interviewees revealed that group chats of DRRM Coordinators, with external agencies and families, and within the school are used as frequent/reliable sources of information.

19.4.3 Disaster Schemas

Some of the responses to the semi-structured interviews revealed key issues concerning the respondents' disaster schemas.

All three respondent schools assessed themselves as moderately knowledgeable (2 to 4 out of 5) of DRRM, with remarks stating their wish for more training opportunities. School A's Interviewee 1 stated that "Training, seminars, and workshops would be helpful for me to be equipped with the knowledge and skills in risk reduction management," and Interviewee 2 "I have enough knowledge but not advanced, and there is still room for learning." School B's Interviewee 4 expressed her willingness to continue learning because "we are not all-knowing about DRR" and "we always forget." School C's Interviewee 7 stated that he needs to develop learning on "actual scenarios" to feel knowledgeable and skillful enough in disaster response.

The three schools have various hazard risks, and the interviewees all have an awareness of these hazards. School A is exposed to rockfalls and landslides, with a river flood risk in the community. However, this risk awareness was overturned by the SDO DRRM Coordinator (Interviewee 3)'s instruction to the School DRRM Coordinator (Interviewee 1) urging her to go to school in a hazardous situation, as he requested the data on the school's evacuation center occupants. School B is aware of the hazards and risks surrounding the school. However, Interviewees 4 and 5 commented that they had not experienced disasters that affected their lives in the past. When their Division DRRM Coordinator (Interviewee 6) advised that soil erosion impacting the school facilities would be a hazard, both Interviewees 4 and 5 agreed that their houses and community are affected by the slowly moving grounds. Their awareness about the soil erosion led the school to identify the majority of their classrooms as unsafe to establish evacuation centers and to consequently lessen the burden on the school personnel in the preparation of the facilities for Typhoon Odette and incoming evacuees. School C's Interviewee 7 referred to the Municipal

DRRM Office's hazard information that there is no flood risk in Cordova and to his home's roof blown away by Typhoon Odette, in addition to earthquake risks. He and his school colleagues had been made aware of the typhoon two weeks prior to its landfall, School C had completed pruning trees and preparing their classrooms for evacuees, and his family had planned evacuation to a safer place (an extended family's home).

Two interviewees from School A highlighted that they had received some kinds of knowledge gained from Typhoon Yolanda in 2013: Interviewee 2 described that the alertness of the general public on December 15 despite the sunny weather was a lesson learned from Yolanda, and her family in Leyte (the most severely affected by Typhoon Yolanda) warned her not to go out because the clear skies were what they had experienced the day before Typhoon Yolanda; and the Interviewee 3 had witnessed coastal residents, particularly fisherfolks, referring to the weather and stating "this is the kind of situation that we are so afraid of. In the evening, this (the typhoon) will be big." Interviewee 3 also noted that the fisherfolks were the first to evacuate in Talisay City. In these cases, the commenters' regards to others' lived experiences of a previous typhoon facilitated their understanding of the potential risk of Typhoon Odette. School C's interviewees reflected on one occasion on December 15 when their colleague told them, "Mind you, if the leaves will not move, we need to prepare because this might be a super typhoon," but the interviewees stated that they "did not listen to that, actually." This indicates that interviewees 4 and 5 did not develop disaster schema based on others' experience.

School A mentioned that their attendance at the PDRA two days prior to the landfall of Typhoon Odette, organized by the Division DRRM Coordinator (Interviewee 3), raised their awareness and level of preparedness. As this school is designated as an evacuation shelter for the community, the PDRA triggered the school personnel's discussion on how the school will be arranged for evacuees. The school was also advised to prune trees for possible wind damage, but this did not occur due to time constraints and the consequential absence of the local government's permission, as well as the unavailability of technical staff to trim branches.

School B's plan for December 17, the next day of Typhoon Odette's landfall, was a Christmas party and team-building among the personnel. Given this and the calm weather observed one and two days prior to the typhoon's arrival, the school personnel focused more on personal preparedness. Interviewees disclosed that "there were two teachers actually that went here in the school because they thought the party was still a go" while the School Head and another teacher visited the homes of the personnel one by one on December 17 after the typhoon's eye passed through.

School C, according to Interviewee 7, commenced their preparedness for Typhoon Odette's arrival two weeks prior. The School Head organized a meeting to discuss the potential hazard and gave the personnel roles such as preparing the classrooms for evacuees and communicating with the learners' families. Interviewee 7 emphasized that his personal preparedness was concurrently ongoing and his family planned for an evacuation once he returned home after completing his duties as the School DRRM Coordinator in the afternoon of December 16.

19.4.4 Internal Processes as Education Personnel

While internal processes leading to action or inaction by the interviewees were not explicitly discussed, the following key concepts relevant to threat perceptions, protective action perceptions, reference group perceptions, fear, and milling, were identified.

Threat perceptions for School A and C took place when the schools' respective "higher-ups" (i.e., SDO and/or District DRRM Coordinators, and School Heads) reached the school personnel with updates on Typhoon Odette. School A became aware of the typhoon threat when the PDRA session was organized across Talisay City. School C became alert of the typhoon two weeks before the landfall. School B was made alert on December 15, one day before the typhoon's landfall, when respective teachers received news updates on the typhoon. Fears were not identified among the respondents concerning the schools' safety, and protective behavior perceptions in relation to the schools were not discussed. The respondents respectively commented on the fear they experienced, though, which led to personal protective behaviors that were evacuation to a safer place.

Reference group perceptions were facilitated through verbal discussions among the school personnel (Schools B and C), and group chats on Messenger (School A). In this scenario, respective schools' internal networks served as social capital for the personnel in taking protective behaviors for their schools. The communications among the school personnel also promoted milling, redefining the situation. School A communicated with the barangay for the opening of evacuation centers on campus, and friends, families and colleagues for the potential impacts of Typhoon Odette. School C engaged in the preparation of the school for evacuees, including providing information to the neighboring residents and guiding learners' families on preparedness. Interviewee 7 described this as collaboration with his colleagues and stated that no teacher was reluctant to help this operation because "we were all after the safety of the people in our community." School B's interviewee 5 reflected on the impact of support around the school in implementing safety initiatives: "We will be encouraged to do something if the people around us will support us," "especially from the higher level," and "the thing that can discourage us (is) if we are left behind to do things on our own without them."

Communications on group chats were no longer usable when phone signals and internet connections were cut off due to strong winds and recovery operations afterwards. This meant that the school personnel's reference groups' perceptions and redefining the situations were no longer practicable. None of the interviewees or their school colleagues went to their schools from the evening of December 16 till the coming morning of December 17.

All the schools interviewed for this research were used as evacuation centers by their neighbors, which the local barangays took charge of. While the levels of the barangay involvements and the time they commenced their operations varied across the three schools, institutional arrangements made between DepEd and local DRRM Offices eased the burden of school personnel in response to Typhoon Odette. Prior preparation of the schools for evacuation center operations and defined roles of school personnel (collecting data of the occupants, communicating the rules of the use of the classrooms, and completing Rapid Assessment of Damage Report (RADaR) with data on the number of the evacuees and immediate damage assessments) meant that the school personnel were not required to be on-site and take charge in responding to evacuees amidst the calamity. Institutional arrangements also include the SDO and/or DRRM Coordinators' information sharing and chain of command via Messenger with School DRRM Coordinators, DepEd instruments (i.e., Department Order no. 033, s. 2021; and Department Memorandum no. 244), and established understanding about the rules of the use of the school facilities. These institutional arrangements can influence the perceptions of protective behaviors: knowing what actions are expected as a School DRRM Coordinator within what timeframe (before the onset of the hazard until the handover of the school facilities to the barangay).

Lastly, School B, where Interviewees 4 and 5 belong, is characterized as the least involved in the school's function as an evacuation center, among the three interview respondent schools. The school did not need to spend much time on the preparation of their classrooms for evacuees as a result of their regular safety inspection of the facilities. This inspection activity was first initiated as part of the school's participation in SEEDS Asia's DRRM Project for the purpose of identifying risks on campus and leading them to improve the learning environment. The school continued this activity and engaged in school improvement, which includes the removal of flower beds along a corridor to smoothen the evacuation of learners; building one additional gate to ensure that a large number of learners could safely evacuate the school grounds; and constructing retention walls to support classroom buildings to mitigate the impact of soil erosion. The school's safety inspection functioned as an advocacy tool for these improvements and led to the decision by the local government that some of their classrooms were not safe for community evacuees. Additionally, other classrooms were also deemed unsuitable for an evacuation center due to their surrounding tall trees; therefore, in Uling Elementary School, only their covered court was assessed as an appropriate evacuation center. This decision precluded the time and efforts of the school personnel and explained the reduced protective behaviors performed by School B.

19.4.5 Situational Facilitators/impediments

Situational facilitators concerning the schools' responses were not identified in the interviews. The school personnel and SDO DRRM Coordinators did not exhibit protective behavior for their schools during the onset of the typhoon. This was largely due to the unnecessity of school personnel's involvement in the operation of evacuation centers on campus.

Impediments included family situations. For instance, interviewee 1 of School A was asked by her husband not to go to school on December 15 because they have a toddler. Interviewee 2, on the other hand, stated that she does not have family responsibilities and had no issue working around school safety concerns with the expected landfall of the typhoon.

19.4.6 Knowledge, Interest, Desire, and Action

Utilizing the KIDA Tree Model, several key aspects of knowledge, interest, desire, and action were identified in these interviews.

Knowledge: School A said knowledge about documentation, including those for evacuation center operation (i.e., records of occupants and tools used in the operation), was the most useful in responding to Typhoon Odette. The school also had prepared their contingency plan for fire incidents which was also useful for typhoon response. School B had equipped themselves with a disaster response manual which was crafted in cooperation with SEEDS Asia as part of their DRRM Project. While this manual was for landslide - the most probable and impactful hazard identified by the school - the basic response actions were the same with those for typhoon. School B's Interviewee 5 stated "The preparation was there, like the one I said about preparing food and flashlights, those were the learnings from the manuals and we applied those parts." School C's Interviewee 7 commented on camp management and basic life support as the most useful training he had received, with a remark that such training should accompany refresher courses so he will remain knowledgeable.

Interest: In those interviews, no particular aspects of interest were identified that the interviewees described as something that raised their "curiosity and wanting to know more about hazards and preparedness."

Desire: Interviewee 3, the SDO DRRM Coordinator, asserted that he wants to instill "what grownups know to do" to the children. Interviewee 7 has voiced twice his willingness to help others in disaster situations, stating "if someone needs my help, I will extend my help" and "(helping neighbors is) not necessarily our role, but initiative of a person to help their neighbors especially if they need help." These remarks present their desire to take action, and strongly demonstrate their readiness to act in order to help others, and are beyond the scope of individual protective behaviors.

19.5 Discussion

This research explored DepEd DRRM Coordinators' protective behaviors concerning their respective schools before and during Typhoon Odette utilizing the working model of Protective Action Decision Making during Disaster by Fujimoto et al. (2019). While quantified variables were not used for verification, uniqueness in information acquisition (external stimuli), disaster schema, and internal processes were observed among the interviewees as schools' DRRM personnel.

19.5.1 Institutional Arrangements

Prior arrangements between DepEd (SDO and school) and local governments (City/Municipality and barangay) contributed to enhanced perceptions of protective behaviors in terms of knowing what actions to be taken as schools, and consequently to situational facilitators during the onset of the typhoon: the school personnel were able to rest assured that the evacuation center operations in their schools were handled by the barangay.

Some interviewees stated that they would prefer decision-making in relation to schools' DRRM concerns being made by their "higher-ups." This relieves the school personnel of the burden of justifying their decision or, ultimately, of internally processing information leading to protective behaviors. DepEd instruments help school personnel take necessary action to protect their schools without having to negotiate with external stakeholders such as barangays and local governments.

19.5.2 Group Chats

Throughout the decision-making by school personnel, group chats consisting of DepEd Division/District DRRM Coordinators and/or with external stakeholders (local DRRM Office and PAGASA) played a significant role. The group chats were the sources of information on the incoming typhoon and warnings, and often facilitated school personnel's precautionary measures. They were also utilized in the communications between the schools and the families of the learners. In the interviews, these group chats were identified as external stimuli (disaster information) and milling.

19.5.3 Local Hazard Knowledge

Interview respondents shared their experiences of local hazard knowledge by older people, people who witnessed Typhoon Yolanda, and fisherfolks. This helped some of them raise their awareness of the potential impact of Typhoon Odette despite the sunny and calm weather the day before the landfall and served as an external stimulus. This also can be interpreted as others' disaster schema contributing to school personnel's perceptions of threats.

19.5.4 Training Needs

This research attempted to identify the relation between SEEDS Asia's previous training and schools' protective behavior. As a result, no connection was identified between the experience of being trained through SEEDS Asia's Project and protective behaviors by school personnel in terms of schools' preparedness and response. All schools stated that they feel the need to improve their DRRM practice and that practical training would address this need. For instance, Talisay City's PDRA raised School A's risk awareness and facilitated plans for action prior to the landfall of Typhoon Odette.

The difference between the schools who had not received training in SEEDS Asia's DRRM Project (School A and C) and who had (School B) was the involvement of the school personnel in the evacuation center operations: School B's personnel were able to focus on their personal safety due to the abovementioned institutional arrangements and their practice of regular safety inspection identifying the school facilities' high risk for their use as an evacuation center. Due to the risk identification on campus and their advocacy to DepEd SDO and the local government, School B's burden of engaging in the operation of an evacuation center in their classrooms was eased.

These reflections by the interviewed schools provide an implication for future training needs: practical and scenario-based training with refresher sessions to build confidence among personnel, and spearheaded risk identification activities for pre-disaster risk mitigation. These training contents should also facilitate better understanding and usage of institutional arrangements, enhanced communications utilizing networks, and attention to local hazard knowledge. Such capacity building could foster a sense of confidence among school personnel, resulting in smooth collaborations across stakeholders to ensure the schools' disaster response and recovery.

To further verify the abovementioned discussion, more samples of interviewees is desired with quantified relevance among the variables for protective behaviors contributing to schools' safety and continued education activities.

19.6 Conclusion

This chapter explored the protective behaviors demonstrated by schools that were stricken by Typhoon Odette in 2021, reflecting on Fujimoto et al.'s protective action decision-making model (2019). As compared to personal protective behaviors, the interviewees' actions for schools' typhoon responses did not involve much of internal processing, and this is considered due to the presence of institutional arrangements within DepEd and with external stakeholders. While SEEDS Asia's training of schools on DRRM prior to the typhoon does not present to be highly relevant to the awareness and actions of the school personnel, their training needs which would develop their disaster schemas were identified for future reference for facilitating their practical knowledge and skills to take protective behavior for their schools.

In conclusion, training sessions that can be repeatedly revisited by personnel, the exploration of locally available knowledge, and risk identification before actual disasters for awareness-raising among personnel, with an emphasis on developing institutional arrangements within DepEd and with external stakeholders were identified as key factors that can encourage school personnel to take proactive and protective actions when a disaster strikes.

References

Dasci Sonmez E, Gokmenoglu T (2023) Understanding the teachers' disaster preparedness beliefs. Intern J Disast Risk Red 85:103511. https://doi.org/10.1016/j.ijdrr.2022.103511

Dela Cruz E, Lema K (2021) Philippines starts to reopen schools after 20-month coronavirus closure. In: Reuters. https://www.reuters.com/world/asia-pacific/philippines-starts-reopen-schools-after-20-month-coronavirus-closure-2021-11-15/. Accessed 27 Dec 2023

Department of Education (2023) DepEd emphasizes bayanihan spirit in simpler Brigada Eskwela 2023 launch. In DepEd website. Available via DepEd. https://www.deped.gov.ph/2023/08/07/deped-emphasizes-bayanihan-spirit-in-simpler-brigada-eskwela-2023-launch/. Accessed 08 Dec 2023

Department of Education (2021) STY-Odette: Consolidated RADaR-1–2:20220126 1200H-1. Philippines

Department of Education Disaster Risk Reduction and Management Service (2022) Typhoon "Odette" Situation Report (SitRep) No. 52 as of March 7, 2022, 1200H (PARTIAL). Philippines

Fishbein M, Ajzen I (1975) Belief, attitude, intention, and behavior: An introduction to theory and research. Reading, MA

Fujimoto S, Kawami F, Kamei T, Tokunaga K, Mitani Y, Tatsuki S (2019) 災害時の防護意思決定構造の理論モデル化とその実証的検討: 大分県 3 市における土砂災害に関する社会調査データへの構造方程式モデリングの適用 (Theoretical model construction and empirical verification of protective action decision making during disaster: Structural equasion modelling of the social survey on landslide disasters in Oita Prefecture). 地域安全学会論文集 (J Soc Saf Sci) 35:305–315

Holden WN, Marshall SJ (2018) Climate change and typhoons in the Philippines: extreme weather events in the anthropocene. Integ Dis Sci Manag 407–421. https://doi.org/10.1016/b978-0-12-812056-9.00024-5

Inal E, Dogan N (2018) Improvement of general disaster preparedness belief scale based on health belief model. Prehosp Disaster Med 33:627–636. https://doi.org/10.1017/s1049023x18001012

International Federation of Red Cross and Red Crescent Societies (2022) Operation update: Philippines – Typhoon Rai (Odette) (MDRPH045). In IFRC website. Available via IFRC. https://www.ifrc.org/appeals?date_from=&date_to=&appeal_code=MDRPH045&text=. Accessed 08 December 2023

Lindell MK, Perry RW (1993) Risk area residents' changing perceptions of volcano hazard at Mt. St. Helens. In: Nemec J, Nigg JM, Siccardi F (eds) Prediction and Perception of Natural Hazards. Advances in Natural and Technological Hazards Research, vol 2. Springer, Dordrecht

Lindell MK, Perry RW (2003) Communicating environmental risk in multiethnic communities. Sage Publications

Lindell MK, Perry RW (2012) The protective action decision model: theoretical modifications and additional evidence. Risk Analy Intern J 32(4):616–632

Mileti DS, Sorensen JH (1990) Communication of emergency public warnings: a social science perspective and state-of-the-art assessment (No. ORNL-6609). Oak Ridge National Lab., TN, USA

Mines and Geosciences Bureau Regional Office (n.d.) Landslide and flood susceptibility map of Cebu Province, Philippines. Available at: https://r7.mgb.gov.ph/geohazard-maps/ (Accessed: 12 January 2024)

Najafi M, Ardalan A, Akbarisari A, Noorbala AA, Elmi H (2017) The theory of planned behavior and disaster preparedness. PLoS Currents, 9.

National Disaster Risk Reduction and management Council (2014) NDRRMC Update: SitRep No. 108 Effects of Typhoon "Yolanda" (Haiyan)

Niesser U (1976) Cognition and reality: Principles and implications of cognitive psychology

Roland HS (1924) Retail advertising and selling: Advertising, merchandise display, sales-planning, salesmanship, turnover and profit-figuring in modern retailing, including "principles of typography as applied to retail advertising" McGraw-Hill Book Company, Inc., USA

SEEDS Asia (15 January 2022) 台風ライ/オデット被災地支援報告書 (Report on the Support for Typhoon Rai/Odette-affected Areas). SEEDS Asia. Available at: https://www.seedsasia.org/philippines/philippines-02-report/11406 (Accessed: 12 January 2024)

Shaw R, Takeuchi Y, Shiwaku K, Gwee QR, Yang B (2009) 1-2-3 of disaster education. Kyoto University, Japan, European Union/United Nations International Strategy for Disaster Reduction (UN/ISDR)

Tatsuki S, Hayashi H, Zoleta-Nantes DB, Banba M, Hasegawa K, Tamura K (2004) The impact of risk perception, disaster schema, resources, intention, attitude, and norms upon risk aversive behavior among Marikina city residents: structural equation modeling with latent variables. In: Proceedings of Asia Conference on Earthquake Engineering March 2004, vol 1, pp 267–272

Vinnell LJ, Milfont TL, McClure J (2021) Why do people prepare for natural hazards? Developing and testing a Theory of Planned Behaviour approach. Curr Res Ecol Soc Psychol 2:100011

Chapter 20
An Approach Towards Enhancing the Capacity Development of Disaster and Climate Risk Education in Pre-service Teacher Training

Tomonori Ichinose

Abstract Following the seismic event of the 2011 Great East Japan Earthquake, Okawa Elementary School in Miyagi Prefecture succumbed to a tsunami, resulting in the unfortunate loss of seventy-four students and ten teachers. Arising from the insights gleaned from this tragic incident, a proposal emerged advocating for the inclusion of Disaster Prevention Education, aligning with the mission of preserving children's lives, as a mandatory component in the curriculum of teacher training institutions. The 2017 Teaching Curriculum Core Curriculum introduced elements like "adequate measures for ensuring school safety" in Japan. While acknowledging the pressing need to address Climate Change, it cannot be definitively asserted that the domains of school safety, climate risk, and Disaster Prevention are inherently interconnected. In this chapter, drawing upon practical experience in teacher training courses and an analysis of curriculum guidelines, we delve into the interrelation between Disaster Prevention Education and Climate Change Education. Additionally, this paper explores the essential skills and competencies educators must possess to realize this objective. Our examination extends to determining whether there presently exists a connection between awareness of Climate Change Education and Disaster Prevention Education in elementary and lower secondary schools. Subsequently, this chapter deliberates on strategies for integrating Climate Change Education and Disaster Prevention Education within teacher training.

Keywords Disaster and climate risk education · Pre-service teacher training · Course guide · Capacity development

T. Ichinose (✉)
Department of Education, National University Corporation, Miyagi University of Education, Sendai, Japan
e-mail: ichinose@staff.miyakyo-u.ac.jp

A. Yildiz and R. Shaw (eds.), *Disaster and Climate Risk Education*, Disaster Risk Reduction, https://doi.org/10.1007/978-981-97-5987-3_20

20.1 Background

In the aftermath of the 2011 Great East Japan Earthquake, Okawa Elementary School in Miyagi Prefecture was inundated by a tsunami, resulting in the lamentable loss of seventy-four students and ten teachers, some of whom perished while others remain missing.

Derived from the lessons learned from this distressing incident, it is recommended that the Ministry of Education, Culture, Sports, Science, and Technology (MEXT), in collaboration with each teacher training university, incorporate education on the environment and Disaster Prevention as a compulsory subject within the structure of basic education or liberal arts education in teacher training programs.

In the 2017 core curriculum for teachers in Japan, the mandatory subjects of "Collaboration between schools and local communities" and "Response to school safety" were introduced. These subjects recognize the potential for parents, families, and local communities to align their goals and address school-related issues, allowing for the accurate and flexible incorporation of local needs into school educational policies and activities.

While the inclusion of these subjects is deemed effective in advancing disaster prevention education, it is noteworthy that the core curriculum of teacher training courses does not explicitly address climate risks in the context of school safety issues.

The incidence of disasters attributable to climate change has been on the rise. According to Munich Re, Geo Risks Research (2019), an analysis of changing or emerging risks from natural hazards, including those arising from climate change and natural climate variability, reveals a noteworthy increase in climatological events such as extreme temperatures, drought, wildfire, meteorological events including tropical storms, extratropical storms, and local windstorms, as well as hydrological events like river floods, flash floods, and storm surges, especially over the past two decades (Fig. 20.1).

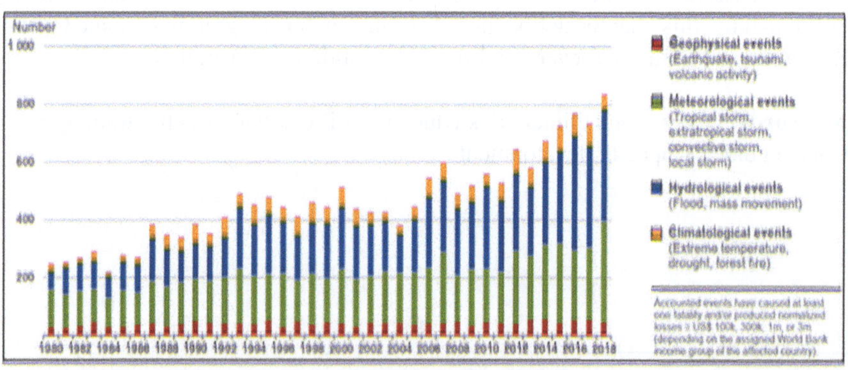

Fig. 20.1 Number of world natural catastrophes (1980–2018). *Source* © 2019 Munich Re, Geo Risks Research, NatCatSERVICE. As of March 2019

The importance of adapting to disasters and climate change has been a subject of deliberation over the past decade. The Intergovernmental Panel on Climate Change (IPCC) reported in 2012 a rise in economic losses attributed to climate- and weather-related extreme events since the latter part of the twentieth century, exhibiting significant variations across regions and years. This increase is primarily ascribed to the heightened exposure of people and economic assets.

ADPC (2013) serves as a guide on integrating Disaster Risk Management (DRM) with Climate Change adaptation, offering advice to DRM practitioners on incorporating DRM into pivotal adaptation processes: policies, adaptation strategies, and adaptation projects.

Schipper et al. (2016) argue that discussions involving disasters, development, and Climate Change, encompassing both adaptation and mitigation, can no longer be considered in isolation. Although these areas require distinct approaches in policy and practice, there is an increasing need for enhanced synergy and overlap to optimize effectiveness in each domain.

As a result, this study seeks to examine the current connection between awareness of Climate Change Education and Disaster Prevention Education in elementary and lower secondary schools. The examination revolves around three key inquiries: First, is there a discernible correlation between Disaster Prevention and Climate Change in school curricula and textbooks? Should such a correlation be identified, what specific elements within school curricula and textbooks are associated with Disaster Prevention and Climate Change? In the event of an absence of connection between Disaster Prevention and Climate Change in school curricula and textbooks, what factors contribute to this lack of correlation? Furthermore, the study will explore strategies to integrate Climate Change issues into the education of disaster risk reduction within pre-service teacher training programs.

20.2 Literature Review

Climate Change Education has progressively advanced through international discourse, epitomized by Article 6 of the United Nations Framework Convention on Climate Change (UNFCCC) ratified on May 9, 1992. The UNFCCC, in collaboration with UNESCO and a consortium of other UN agencies, advocates for Action for Climate Empowerment (ACE), emphasizing six priority domains, including education and training. Article 12 of the Paris Agreement, formulated during COP 20 in Lima in December 2014, and the Doha Work Program on Article 6 of the Convention, leading to the comprehensive review of the Doha Work Program in 2020, further enrich the overarching framework. Under these agreements, notable initiatives for Climate Change Education (CCE) teaching practices have emerged both locally and internationally, as illustrated in Table 20.1.

Conducting a literature review focused on the capacity development of Disaster and Climate Risk Education in pre-service teacher training, Mavuso et al. (2022)

Table 20.1 Representative CCE International practices

UN CC: Learn	2011^	Thirteen National Climate Change Learning Strategies have been initiated, accompanied by a Regional Climate Change Learning Strategy specific to Central America. More than 25 learning initiatives have been executed
UNESCO Getting Climate Change Ready	2017^2019	Fifty-five schools across 12 countries, all of schools are affiliated with UNESCO's Associated Schools Project Network (ASP net)
OECD Global teaching insight climate action	2021^	In total, approximately 850 educators actively participated in this endeavor, with involvement from over 6,500 visitors spanning 157 countries

employed an approach exploring Climate Change Education and strategies for educators to integrate it into classroom lessons. The study delineates teaching strategies encompassing transdisciplinary, critical thinking, and problem-based approaches to facilitate effective teaching and learning of Climate Change Education. This research serves as a resource for policymakers, aiding the design of pedagogical approaches that align with the curriculum and enhance local relevance.

In a study by Majid et al. (2023), data from 71 preservice teachers across three Initial Teacher Education institutions in England revealed a lack of key subject knowledge in Climate Change and Sustainability Education (CCSE). The findings underscore the need for further efforts to embed CCSE in Initial Teacher Education courses across England. The suggested CCSE framework could serve as a crucial national document to guide ITE institutions in conceptualizing CCSE teaching and learning within their programs.

Rushton et al. (2023) discovered that young people primarily focus on content knowledge about Climate Change, including global warming and associated negative impacts. Climate Change and environmental education have a limited presence in education policy-making in England, as evidenced by their absence in inspection frameworks and recent teacher education policies. The secondary national curriculum includes Climate Change and environmental topics in subjects like design and technology, geography, and science.

In discussing curriculum development, Catallo et al. (2022) underscored the significance of earnestly addressing Climate Change and ecological crises to facilitate students' comprehension and engagement with these matters. In their analysis of the updated national curriculum (2014/2017) of England as a reference point, they highlighted disciplines such as Science, Geography, Design and Technology (D&T), Art and Design, History, English, Religious Studies, Citizenship, and Personal, Social, Health, and Economic Education (PSHE).

Opuni-Frimpong (2022) demonstrated that teachers' readiness to educate about Climate Change was influenced by the subjects they taught in Ghana. Non-science-based subjects provided limited information on Climate Change. The study suggests an interdisciplinary approach to teaching Climate Change and emphasizes the potential benefits of in-service training for teachers.

Based on the aforementioned research, it is apparent that Climate Change Education has not been integrated into the Initial Teacher Training Core Content Framework. To address this, it is recommended that Climate Change be introduced across subjects, including design and technology, geography, and science. Transdisciplinary, critical thinking and problem-based approaches can be employed for effective teaching and learning in Climate Change Education.

In relation to Disaster Risk Reduction Education, Selby and Kagawa (2015) conducted a survey across 30 countries to identify the integration of disaster risk reduction components within their educational frameworks. The study highlights the prevalence of subjects mentioned in disaster risk reduction curricula. Findings reveal that Natural Sciences are frequently referenced, alongside Social Sciences/Studies, Geography, and Language Studies. Türker and Sözcü (2021) aimed to determine the natural disaster literacy levels of pre-service teachers specializing in natural geography in Turkiye. Results showed high general natural disaster literacy but moderate levels in the behavioral dimension. Recommendations include incorporating subjects and practices on natural disasters at all education levels and enhancing information on natural disasters in geography courses.

Şeyihoğlu et al. (2021) demonstrated that the Interdisciplinary Disaster Education Program (IDEP) statistically improved teachers' natural disaster literacy in Turkiye. Different instructional activities related to natural disasters contributed significantly to professional and personal knowledge, particularly in disaster management.

Kawasaki et al. (2022) emphasized the importance of developed training to enhance aspiring Japanese teachers' understanding of their responsibilities after a disaster. Despite the Ministry of Education, Culture, Sports, Science and Technology (MEXT) emphasizing the need for such subjects, a beginner-level training plan is urgently required to raise awareness among aspiring teachers about post-disaster school situations.

While previous research highlighted the significance of Disaster Risk issues and effective methods, such as the Interdisciplinary Disaster Education Program, no documented connection between Disaster Prevention and Climate Change was identified. This underscores the need for discussions in the realms of school education and teacher training to encompass both Disaster Prevention and Climate Change.

20.3 Research Question and Methods

Japan is considered an advanced country in integrating Disaster Risk Reduction (DRR) into school curricula, with the inclusion of Disaster Prevention as a compulsory subject in the 2017 Core Curriculum for Teachers. This study focuses on exploring the incorporation of Climate Change issues into Disaster Risk Reduction Education, particularly in the context of the Japanese DRR curriculum. Given that pre-service teacher training in Japan primarily focuses on compulsory education, specifically at the elementary and lower secondary levels, this research concentrates on investigating the curriculum and textbooks for these educational stages.

The research questions are formulated as follows:

RQ1: Is there a correlation between Disaster Prevention and Climate Change in school curricula and textbooks?

RQ2: If such a correlation exists, what specific elements are related to Disaster Prevention and Climate Change in school curricula and textbooks?

RQ3: In the absence of a connection between Disaster Prevention and Climate Change in school curricula and textbooks, what are the reasons for this lack of correlation?

To address these questions, the research utilizes the following materials:

- Japanese curriculum course guide revised in 2017, as proposed by the Ministry of Education, Culture, Sports, Science and Technology (MEXT)
- Textbooks for elementary and lower secondary schools from three different publishers

The research methodology comprises the following steps: Firstly, conducting a search for terms associated with Disaster Prevention and Climate Change within the national curriculum guidelines. Subsequently, scrutinizing the treatment of Disaster Prevention and Climate Change as study units. Secondly, evaluating how Disaster Risk Reduction education and Climate Change Education are presented as study units in textbooks. The objective is to ascertain whether there is an interconnection between Disaster Risk Reduction education and Climate Change Education.

20.4 Findings

Table 20.2 illustrates the descriptions of Disaster Prevention in the Couse Guide of the MEXT and the disasters referred to in the textbook. Table 20.3 illustrates the descriptions of Climate Change issues in the course guide and textbooks.

Within Japanese elementary school curricula, subjects such as Social Studies, Natural Science, and Special Activities are integrated. Geography, Civics, Natural Science, Health, Home Economics, and Special Activities are part of the lower secondary curriculum. As Selby and Kagawa (2015) suggested, Disaster Prevention is intricately linked to Social Science and Natural Science. The elementary curriculum intersects with the recognition of local community and national landscape topics in social studies, as well as running water and volcanic activities in natural science. In the lower secondary curriculum, Disaster Prevention is integrated with the investigation of the local area in Geography, Modern Society in Civics, and Nature and humans in Natural Science. Aside from Social Science and Natural Science subjects, issues related to disasters are addressed in Health and Home Economics, focusing on injury prevention and safe living.

Regarding Climate Change issues, it is discussed in Social Science as an international cooperation matter. However, the combination of Disaster Prevention and

Table 20.2 Disaster related descriptions

Subject and grade	Description of course guide	Disasters referred in the textbook
Elementary social studies 4th grade	2(3) Disaster Prevention Activities: There is an established emergency response system facilitated through collaboration among relevant organizations Recognition of the collaborative efforts of local organizations and individuals in addressing various aspects of natural disasters and their preparations for future occurrences is essential	Earthquake Tsunami Wind and flood damage Volcanic disaster Snow damage
Elementary social studies 5th grade	2(5) Japan's Natural Environment and People's Lives: Natural disasters are linked to the inherent natural conditions of the country. Both national and prefectural governments are engaged in diverse initiatives and projects aimed at preserving the nation and safeguarding lives from such disasters	Earthquake Tsunami Wind and flood Volcanic disaster Snow damage
Elementary natural science 5th grade	2.Life and Earth: (3) Running Water: The velocity and volume of water flow vary based on precipitation patterns, leading to significant alterations in the appearance of the land due to increased water	Flood
Elementary natural science 6th grade	2.Life and Earth: (4) Land Formation and Change: Students will gain the ability to observe and explore the land and its contents, investigate the structure and formation of the land, and formulate insights into the configuration and transformations of the land	Volcanic disaster
Elementary special activities	2(3) Health Safety/Physical Education Events: Promoting the healthy development of the mind and body, maintaining and enhancing health, learning secure behavior, fostering disciplined group behavior for self-protection from incidents, accidents, disasters, etc., instilling familiarity with exercise, and cultivating a sense of responsibility	Disaster prevention drills
Lower secondary geography	2(c) Different Regions in Japan: Comprehend the distinct features of Japan's natural environment, considering the topography, climate, the nation's ocean-surrounded landscape, and factors such as natural disasters and corresponding prevention strategies	Local disaster issues

(continued)

Table 20.2 (continued)

Subject and grade	Description of course guide	Disasters referred in the textbook
Lower secondary natural science [Second field]	(4) Weather and Its Changes: Blessings of Nature and Weather Disasters: Explore the benefits and meteorological adversities resulting from weather phenomena. Understand these aspects in the context of changes and Japan's weather	Flood
Lower secondary health	2(3) Through activities designed to identify and address issues related to injury prevention, students should acquire the following skills: A. Recognize that injuries from traffic accidents, natural disasters, etc., are influenced by both human and environmental factors C. Acknowledge that injuries from natural disasters extend beyond the event itself and may result from secondary disasters. Preparedness and safe evacuation can prevent many injuries associated with natural disasters	Injury caused by disaster
Lower secondary home economics	(6) Housing Functions and Safe Living: A. Grasp the correlation between family life and living space, understanding the fundamental functions of a house B. Contemplate and strategize how to organize a living space with the safety of the family in mind, particularly in preparation for natural disasters	Disaster prevention
Lower secondary health	(3) Health and Safety/Physical Education Events: Promote the holistic development of the mind and body, enhance health, instill safe behavior and disciplined group conduct for self-protection from incidents, accidents, disasters, etc. Cultivate a positive attitude toward exercise, nurture a sense of responsibility and solidarity, and aim to improve physical strength	Disaster prevention drills

Climate Change is only found in Natural Science, specifically within weather, typhoons, and rainfall topics (Fig. 20.2).

These findings indicate that Climate Change issues are categorized under Meteorology. There is a challenge for students and teachers to establish connections between Disaster Risk and Climate Changes in other topics without meteorology. While the Natural Science section, particularly "Nature and its blessings and disasters" (Life and Earth in Elementary), provides an excellent opportunity to explore the familiar natural environment and local natural disasters, it is advisable to systematically include Disaster Risk and Climate Change issues within these units.

In response to Research Question 1, a correlation is identified between Disaster Prevention and Climate Change in the school curricula and textbook within the

Table 20.3 Climate change-related descriptions

Subject and grade	Description of course guide	Climate change issues
Elementary social studies 6th grade	2(3) Japan's role in the globalizing world Concentrating on collaboration and cooperation to address global challenges, we will examine the efforts of the United Nations and Japan's contributions to international cooperation. Additionally, we will explore Japan's role in the global community	Sustainability SDGs Climate Change issues
Elementary natural science 6th grade	2 B. Life and Earth Changes in Land Formation: It is essential to address alterations in weather patterns resulting from the trajectories of typhoons, the correlation between typhoons and precipitation, and the associated natural disasters	Flood Climate Change issues
Lower secondary civics	D. Issues facing us and the international community (1) Challenges confronting both our society and the global community: (a) Advancing global peace and human well-being: Recognizing the significance of economic and technical collaboration in addressing issues like global environmental concerns, resource and energy management, and poverty	Sustainability SDGs Climate Change issues
Lower secondary natural science [Second field]	2 Nature and Humanity: B. Natural Phenomena and Disasters: (a) Delving into the benefits and adversities brought about by nature, approaching them from various perspectives, and contemplating the intricate relationship between nature and humans (b) Engaging in observations and experiments to explore the local natural environment and occurrences of natural disasters Scientifically evaluating and deciding on strategies concerning the preservation of the natural environment and the utilization of science and technology	Flood Climate Change issues

domain of Natural Science. Regarding Research Question 2, Meteorology emerges as a specific element within Natural Science that serves as a link between Disaster Prevention and Climate Change in school curricula and textbooks. Turning to Research Question 3, two potential explanations for the absence of correlation are considered. Firstly, the prevalent imagery of disasters often focuses on local issues, creating a disconnect from global phenomena like Climate Change. Consequently, the topic of Climate Change may be addressed in the section of international collaboration within the realm of social science. Secondly, the prevailing imagery of disasters

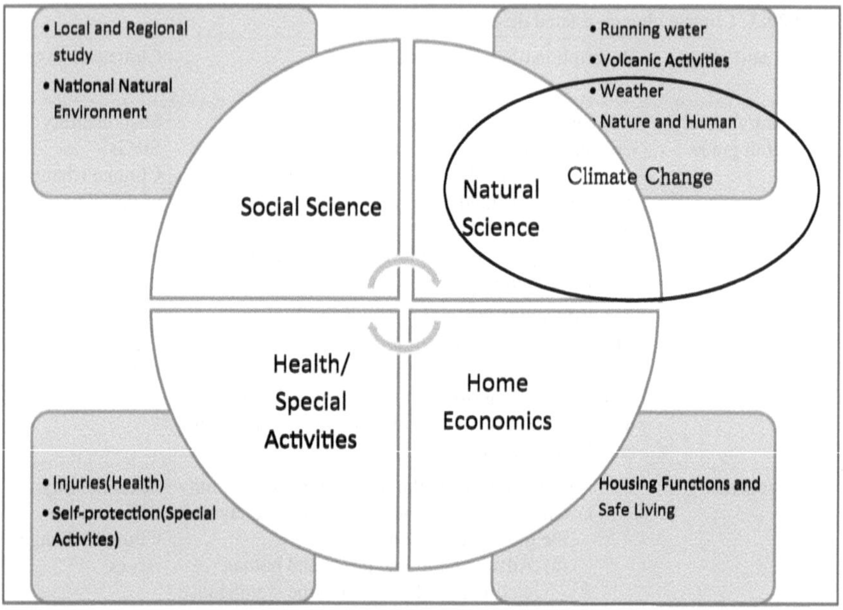

Fig. 20.2 Overlapping topics of disaster risk and climate change in Japanese compulsory education curriculum

is typically associated with meteorological events, making it challenging to establish connections with geophysical events such as earthquakes and tsunamis.

20.5 Discussion

Miyagi University of Education, serving as a pre-service and in-service teacher training institution, is proactively addressing this educational challenge. In alignment with the 2017 Core Curriculum for Teachers, the university is actively incorporating Disaster Risk and Climate Changes across different subjects, as outlined in Table 20.4, within its pre-service teacher training programs.

To assess the comprehension of Climate Change Education (CCE) among pre-service teachers, the author tasked 100 pre-service teachers with creating CCE lesson plans for practical application in schools (Fig. 20.3). The findings indicated that pre-service teachers predominantly centered their attention on elements like litter management, waste reduction, recycling, energy conservation, and environmentally-friendly practices. Particularly noteworthy was the emphasis on energy conservation, with a noticeable inclination towards directly addressing Climate Change in their teaching approaches. However, Fig. 20.2 illustrates that a mere 3 percent of the teachers acknowledged the interrelation between Climate Change issues and Disaster Risks.

Table 20.4 Disaster risks and climate change-related subject in Miyagi university of education

1. Subject "Disaster Risk Reduction in School" First-year students for the teacher training course, Compulsory, 2 credits from 2017
2. Subject "Integrated study time in School" Second-year students for the teacher training course, Compulsory, 2 credits from 2021
3. Subject "International Understanding Education" First to Third-year students for the teacher training course, Elective, 2 credits from 2015
4. Subject "Local and Global collaboration based on SDGs." First-year students for the in-service teacher training course, Compulsory, 20 h from 2020

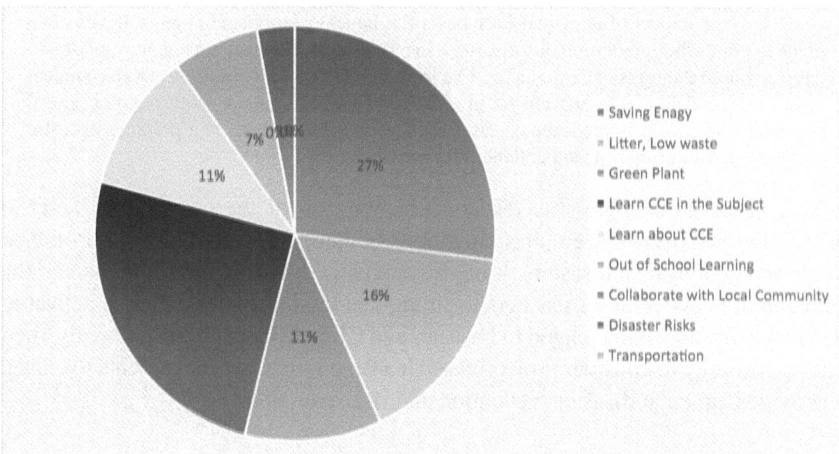

Fig. 20.3 Recognition of the pre-service teachers' learning contents of climate change education (N = 100)

To foster systematic contemplation of Disaster Prevention and Climate Change within the school curricula among students in pre-service teacher training courses, two distinct approaches are employed. The first approach involves encouraging a systematic examination of both local disasters, such as earthquakes and tsunamis, and global disasters arising from climate change. The second approach emphasizes the systematic linkage of meteorological events, earthquakes/tsunamis, and challenging connections with geophysical disasters like earthquakes/tsunamis, as well as climatological disasters such as storms, floods, forest fire, and extreme temperatures. Additionally, student teachers engage in discussions concerning the definition of disasters as outlined by the Sendai Framework for Disaster Risk Reduction (SFDRR).

As a result of this initiative, teachers exhibited heightened awareness, extending their recognition of Climate Change issues beyond typhoon-related floods to encompass other disasters like droughts, forest fires, landslides, and snow damage. Subsequent to the "Disaster Risk Reduction in School" session, students conveyed the following observations:

> In Miyagi Prefecture, disasters typically evoke thoughts of the Great East Japan Earthquake. However, I was struck by overseas forest fires. Observing animals in the forest enduring severe burns and witnessing the wind spread the damage made me realize my own vulnerability. It is crucial to contemplate actions for both domestic and international disaster scenarios and engage in preventive activities. Additionally, considering various disaster types and expanding our preparedness is imperative.
>
> Due to the impact of the Great East Japan Earthquake, earthquakes and tsunamis often come to mind when contemplating disasters. In past classes, the term 'emergency' automatically conjured thoughts of earthquakes. I've learned that disasters occur universally, emphasizing the need for increased efforts in prevention and mitigation. With ongoing global warming, previously unprecedented disasters are emerging. There is a pressing need for global information sharing and collaborative measures.

This observation highlights teachers' recognition of the connections between local and global disasters, emphasizing the systematic integration of meteorological events and geophysical disasters through systematic thinking. To further bolster this approach, it is essential for teachers to strengthen lesson studies aimed at activating the knowledge and skills related to Disaster and Climate Change. Additionally, there is an opportunity to consider project-based learning initiatives for students to engage in activities aimed at disaster prevention and risk reduction.

20.6 Conclusion

In the 2017 Core Curriculum for Teachers in Japan, the compulsory subject of "Response to school safety" was introduced. Despite this, there is a notable lack of awareness regarding Climate Change risks among both students and teachers. This research delves into the existing correlation between the awareness of Climate Change Education and Disaster Prevention Education in elementary and lower secondary schools in the Japanese curriculum. The results suggest that Climate Change issues are classified under Meteorology and are largely disconnected from Disaster Prevention.

The suggestion is to integrate Climate Change issues into discussions on disaster risks within various subjects. This includes incorporating them into conversations about local community and national landscape topics in Social Studies, as well as addressing running water and volcanic activities in Natural Science. Moreover, Catallo et al. (2022) underscored the significance of integrating Climate Change into subjects such as Language, Technology, Art and Design, Moral Studies, Citizenship, and Interdisciplinary Subject.

This chapter has introduced an approach aimed at enhancing the capacity of teachers, specifically through the utilization of systematic and creative thinking. In the context of disaster considerations, the prevailing image within the mindset of teachers and students often focuses solely on natural phenomena, neglecting the association with man-made disasters resulting from Climate Change. To overcome these conceptual barriers within the teacher training curriculum, the incorporation of both system thinking and creative thinking becomes imperative.

Facilitating Climate Change Education necessitates adopting a perspective that encompasses global disasters. Ongoing initiatives at the global and regional levels, such as UN, UNESCO and OECD, underscore the importance of engaging with these networks for more effective participation in Climate Change Education.

As per Sendai Framework for Disaster Risk Reduction (SFDRR) 15, the current framework is designed to encompass risks from small-scale and large-scale, frequent and infrequent, sudden and slow-onset disasters caused by natural or man-made hazards. This includes associated environmental, technological, and biological hazards and risks. SFDRR aims to guide the comprehensive management of disaster risk across all levels of development and within and across various sectors.

This approach seeks to eliminate conceptual barriers between natural and manmade disasters, as well as distinctions between local and global disasters.

References

Asian Disaster Preparedness Centre (ADPC) (2013) Integrating disaster risk management into Climate Change adaptation, Disaster Risk Management Practitioner's Handbook Series, Asian Disaster Preparedness Centre, Bangkok, p. 114 (accessed on 30th November 2023)

Catallo A, Lee E, Vare P (2022) Curriculum for a changing climate: a track changes review of the national curriculum for England; Teach the future: London, UK.

Intergovernmental Panel on Climate Change (2012) Managing the risks of extreme events and disasters to advance climate change adaptation. A special report of the intergovernmental panel on climate change. http://www.ipccwg2.gov/SREX/ (accessed on 30th November 2023)

Kawasaki H, Yamasaki S, Yamakido M, Murata Y (2022) Introductory disaster training for aspiring teachers: a pilot study. Sustainability 14:3492. https://doi.org/10.3390/su14063492 (accessed on 30th November 2023)

Mavuso MP, Khalo X, Kafu-Qavane BP, Olawumi KB (2022) Strategies Used by Secondary Teachers integrating Climate Change Education in their Lessons: Toward A Framework for Combating Climate Change through Education. Special Issue: Human, Nature and Society, E-Bangi J 19–3, Bangi

Majid N (2023) Reconceptualising preservice teachers' subject knowledge in climate change and sustainability education: a framework for initial teacher education from England, UK. Sustainability 2023, 15, 12237. https://doi.org/10.3390/su151612237 (accessed on 30th November 2023)

Munich Re (2019) Geo Risks Research, NatCatSERVICE. As of March 2019. https://www.iii.org/graph-archive/96425 (accessed on 30th November 2023)

Opuni-Frimpong NY, Essel HB, Opuni-Frimpong E, Obeng EA (2022) Sustainable development goal for education: teachers' perspectives on climate change education in senior high schools (SHS). Sustainability 14:8086. https://doi.org/10.3390/su14138086 (accessed on 30th November 2023)

Rushton EAC, Sharp S, Kitson A, Walshe N (2023) Reflecting on climate change education priorities in secondary schools in England: moving beyond learning about climate change to the emotions of living with climate change. Sustainability 15:6497. https://doi.org/10.3390/su15086497 (accessed on 30th November 2023)
Selby D, Kagawa F (2015) Disaster risk reduction in school curricula: case studies from thirty countries. United Nations Children Fund UNICEF.
Şeyihoğlu A, Kartal A, Tekbiyik A, Sezen VG, Birinci KK (2021) The design and implementation of a teacher training program for improving teachers' disaster literacy: Interdisciplinary disaster education program (IDEP). Probl Educ 21st Cent 79(5):781–803. https://www.scientiasocialis.lt/pec/node/1485 (accessed on 30th November 2023)
Schipper ELFF, Thomalla G, Vulturius M, Davis KJ (2016) Linking disaster risk reduction, climate change and development. Intern J Disast Resil Built Environ 7(2):Special Issue: UN Global Assessment Report 2015
The Ministry of Education, Culture, Sports, Science and Technology (MEXT) (2017) Teacher Training Core curriculum (in Japanese). https://www.mext.go.jp/b_menu/shingi/chousa/shotou/126/houkoku/1398442.htm (accessed on 30th November 2023)
Türker A, Sözcü U (2021) Examining natural disaster literacy levels of pre-service geography teachers. J Pedag Res 5(2):207-221. https://doi.org/10.33902/JPR.2021270164 (accessed on 30th November 2023)
UNESCO (2016) Getting climate-ready: A guide for schools on climate action. UNESCO, Paris
UNESCO, UNFCCC (2016) Action for climate empowerment: Guidelines for accelerating solutions through education, training, and public awareness. Paris: UNESCO, & Bonn: UNFCCC
UNESCO (2021) Getting every school climate-ready: How countries are integrating climate change issues in education. UNESCO, Paris
UNISDR (2015) Sendai framework for disaster risk reduction 2015–2030

Part V
Technology

Chapter 21
Using Virtual Reality and Augmented Reality for Enhancing School Education

Tomoki Itamiya

Abstract Virtual reality (VR) and augmented reality (AR) technologies enable people who have never been affected by a disaster to experience it as if it were real and to take concrete countermeasures. Using smartphones, tablets, and inexpensive head-mounted displays makes it possible at a relatively low cost. Virtual reality is immersive and particularly useful for experiencing earthquakes. However, creating disaster situations with computer graphics is costly and challenging to produce in a large variety. Augmented reality is cost-effective because it synthesises and displays the disaster situation at the location of the experience, including floods and fires. Disaster risk reduction education initiatives using VR and AR have been gaining momentum since around 2014; AR has also been used in disaster drills for all children in primary schools, showing its usefulness. If the issues of introduction and maintenance costs can be resolved, they are expected to become an indispensable part of disaster education in the future.

Keywords Augmented reality · Virtual reality · Evacuation drills · Training · School

21.1 Introduction

This section includes virtual reality and augmented reality for disaster management in Japan and its future development. Virtual reality (VR) allows users to experience reality in the virtual world. They do it by wearing a unique head-mounted display to see the all-around images created by CG or 360-degree cameras. The images are used in VR meetings and training. Avatars are displayed in the virtual world, on VR campuses, and in offices using similar technology. On the other hand, augmented reality (AR) allows users to experience a virtual reality overlaying the real world. The real world is "augmented" by data and images that become virtual entities. This can be explored through a smartphone or head-mounted display. A user can

T. Itamiya (✉)
Division of Dental Education, School of Dentistry, Kanagawa Dental University, Yokosuka, Japan
e-mail: itamiya@kdu.ac.jp

A. Yildiz and R. Shaw (eds.), *Disaster and Climate Risk Education*, Disaster Risk Reduction, https://doi.org/10.1007/978-981-97-5987-3_21

experience the feeling of being at the scene of a disaster from the comfort of one's home through images of the natural world and information from the virtual world. Similarly, mixed reality (MR) is a technology that blends real and virtual worlds. The AR tries to "augment" reality by displaying information from the virtual world in the real world, while MR aims to "merge" the real and virtual worlds more closely. MR does "fieldwork" technologies such as overlaying 3D images of disaster predictions on the actual construction site or recording the progress of a construction site using a 3D camera to check it remotely. Recently, these technologies have spread rapidly.

The first usage of VR in disaster management was as an experiment to alleviate PTSD among US Army soldiers in the aftermath of the 9/11 incident (Difede et al. 2007). Further, Ukawa et al. (2014) developed a smartphone tsunami AR app as a teaching tool in disaster preparedness education. The application simulates the height of the tsunami during the Great East Japan Earthquake site. It shows the icons and numbers in real time, and the Computer Graphics is used to represent the tsunami. The University of Tokyo's Institute of Industrial Science, Takaaki Kato Laboratory, and Cad Center Corporation (2014) have developed an application for visualizing disaster prevention information using a smartphone. This is based on the GPS location information of the device, and a CG image of the water surface at the assumed flood level is superimposed on the real-time image of the smartphone camera. This allows the user to understand the danger of the current location in three dimensions. Itamiya et al. (2018) developed an AR application Disaster Scope that uses a smartphone and paper goggles to create an immersive experience of a three-dimensional CG display. It superimposes floodwaters and smoke over the entire landscape.

Disaster Scope is a mixed-reality application that utilizes a smartphone and a tablet or a Head Mounted Display to display CG images of floods, smoke from fires, and other disaster situations superimposed on natural landscapes to generate an immersive experience. A smartphone and a tablet or a Head Mounted Display equipped with a sensor can accurately determine the height above the foreground and distinguish the objects around it. The flooding of a low water level of 1 m or less and the filling of a room with smoke can be achieved realistically. The system has been shown to simulate a real sense of crisis that promotes crisis awareness among students during elementary and junior high school disaster prevention events.

21.2 Virtual Reality for Disaster Risk Reduction Education

Virtual reality (VR) creates an immersive experience. For instance, aircraft pilots can use it during training and in manufacturing for design. On the other hand, traditional VR equipment is so costly that few people can experience it. However, recent developments make it more accessible due to its lower cost. Thus, this technology offers an alternative teaching tool to learn about disasters.

Fig. 21.1 A facade of a VR Tsunami driving simulator

21.2.1 Tsunamis and Flash Floods if You Drive a Car

To evacuate as quickly as possible is important in the event of a tsunami. However, evacuating by car is risky, and many people are not aware of this risk. They cannot properly conceptualize a disaster situation. A tsunami driving simulator using Oculus Rift was developed. It is a virtual tsunami disaster experience system. This is a low-cost head-mounted display (HMD). The technology was developed by our team. The system provides immersive experiences of different situations, such as being hit by water during flash floods and tsunamis while driving a car. A person wearing a head-mounted display can imagine the disaster event in the actual location where it could happen around Japan. They can see a flood and the inside of a car, which is close to reality, through three-dimensional computer graphics (3D-CG) using digital images. People have used the system in their disaster prevention exhibitions organized by their local governments. This learning experience is useful for disaster preparedness education to raise awareness about tsunamis. Figures 21.1, 21.2 and 21.3 show the experience of using the VR tsunami driving simulator.

21.2.2 Virtual Reality Tsunami Driving Simulator Evaluation

Figure 21.4 shows the comparison of using VR, the movie and the map. The participants experienced a VR tsunami driving simulator. A movie was played showing a real flood situation filmed in a car and they looked at the paper hazard map. Then, they answered a questionnaire after each session. The question was: "*Did the experience make you think of preparing during disasters?*" The experience was randomized to avoid biased results.

Fig. 21.2 A VR scene of Tsunami driving simulator

Fig. 21.3 A Tsunami driving simulator VR scene

21.2.3 Earthquake

The VR Earthquake Experience application was developed to educate people who do not have experience with a major earthquake. It aims to teach them about earthquakes. The information is based on the earthquake data released by the Japan Meteorological Agency. It can be easily used for disaster preparedness events such as evacuation drills. They can experience earthquakes by launching the application and using the controller to select the location where the past massive earthquake occurred, and its location in Japan is displayed in the VR space. The application can be used by a

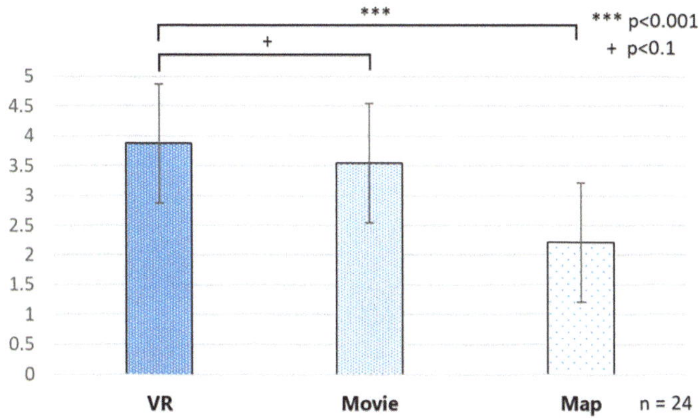

Fig. 21.4 Comparing VR and movie, map

large number of people simultaneously at a low cost. They can use Oculus Go. This is a standalone head-mounted display. This can work without the need to use a PC. Earthquake ground motions can be performed at the same location to make them distinguish different shaking motions during earthquakes. The experience of this application provides an opportunity to rethink our perspectives during evacuation and everyday activities. The application was used for disaster preparedness seminars at Nasu Junior High School in Nasu City in Tochigi Prefecture, Japan. 500 junior high school students expressed that they acquired great knowledge with the experience. Figures 21.5, 21.6, and 21.7 show the experience of earthquake VR.

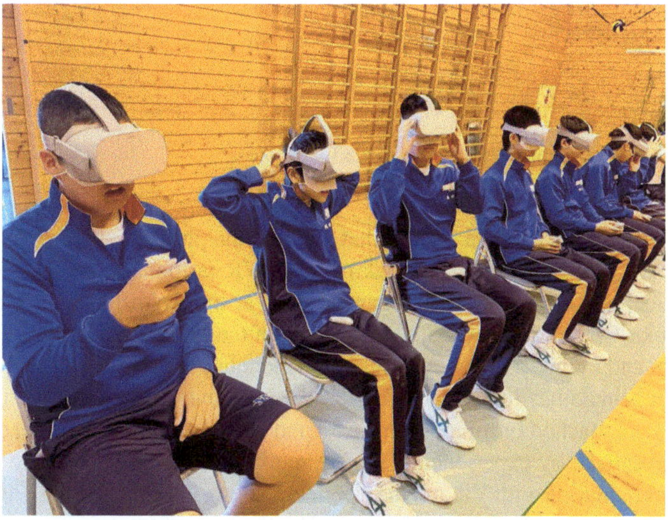

Fig. 21.5 Junior high school students exploring Earthquake VR

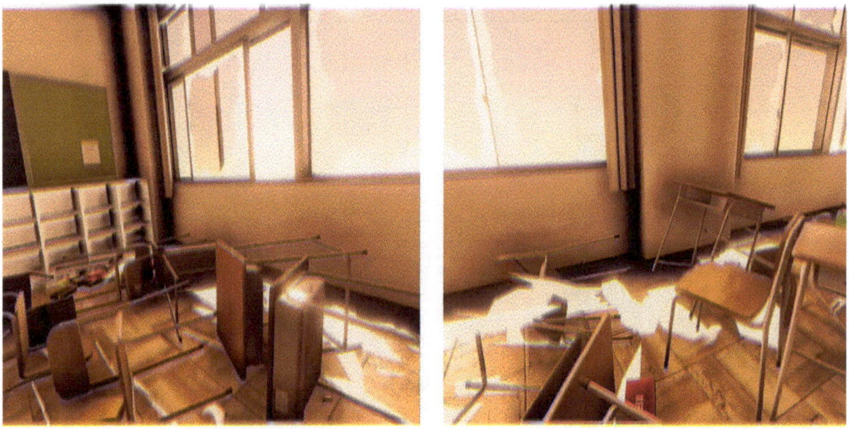

Fig. 21.6 A classroom scene of earthquake in the VR

Fig. 21.7 A kitchen scene of earthquake in the VR

We have also developed a system that can link the physical shaking of the earthquake experience vehicle with the shaking in the VR earthquake experience content, and practical use of this system began in 2023 at the Yokosuka Fire Department in Yokosuka City, Kanagawa, Japan. Similar initiatives are being taken by other companies and municipalities. Although experiencing the shaking in an earthquake simulation vehicle has long been used in disaster prevention education, the risk of injury from unattached furniture falling over and the risk of blocked evacuation routes cannot be realized by experiencing an earthquake simulation vehicle alone. Therefore, the combination with VR and the physical shaking is very effective. Figures 21.8 and 21.9 show the earthquake experience vehicle.

Fig. 21.8 The earthquake experience vehicle in Yokosuka City

Fig. 21.9 Earthquake VR in the earthquake experience vehicle

21.3 Augmented Reality for Disaster Risk Reduction Education

Augmented reality (AR) can overlay computer graphics onto a real landscape and this is what makes it different from VR. It has become popular due to the innovations being developed with smartphones. Its usage ranges from giving directions and work instructions as well as in disaster preparedness education. This technology has been utilized in recent years in Japan where natural disasters occur frequently.

Some people do not have enough awareness to evacuate quickly and safely, particularly in the Tohoku earthquake in March 2011 and the rain in Western Japan in July 2018. Numerous studies have been conducted on flood simulation (National Research Council 2000; U.S. Army 1992). However, ordinary people and young people find it hard to understand. Disaster preparedness seminars utilize hazard maps, and pictures of previous disasters are conducted to teach them of the potential risks. These activities are being held in primary and secondary schools and local government.

The seminar entails working on a task to learn about the risk. For instance, participants locate a house on the hazard map. Next, they look for the flood depth of one's homes and schools from the map. Then, they conceptualize the disaster situation. However, these learning tasks are difficult for a young student to accomplish. Thus, an alternative teaching method is necessary to enhance students' disaster preparedness skills. We have developed Disaster Scope. It is an augmented reality application for smartphone/tablet users. The users can see the danger of flooding even if the water level is less than 1 m. They can also experience how smoke fills a room in the event of a fire. This application requires only a smartphone/tablet and an inexpensive paper headset or a Head Mounted Display. It is easy to navigate and can be used by a lot of people simultaneously.

21.3.1 Application of Tsunami and Flash Floods

Computer Graphics (CG) floods or fire smoke can be displayed over the real scenery video captured by the device's camera when the Disaster Scope application is installed on a smartphone or a tablet or a Head Mounted Display(HMD). Disaster Scope is an AR disaster simulation application developed by Itamiya since 2015, characterised by its realistic CG representation that closely interferes with real space, and its ease of use that also works on smartphones. A paper headset on their smartphone should be worn by the users to have an immersive experience. They can have an interesting experience, and other people can participate simultaneously because setting it up is easy.

The surface of CG water is shown at a constant height, which can be adjusted at 10 cm, where a real-time occlusion can be done. The water facade is not displayed behind the real objects and above the set water height. It can also show the flood current and the moving objects that accompany it. It can detect the collision between CG debris and real-world objects. CG debris will bounce once it hits the real wall and people. Three types of flow speed (stable, 0.5 m/s, 1.5 m/s) can be set. This application can be connected to the server of the Japanese Ministry of Land, Infrastructure, Transport, and Tourism (MLIT). It can obtain the estimated flood depth of the current location. The user can change the water surface color from clear to brown or muddy, like in real floods. Figures 21.10, 21.11, 21.12, 21.13, 21.14, 21.15, and 21.16 show the AR experience of tsunamis and flash floods.

Fig. 21.10 Elementary school students exploring AR view of tsunami and flash floods

Fig. 21.11 A tsunami situation with floating debris AR view

Underwater expression is displayed in the smartphone when the water level is higher than its height. The water view becomes brown and cloudy. The base of the water surface can be viewed when you tilt it up. The augmented reality (AR) depth gauge and occlusion display make the height easy to understand even above the water level.

Fig. 21.12 A tsunami situation with moving debris AR view

Fig. 21.13 A tsunami situation with the AR water level gauge view

21.3.2 Heavy Rain and Storm Application

Rains and storms can be simulated in Disaster Scope. Changing the wind speed and hourly rainfall can be done by hitting the icons (shown in Fig. 21.15) on the screen. It can also display moving objects caused by disasters.

Fig. 21.14 The AR view of a clear water

Fig. 21.15 The display of interface buttons

Fig. 21.16 The underwater expression AR View

Fig. 21.17 A heavy rain and storm AR view

Fig. 21.18 A rain and storm with moving objects AR view

21.3.3 Application of Fire and Smoke

A situation is created when a fire breaks out and smoke fills the room indoors in the Fire Smoke Experience app. A smoke created using CG is overlaid on the picture of the real landscape once the user launches the application while wearing paper goggles.

The smoke in CG is dense during the standing state. This is due to the height position of the smartphone or the HMD from the ground can be accurately detected from the 3D depth sensor or the six degrees of freedom (6DoF). There is low visibility, but when the user squat, the smoke becomes thin that makes it convenient to see. Thus, a user can feel that there is little smoke near the ground. The user can see the neutral zone, which is the boundary between smoke and fresh air. The location for this application is estimated to be an accessible room with a ceiling height of

Fig. 21.19 Elementary school students exploring Fire and Smoke AR

approximately 2.8 m. The eye level of the user using the application in a standing position should be around 1.4 m from the floor or 1.4 m from the ceiling. The starting position of the neutral zone measures 0.5 m or 2.3 m from the floor and ceiling, respectively. The travel time of the smoke is determined by the combustible material and the ceiling height. The distance from the ceiling should be measured to assess the concentration of the smoke. The distance from the floor is also easier to obtain rather than from the ceiling due to the design specification of the application. The placement position of the smoke is based on the height position from the floor. This setting can be changed by the user. The combustion point can be set anywhere. The surface area of the smoke can be set to a shape that matches the actual interior of a building. The smoke display of this application is based on the author's video of the actual fire training room of the Kyoto City Firefighting Activity General Centre in Minami-Ku at Kyoto City under the guidance of the city firefighters. Figures 21.19, 21.20, 21.21 and 21.22 show the AR experience of fire and smoke.

21.3.4 Evaluation Scheme

We used the technology in the evacuation drills for 334 students at the 7th Elementary School and the Takayama Elementary School in Mitaka City in Tokyo, Japan in 2020. Students of the 7th Elementary School experienced it and the 5th-grade students of Takayama Elementary School experienced it in one day. Five sets of tablets were prepared for the tsunami flash flood application and five head-mounted displays for the smoke experience application. Five students in each group were led by teachers.

Fig. 21.20 Fire and smoke app in AR

Fig. 21.21 Fire and smoke app in AR with a low posture

Fig. 21.22 Fire and smoke app in AR with a standing state

The drill started with students experiencing a fire smoke in the hallway of the school building. Then, they transferred to the school ground and tried the flood caused by the tsunami situation. Designated timekeepers timed for 3 min per session. A survey was administered to the students, and a class led by a disaster risk reduction specialist was conducted after the session.

21.3.5 Findings

The free response questionnaire survey shows the different perspectives of the students. They mentioned "In the smoke situation, smoke is everywhere. Everything was blurred." "Water is reaching my neck", "It seems that I can feel the emotions of the people who died in the tsunami". "This activity is a good exposure. I want to act calmly during a disaster". "I was more afraid than the usual evacuation drills." "Drills were essential." "It is important to prepare for the disaster every day." "I understand it." The primary school principal mentioned that "There was an effect of thinking of oneself, thinking of ways to evacuate, this is good to use here and can be shared with other schools." At one item "Have you thought about preparing for a disaster by experiencing floods and smoke?" 70% of the 7th Elementary School and Takayama Elementary School in Mitaka City in Tokyo, Japan answered, "Very much" and 24% answered, "Quite a lot". Furthermore, student assessments were analyzed for each grade level from 1st grade, which is around 7 years old, up to 6th grade, which is 12 years old, at the 7th Elementary School in Mitaka City in Tokyo, Japan. We also practiced in many schools around Japan, where 12,000 students participated in learning the AR application. Figures 21.23 and 21.24 show the results of the AR app 1 and AR app 2 surveys, respectively.

A comparison survey was administered by 31 participants in evacuation drills and disaster prevention events in Nishio City and Toyohashi City in Aichi. The evaluation experiment that was set up has two conditions. Two kinds of tsunami hazard maps published by the said cities were constructed for the first one while the AR application to indicate flood height up to 1.2 m above the ground is put for the second one. Then, a hazard map about tsunami that has a circle marked of a 1.2-m above-ground inundation point was designated and discussed to the participants. Table 21.1 shows the AR app and Hazard map Evaluation Item.

Figure 21.25 shows the comparison between the AR app and the Hazard map. The questionnaire survey results show that the use of AR applications will benefit the improvement of awareness among citizens and young students.

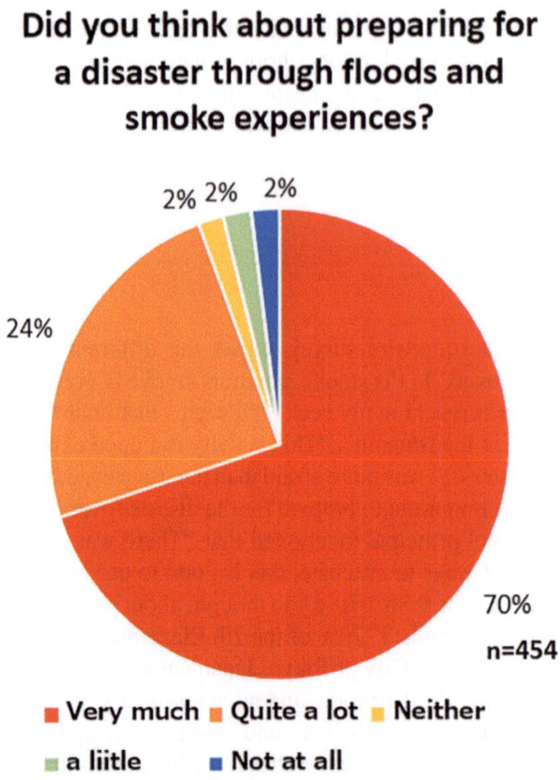

Fig. 21.23 The questionnaire survey result of the AR app 1

21.4 Discussion

VR and AR are effective tools for disaster preparedness education, but each has its pros and cons. Users can have interactive operations in VR and attain good experience and immersion but, the usage of VR requires the use of CG to create the experience site, and the costs needed are expensive. The average cost to create a new experience site desired by the user is around 10,000 USD and the average time for a user to experience VR is around 3–5 min. This limited time makes it difficult to make all the people in a disaster prevention event to explore the device. Also, hiring staff to operate and guide the users are needed, which makes the operation costly.

Similarly, developers should also consider the frame rates. When the HMD screen update speed is slow, it makes the users feel uncomfortable using VR and if the 3D-CG model has a complicated design, the frame rates tend to slow down. On the other hand, AR can overlay a CG onto a real landscape. There is no need for the user to create a CG model of the experience site like in VR, and it is cheaper than using VR. The improvements in smartphones enable the creation of AR expressions

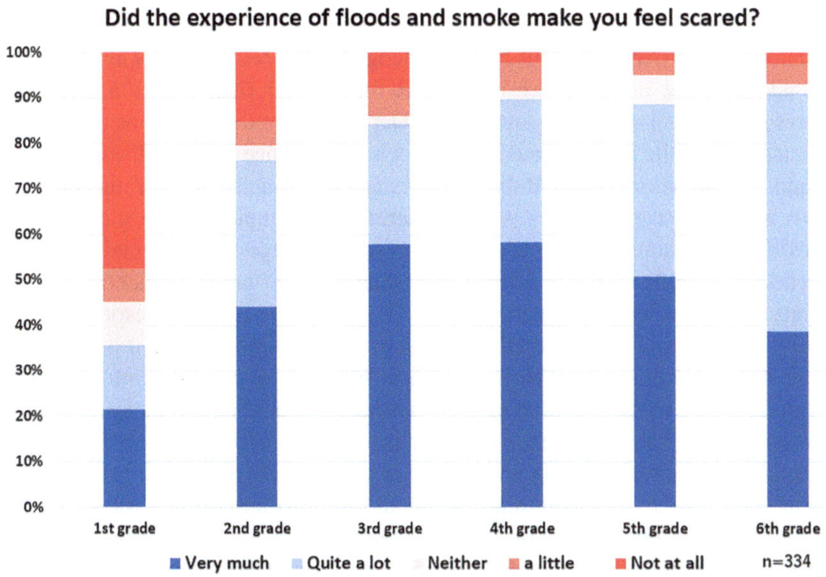

Fig. 21.24 The questionnaire survey result of the AR app 2

Table 21.1 An AR app and the Hazard map evaluation item

	Questions	Evaluation
Q1	Can you accurately imagine the depth of the flood?	5 stages from 1 to 5
Q2	Did you feel a sense of crisis due to flood depth?	5 stages from 1 to 5
Q3	In the event of an earthquake or heavy rain, will you think about evacuating yourself by thinking of tsunamis and flooding?	5 stages from 1 to 5

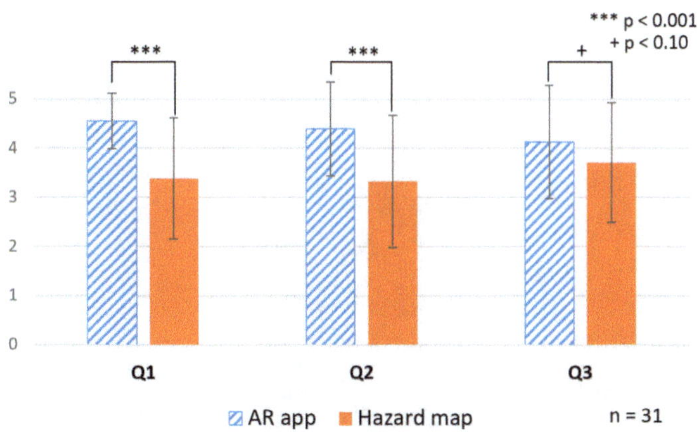

Fig. 21.25 Comparison of the AR app and the Hazard map

such as realistic disasters but the rapid development in AR requires developers to continuously improve their skills and knowledge for the best practices available.

In addition, the effective use of VR/AR for disaster risk reduction education requires operational skills and know-how. Depending on the school or region, the characteristics of the students and residents who experience VR/AR disaster education and the educational content that teachers and community disaster management experts want to convey will vary widely. Therefore, the implementing staff needs to be flexible and adaptable. The results of the survey suggest that special ingenuity is needed to provide effective educational benefits to 1st-grade elementary school students. A certain level of experience and IT skills are required to respond appropriately. In addition, for VR/AR education to be widely disseminated, it is important that the necessary operational skills and know-how can be compiled into a manual and implemented without burdening busy teachers and administrative agency staff.

Most of the companies currently producing and selling VR/AR content for disaster risk reduction education are not specialized in the field. As a result, it is difficult to have in-depth discussions with teachers about the most appropriate ways to work in the educational field and to respond immediately to software improvements. In this context, AR Bosai, a general registered association in Tokyo, was established in 2020, specializing in the production of VR/AR content for disaster prevention and the implementation of disaster risk reduction education. AR Bosai's content creators sometimes conduct disaster prevention training in schools, enabling them to achieve optimal educational effects. It is hoped that the sector will develop further in the future as more companies specialize in this field.

VR/AR can also be useful for climate change education. In particular, the AR flood experience app can be directly applied to experiential education on sea level rise. An AR sea level rise experience app developed by Itamiya is on permanent display at the Miraikan—The National Museum of Emerging Science and Innovation in Tokyo from 2023. Figure 21.26 shows the sea level rise AR educational content at the Miraikan.

Fig. 21.26 Sea level rise AR educational content at the Miraikan, Tokyo

21.5 Conclusion

The study has shown that AR and VR technologies are useful tools in teaching disaster risk reduction. Devices such as eyeglass-type devices have a potential source of information after smartphones for mainstream learning (Kelly 2016). The proliferation of AR using these devices and 5G can make real-time disaster prediction possible and be displayed on the glasses. Thus, this can help increase people's awareness about disasters and promote early evacuation, which can minimize the rise of disaster victims.

References

AR Bosai (Reference date: 11.12.2023). https://arbosai.org/

Difede J, Cukor J, Jayasinghe N, Patt I, Jedel S, Spielman L, Giosan C, Hoffman HG (2007) Virtual reality exposure therapy for the treatment of posttraumatic stress disorder following September 11, 2001. J Clin Psychiatry 68(11):1639–1647

Itamiya T, Yoshimura Y (2018) Development of immersive disaster experience smartphone-application "Disaster Scope" and utilization in evacuation drill. J Disast Inform Stud 16–2:191–198

Kato T, Kato T (2014) Location (2014), Location-based Disaster Prevention Map Application "Ten Sai! Manabu-kun," 2014 Information and Communication White Paper, p 260

Kelly K (2016) The inevitable: understanding the 12 technological forces that will shape our future. Penguin Books

Miraikan, Planetary Crisis (Reference date: 11.12.2023) https://www.miraikan.jst.go.jp/en/exhibitions/world/planetarycrisis/

National Research Council (2000) Risk analysis and uncertainty in flood damage reduction studies. National Academy Press, Washington, D.C.

Ukawa Y, Fukuchi A, Kuriki (2014) Development of a Tsunami AR application for disaster prevention education using smartphones. J Environ Educ Miyagi University of Education 16:7-12

U.S. Army Corps of Engineers Institute for Water Resources (1992) Guideline for Risk and Uncertainty Analysis in Water Resources Planning, Volume I: Principles with Technical Appendices, IWR Report, 92-R-1

Chapter 22
Community Education and Engagement for Blue-Green Infrastructure (BGI) Projects: Insights from Assam, India

Bhola Saha, Dikshya Saikia, Amrita Sabhapandit, Tim Marjoribanks, Lee Bosher, Mahua Mukherjee, Siva Ram Edupuganti, and Atul Kumar

Abstract The use of blue-green infrastructure (BGI) for addressing problems of urban flood risk is increasingly being encouraged. However, effective and appropriate implementation of BGI requires local community involvement. This research explores attitudes towards BGI within communities in Guwahati, Assam. A participatory case study method was used to identify the role of community education and engagement in enabling BGI implementation. Through community workshops, physical models and creative media were used to help educate the communities regarding BGI solutions. Feedback and insights into local knowledge were then used to develop community-led BGI solutions. This study provides important evidence of the benefits of understanding diverse stakeholder needs and in utilising a co-creation process that can contribute towards removing potential social barriers and facilitate the design of scientifically informed, locally appropriate solutions to urban flood risk.

Keywords Nature-based solutions · Blue-green infrastructure · Community engagement · Community education · Community resilience

---

B. Saha · A. Sabhapandit
Royal School of Architecture, The Assam Royal Global University, Guwahati, Assam 781035, India

D. Saikia
Jogananda Deva Satradhikar Goswami College, Bokakhat, Assam 785612, India

T. Marjoribanks
Water Engineering and Development Centre, School of Architecture, Building and Civil Engineering, Loughborough University, Loughborough LE11 3TU, England

L. Bosher (✉)
School of Business, University of Leicester, Leicester LE2 1RQ, England
e-mail: lsb23@leicester.ac.uk

M. Mukherjee · S. R. Edupuganti · A. Kumar
Department of Architecture and Planning, Indian Institute of Technology Roorkee, Roorkee, India

A. Yildiz and R. Shaw (eds.), *Disaster and Climate Risk Education*, Disaster Risk Reduction, https://doi.org/10.1007/978-981-97-5987-3_22

22.1 Introduction

Many low- and middle-income countries across the world are experiencing rapid urbanisation, which requires extensive investment in supporting infrastructure while also raising concerns about the negative impacts of such development on the natural environment. Some of the environmental degradation experienced can include precious agricultural land and dwindling wetlands being built upon for construction (Konadu and Fenner 2017). Typically, urban development over the past few decades has included the construction of solid/concrete pipe systems that are used for drainage and water management. However, it is noted that such systems are technocratic, and 'top-down' and can lead to a technical lock-in and thus reduce flexibility to modify the system performance in light of climate and development uncertainties (Konadu and Fenner 2017). It has been suggested that blue-green infrastructure (BGI) solutions can be more flexible and more sustainable and are often delivered at lower cost while generating a wide variety of benefits in addition to flood protection (Ashley et al. 2018; Konadu and Fenner 2017; Kumar et al. 2023). This chapter reports upon research conducted with a broad range of local stakeholders in Guwahati, Assam, India, with the aim of better understanding who the key stakeholders are and what their key concerns are and to utilise their local knowledge in the co-design of potential BGI measures.

22.1.1 Blue Green Infrastructure

Blue-green infrastructure (BGI) is a nature-based solution for mitigating the impacts of urbanisation and climate change on the built and natural environment. BGI is defined as an 'interconnected [strategically planned] network of natural and semi-natural elements capable of providing multiple functions and ecosystem services' (Hamel and Tan 2022). Accordingly, BGI can include green spaces, riparian and coastal vegetation, trees, wetlands, basins and engineered systems such as bio-retention ponds or green roofs (Mishra et al. 2021). In addition to providing a means for coping with extreme weather conditions/floods (Sánchez and Govindarajulu 2023), BGI has become a popular solution because of the multiple ecosystem services and benefits it can produce. For example, BGI can lead to water quality improvement, climate change mitigation, temperature reduction, habitat improvement, and carbon sequestration (Ashley et al. 2018; Mukherjee et al. 2022). Beyond physical and ecosystem benefits, BGI can produce social benefits too, such as energy-saving, reduction in noise pollution (Liu et al. 2021), and improvements in health and well-being both directly and also through increasing recreational activities like jogging, picnicking, observing nature, bench sitting etc. (Sadeghi and Bin 2023). Finally, community cohesion through BGI implementation can result in increased interaction between residents and lead to the development of informal relationships among neighbours that promote a cohesive human society (Mills and Waite 2018).

Successful implementation of BGI for flood risk management requires community participation and an understanding of the local context (Staddon et al. 2018; Drosou et al. 2019). In the global South, understanding local factors like climate and its changes, long-term maintenance of possible solutions, integration with grey infrastructure and considerations of BGI becomes very important and are generally not studied in the Global North which leaves a knowledge gap about the potential benefits of BGI (Hamel and Tan 2022). Given the ecological similarities within countries, mainly climate and vegetation, there are ample opportunities for knowledge transfer within the Southeast Asian ecological context (Hamel and Tan 2022). However, a paucity of studies in the Indian context leaves stakeholders with a limited understanding of how BGI can be instrumental in reshaping urban systems (Sánchez and Govindarajulu 2023).

22.1.2 Working with Local Stakeholders

To be sustainable, BGI cannot be limited to the efforts that need to be performed by national, regional, or municipal governments; instead, the community must be involved in new stewardship practices (Soetanto et al. 2022). Community education and engagement is necessary to bring about behavioural changes in the community, and this is frequently accomplished by highlighting the possible advantages and inspiring communities to embrace and value BGI (Lamond and Everett 2019). It has been found that community engagement is an efficient strategy for promoting well-being and reducing inequalities (Attree et al. 2011). In addition to this, community members experience positive benefits, in terms of self-confidence, and self-esteem; which results in better physical and emotional health, and improved social relationships and individual empowerment (Attree et al. 2011). Thus, the involvement of the community as an active stakeholder is a prudent decision in the discourse of BGI as well as broader disaster risk management (Drosou et al. 2019).

Functionally, effective environmental management using BGI requires informed engagement from relevant practitioners, governmental agencies and local-level representatives (i.e., community groups) during the planning, design, installation and maintenance stages (Bosher and Chmutina 2017). Hence, there is a need to establish a multidisciplinary, research-informed approach to BGI which addresses a myriad of influencing factors, is technically desirable, and is culturally feasible within the local context (UK-GBC 2015; Bozovic et al. 2017). Therefore, community education and relevant (non-tokenistic) engagement are considered vital to the success of BGI. Accordingly, this chapter reflects upon the process and most relevant approaches to working with local stakeholders so that their local knowledge can be utilised to help co-create BGI solutions that are appropriate to their needs and local context.

22.2 Methodology

22.2.1 SEMP: Stakeholder Engagement Management Plan

For the development of a coherent BGI strategy, it is necessary to co-produce outputs that can operate within 'top-down' derived processes and frameworks (working with city-level authorities) while at the same time ensuring 'bottom-up' capacities (working directly with the community), that can be nurtured for structural and non-structural BGI solutions (Drosou et al. 2019). 'Stakeholder engagement' refers to the method by which a project or organisation involves those who could be impacted by the decisions being taken or can influence how those decisions are carried out. From the early stages of the case study, stakeholders[1] may be informed about the stakeholder engagement workshops that aim to address key information and coordination requirements of the key stakeholders. In addition to maintaining close collaboration among partners, the workshops should be designed to enable the initial dissemination of findings and refinement of the later impact-focused outputs. During the workshops, stakeholders cannot be treated as passive recipients of 'research' but should be brought into the idea generation process itself, co-producing outcomes.

A Stakeholder Engagement Management Plan (SEMP)[2] can be established that provides a framework for the following stages:

1. **General engagement, stakeholder identification and mapping exercise**, including explaining/demonstrating what BGI is (i.e., using models and storytelling) and to ensure that a suitably relevant and broad range of stakeholders were being included in the study;
2. **Diagnostic phase (1)**: Research activities that will build on existing work and networks held by local academic partners and assess key problems across governance scales, identifying the needs of stakeholders and communities with the aim of establishing a researcher–community–stakeholder partnership with responsibility for the co-production of the research project outputs.
3. **Creative phase (2)**: Research activities, including workshops designed to facilitate creative thinking and the co-production of suitable BGI measures (from a broad range of stakeholder needs and perspectives).
4. **Legacy phase (3)**: Multi-stakeholder scenario-based workshops towards the end of the project will co-design and evaluate outputs (i.e. papers, posters, drawings, comics, policy briefs) and validate the BGI solutions that might be adopted by the local stakeholders.

This type of strategic intervention, as shown in Fig. 22.1, can be implemented at the local level so that an effective method of addressing the engagement of stakeholders and co-production of solutions for resolving urban climate issues through BGI can

[1] For this study, a 'Stakeholder' was anybody who can affect or is affected by an organisation, strategy or project. They can be internal or external and they can be at senior or junior levels.

[2] The framework for the Stakeholder Engagement Management Plan (SEMP) was adapted from the work of Busque (2016).

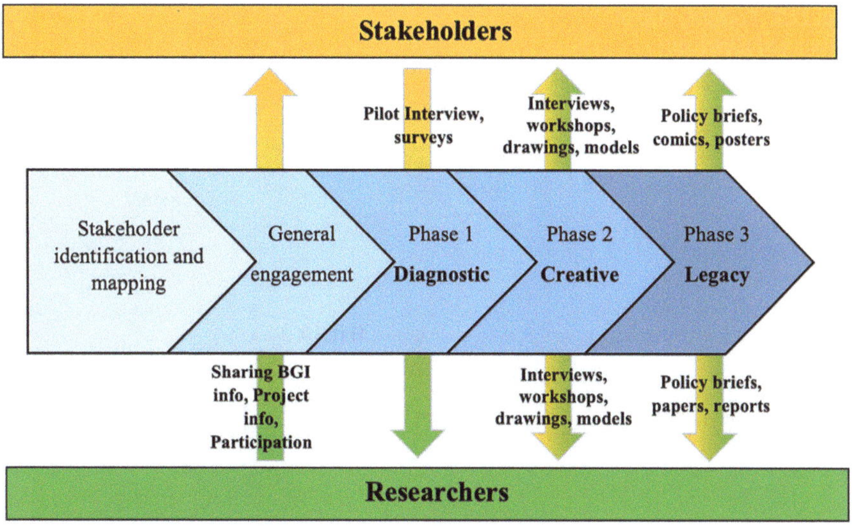

Fig. 22.1 Linear diagram representation for developing a SEMP framework

be identified. Such intervention, when brought into action, will enhance community education and community resilience.

22.2.2 *Case Study Locations*

Considered to be the gateway to North-east India, Guwahati is the largest city in the region (Fig. 22.2) and is a major commercial and educational centre in eastern India. Present-day Guwahati is a cosmopolitan city that is developing at a rapid speed. By 2011, Guwahati had expanded from a small, marshy village with 12,000 residents in 1911 to a sizeable urban mass with around 1.26 million residents, who are accommodated in 60 wards within 328 km^2 under the Guwahati Metropolitan Area (Hemani and Das 2016). The Guwahati Metropolitan Development Authority (GMDA) has initiated four water supply projects as externally aided projects for 27.13 billion Indian Rupees (Pawe and Saikia 2020). Guwahati is located on a plain that is surrounded by 18 hills and 5 broad natural drainage basins; Deepor beel (about 144 sq.km.), Bharalu (about 100 sq.km.), Silsako beel (about 92 sq.km.), Kalmoni basin (66.5 sq.km.) and Foreshore basin. But, having such huge natural drainage basins, urban flooding is still an issue in Guwahati. It is primarily caused by significant runoff generation from the mountainous terrain (due to deforestation) and sediment yield.

Just like in other Indian cities, urban sprawl and intensification have exacerbated the consequences of climate change in Guwahati through the reduction in green cover and water bodies compared to rural areas (TERI 2013). Here, urban growth has been

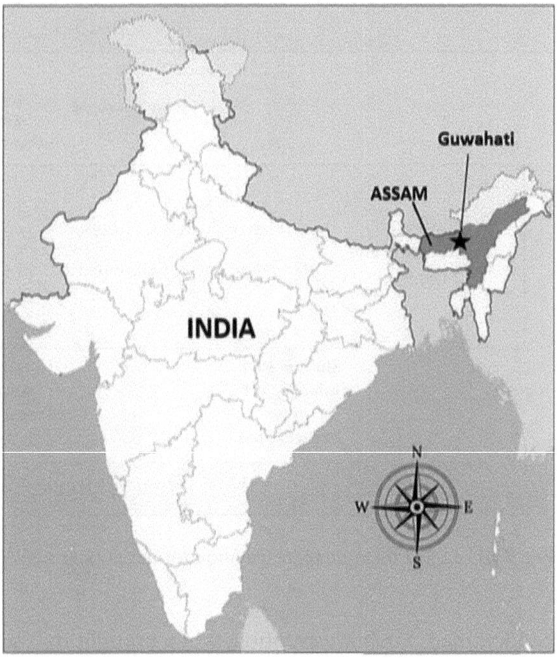

Fig. 22.2 Location map of Assam and Guwahati

rapid, unplanned and organic; and the ecology and environment have suffered greatly as a result of changes in land-use patterns brought on by unchecked development activities. for instance, encroachment and unplanned urban extension are putting Deepor Beel, one of the Ramsar Notified wetlands, (which is surrounded by the city), in danger (Fig. 22.3). The Basistha-Bahini-Bharalu river system has become a network of sewage channels or dead rivers, contributing to an increased incidence of urban flooding in the city (CEPT 2011; TERI 2013). Bharulu is the most flood-prone area in the region because most of the drains fall on the upstream side of Bharalu (CEPT 2011; TERI 2013) and with that, heavy siltation and dumping of garbage blocks drainage channels, which worsens the flooding scenarios along Bharulu.

For this case study, potential urban flooding hotspots were identified based on desk-based research. All identified potential hotspot sites were visited, and cross-verification of the desk-based research was conducted (Fig. 22.3). Out of the many sites under discussion and review, two sites were selected for this study based on various parameters, such as their dynamics, area of influence, urban growth, and flood-related problems faced by the community members. The sites were located in Lokhra and Deepor Beel Bird Watch Tower.

***Site 1- Lokhra*:** Lokhra is one of the major residential localities in southern Guwahati. The site selected is a linear stretch along the Basistha River, which bifurcates the site into north and south zones (Fig. 22.4). From the initial public response, it was understood that Lokhra usually faces the problem of flash floods and waterlogging

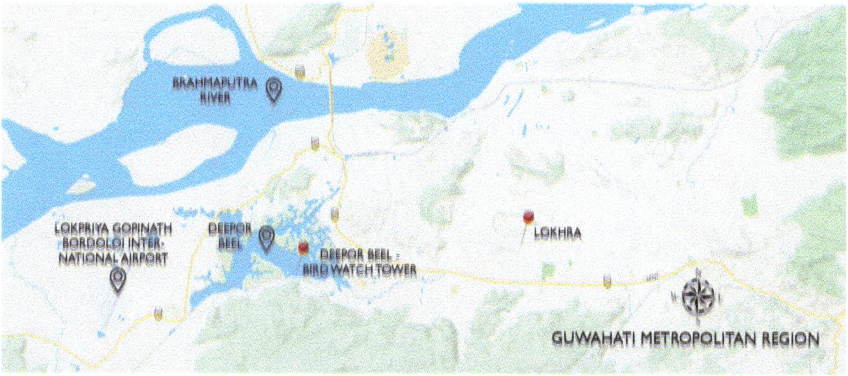

Fig. 22.3 Guwahati and case study locations. *Source* Google maps

during monsoon season and has been a recent victim of grey infrastructure development, which has led to the decrease in green cover on hills around the site. Pressurised by overpopulation and waste dumping activities, the health and natural state of the river have been highly damaged. As per the respondents, the river overflows at times and creates problems for the inhabitants nearby.

***Site 2- Deepor Beel Bird Watch Tower Site*:** The Deepor Beel is a large and prominent floodplain lake located in the southwest of Guwahati city in Assam- (Fig. 22.5). Every day urban waste from the Guwahati Municipal Corporation is dumped in the Boragaon area located at the periphery of the Beel. Further, the polluted rivulet Bharulu flows to Deepor Beel and is aggravated by human settlements with industrial growth in the surrounding areas. To add to the ecological degradation, a railway track passes through the Beel on the southern side and cuts through an elephant corridor, disturbing the flow of water to the beel. The exact size of the Beel is contentious: Guwahati Smart City Projects Limited claims that the notified area of the Beel is approximately 6.89 sq. km, and the current total wetland area is approximately 13–15 sq. km.

22.2.3 SEMP Development

In Guwahati City, flood-prone areas were taken into consideration for the selection of probable sites. For the work, both quantitative as well as qualitative types of study were conducted; and primary and secondary data were also collected. For this descriptive research work, stratified random sampling was done with a sample size of 440 from the community and 40 from administrative personnel. A questionnaire and schedule method were used for data collection; Microsoft Excel and SPSS were used as tools for data analysis. Participation in the study was voluntary and all participants provided written informed consent before participating in the study. The study was submitted for ethical approval at the local lead institution, IIT Roorkee.

Fig. 22.4 Site morphology, catchment areas and surface flow direction of Lokhra site

The project started with the identification of the stakeholders. Provisional stakeholder mapping was done to help understand each stakeholder according to their power and influence. Engagement included all identified, but those with high interest and/ or power were a priority (Fig. 22.6). The levels of influence on the project/outputs have been categorised in the following ways: ***1***—high power and high interest; ***2***—low power and high interest; ***3***—high power and low interest; ***4***—low power and low interest (Bosher et al. 2007).

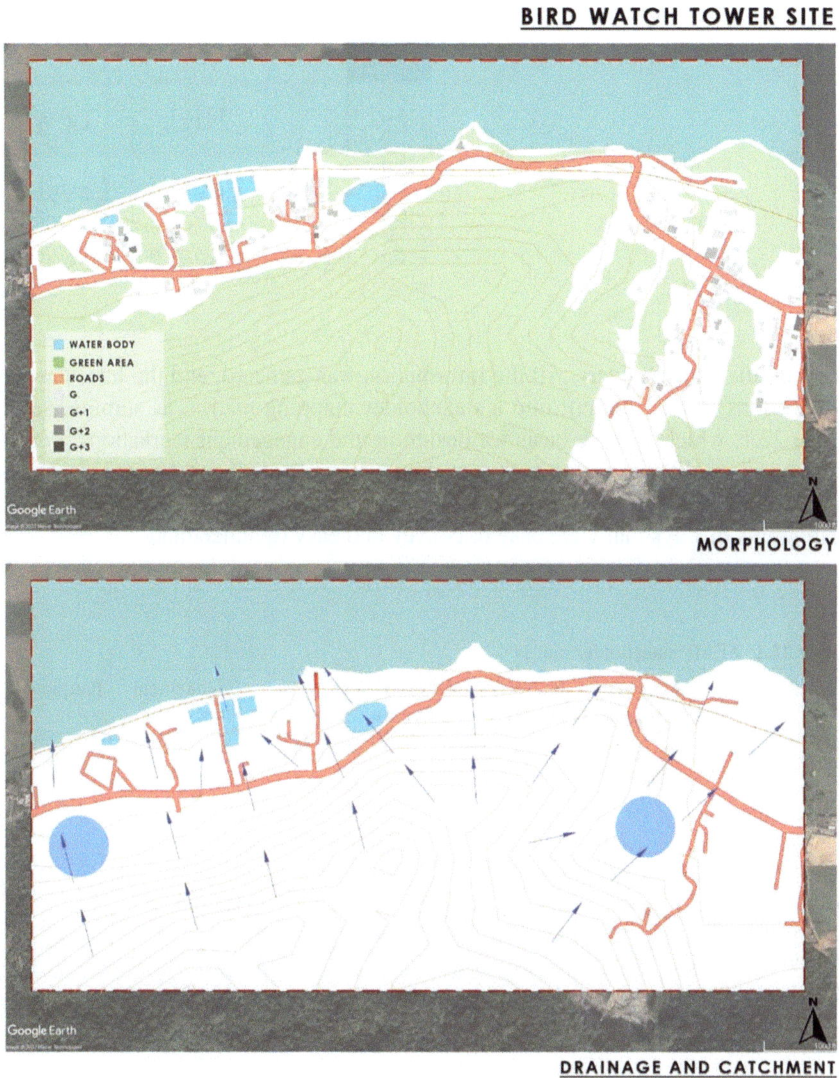

Fig. 22.5 Site morphology, catchment areas and surface flow direction for Deepor Beel bird watch tower site

Later, based on the stakeholder engagement activity planning, public consultations and communications methods were identified to ensure that suitable communication and engagement methods were deployed for the specific target audiences. The engagement methods that were identified included one-to-one meetings, group meetings, workshops, and participatory activities, including demonstrations using models and posters. Developing the questionnaires for both the community members and administrative personnel was equally important for gathering information from

Fig. 22.6 Stakeholder power/interest matrix

		POWER	
INTEREST		High	Low
	High	Level 1	Level 2
	Low	Level 3	Level 4

the identified stakeholders. All the information was gathered, and the interviews or connections made were put into a stakeholder mapping matrix, as summarised in Table 22.1, which was the basis for designing the engagement workshops. For the interviews, qualitative and quantitative data were collected, and for site visits, that information and data were turned into graphics, maps, images, graphs, charts and histograms in ways that were easy to convey and easy to understand.

Table 22.1 SEMP stakeholder matrix

Stakeholders	Area(s) of influence	Phase(s) of project	Level of influence	Key contacts	Engagement tools	Frequency
Government				Anonymised		
Guwahati metropolitan development authority	1	All	1		f2f, policy	m
Pollution control board Assam	1	1,3	1		f2f, policy	m
Guwahati development department	1	All	1		f2f, policy	m
Guwahati municipal corporation	1	1,2,4	1		f2f, policy	m
Housing & Urban development corporation Ltd	2	3	3		f2f, policy	q
Water and sanitation support organization, PHED	1	1,3	1		f2f, policy	q

(continued)

Table 22.1 (continued)

Stakeholders	Area(s) of influence	Phase(s) of project	Level of influence	Key contacts	Engagement tools	Frequency
Assam chamber of commerce	3	1	4		f2f, policy	A
MSME development institute	4	1	4		f2f, policy	A
Assam tourism development corporation	2	All	2		f2f, policy	A
Assam state disaster management authority	1	1,3,4	1		f2f, policy	q
Assam engineering college	3	1,4	4		f2f, policy, w/s	A
Forest department	2	1,3,4	1		f2f, policy	A
Guwahati revenue circle	2	3,4	3		f2f, policy	q
Guwahati smart city limited	2	All	1		f2f, policy	m, A
Wildlife department	2	1,4	1		f2f, policy	A
Water resource department & irrigation dept	1	1,3	1		f2f, policy	q
Brahmaputra water board	1	1,3	1		f2f, policy	q
Non-governmental organizations						
Save Guwahati build Guwahati	1	1	2		All	q
Agriculture and rural tourism	2	1	3		f2f, policy, email	A
Research and academic organizations						
Center for environment education North East	3	1,2	4		All	A
Department of environmental studies, Gauhati Uni	3	1,2	4		All	A

(continued)

Table 22.1 (continued)

Stakeholders	Area(s) of influence	Phase(s) of project	Level of influence	Key contacts	Engagement tools	Frequency
Private sector						
Indian Institutes of Architects, Guwahati Chapter	3	All	2		All	q, A
Architects, Planners, Entrepreneurs and Professionals	3	All	2		All	q, A
Local associations						
PHD Chamber of Commerce and Industry NE	4	1	4		W/s, post, fdg, email	A
End Camp Adventure	4	1	4		e	A
Ecological	4	1	4		e	A
Community groups						
Fishermen of Deepor beel	1	All	2		All	w, m
Villagers (Karbi & Kalita in Deepor Beel)	1	All	2		All	w, m
Karbi community at Lokhra	1	All	2		All	w, m
Jyoti sangha	2	All	2		All	w, m
Residents of Anil Nagar—Nabin Nagar	1	All	2		All	w, m
Others						
News channel: Pratidin Times	4	1,2,4	3		F2f, fdg, leaf	q, A
Media channel: Dy 365	4	1,2,4	3		F2f, fdg, leaf	q, A

(continued)

Table 22.1 (continued)

Stakeholders	Area(s) of influence	Phase(s) of project	Level of influence	Key contacts	Engagement tools	Frequency
Notes						
Areas of influence (Case study specific)						
Zone 1: Major—Stakeholders directly affected by or involved with the 4 **case study sites in/ around Guwahati**						
Zone 2: Moderate—Stakeholders that have a say (directly and indirectly) in how water/ environmental resources are managed in **Guwahati** and people that are affected by floods and environmental issues related to NbS in the city						
Zone 3: Minor—Stakeholders that have an indirect say in how water/environmental resources are managed in Guwahati and those that are affected by floods and environmental issues related to NbS in the **environs around the city**						
Zone 4: Negligible—Stakeholders in the **region** (upstream and downstream) that may have inputs to decision-making on an occasional/incidental basis						
Phase(s) of project						
1—Preliminary concept/idea						
2—Pre-construction design and planning (and redevelopment)						
3—Construction (and re-construction)						
4—Post-completion (and operation)						
Level of influence on the project/outputs (suggestion)						
1—high power and high interest						
2—low power and high interest						
3—high power and low interest						
4—low power and low interest						
Key contacts						
Names of relevant stakeholders/key contacts were provided but have been anonymized for this publication						
Engagement tools						
Includes Face-to-face interviews (f2f), emails €, policy briefs (policy), posters (post), leaflets (leaf), workshops (w/s), focus groups (fdg), community events (commeve), etc						
Frequency						
To be refined once key contacts are identified—may consider weekly (w), monthly (m), quarterly (q), and annual (A) engagement where relevant						

The development of the SEMP matrix (Table 22.1) enabled the research team to identify the key stakeholders, how they could be of benefit to the study and how the study could be helpful to them. A summary of key stakeholder roles is provided in Table 22.2.

Table 22.2 Stakeholders' role in BGI scenarios development

Stakeholders	Roles of stakeholders in BGI development
Government	
Guwahati metropolitan development authority	Share past experience and data about flooding in the urban areas and existing and future drainage plans to address urban flooding. Share insights related to urban development plans and plans for the development of green spaces and advise on potential BGI and the possibility of implementing those measures in Guwahati. Identify data gaps and needs related to flood management
Pollution control board Assam	Advise on where and how to link BGI measures to the City Action Plan to Respond to Climate change. Advise on the possibility of implementing BGI measures in their sector
Guwahati development department	Contribute ideas about potential BGI in their sector. Share relevant information about the city land use plan
Guwahati municipal corporation	Contribute ideas about potential BGI. Advise on the opportunities and challenges in implementing BGIs in Guwahati
Housing & urban development corporation Ltd	Contribute ideas about potential BGI in Guwahati
Water and sanitation support organization, PHED	Contribute ideas about potential BGI in Guwahati
Assam chamber of commerce	Share past experience and data about flooding and existing and future drainage plans to address urban flooding in the district; Advise on potential BGI and locations where these measures can be applied and the possibility of implementing BGI measures in the district and analysis of the economic impact of these
Micro, small and medium enterprises (MSME) development institute	Share past experience and data about flooding and existing and future drainage plans to address urban flooding in the district; Advise on potential BGI and locations where these measures can be applied and the possibility of implementing BGI measures in the district. Analysing the contribution of industries in the scenario of BGI
Assam tourism development corporation	Advise on potential BGI and the possibility of implementing BGI measures in Guwahati. Identify data gaps and needs related to flood management
Assam state disaster management authority	Shares knowledge and data related to flooding in Guwahati. Share BGI-related measures that have been applied earlier, if any
Assam engineering college	Contribute ideas about potential BGI in Guwahati. Advise on the possibility of implementing BGI measures in their sector and its impact analysis

(continued)

Table 22.2 (continued)

Stakeholders	Roles of stakeholders in BGI development
Forest department	Sharing past experience and data about flooding in the city and existing plans to address flooding. Contribute ideas on BGI related to flood risk management/reduction actions of the city
Guwahati revenue circle	Shares knowledge and experience in urban resilience building. Share insights about the city resilience strategy and the role of BGI in building flood resilience. Contribute ideas related to potential NbS linked to floods in the city
Guwahati smart city limited	Sharing knowledge and data about past and future climate change and its impacts on flooding Sharing knowledge about adaptation measures related to flooding. Advise on potential BGI and the possibility of implementing BGI measures in Guwahati
Wildlife department	Contribute ideas about potential BGI for Guwahati and the impact on the forestland wildlife sector
Water resource department & irrigation department	Contribute ideas about potential BGI in Guwahati
Brahmaputra Water Board	Contribute ideas about potential BGI in Guwahati
Non-governmental organizations	
Save Guwahati build Guwahati	Share knowledge/findings and experience related to flood management. Share ideas about potential BGIs for the city of Guwahati
Agriculture and Rural Tourism	
Research and academic organizations	
Center for environment education north east	Share knowledge, research findings/results and experience related to flood management and climate change in Guwahati. Share ideas about potential BGI for Guwahati
Department of environmental studies, Gauhati university	
Private sector	
Indian institutes of architects, Guwahati chapter	Share ideas about potential BGI for Guwahati and give inputs in terms of building layouts, city design and city planning

(continued)

Table 22.2 (continued)

Stakeholders	Roles of stakeholders in BGI development
Architects, planners, entrepreneurs and professionals related to NBS	
Local associations	
PHD chamber of commerce and industry NE	Share ideas about potential BGI for Guwahati and help in understanding the economic angle associated with it
End camp adventure	Share ideas about potential BGI for Guwahati and help in understanding the economic angle associated with it. They can help in conducting awareness programs related to BGI
Community groups	
Fishermen of deepor beel	Share their experience and challenges in dealing with flooding at the community level and their expectations in relation to BGI. Also, contributing towards ideas/solution generation and potential implementation at the community level
Villagers (Karbi & Kalita in Deepor Beel)	
Karbi community at Lokhra	
Jyoti sangha	
Residents of Anil Nagar-Nabin Nagar	
Others	
News channel: Pratidin times	Communicate BGI approach and working with the local stakeholders and the potential contribution of BGI to addressing flooding and other concerns to the local audience in Guwahati
Media channel: Dy 365	

22.2.4 Workshops and Creative Media

Following initial stakeholder mapping and engagement, focus group discussions took place over two workshops that were held in the Deepor Beel Bird Watch Tower and Lokhra areas. The workshops aimed to enable discussion amongst the local community members on the perspective of blue-green infrastructure. Initially, a survey was conducted, capturing all the age groups of the society as well as their

Fig. 22.7 Explanation of BGI through live models

professions. Then, a community workshop was set up after the completion of the questionnaire survey. During the workshop, models and information sheets were utilised to explain and demonstrate local issues and potential BGI approaches to the community. The discussion focused on the interaction with the community members regarding the problems faced by them and also on making them familiar with BGI through information sheets and live model demonstrations.

The workshop started with an introduction, followed by an explanation of the infographic sheet to the community. Then, a demonstration using physical models (see an example in Fig. 22.7) was conducted to showcase the issues at each site. After the demonstration, the participants were divided into three groups, which contained a mixture of community members, students from educational institutes and project team members as facilitators. They interacted and shared their viewpoints, which facilitated discussions regarding the enablers and barriers of BGI. The threats and opportunities in the area, along with the challenges, were also discussed.

Finally, potential solutions were discussed. Here, different individuals came up with their household and community-level solutions. This discussion was conducted in the native language of Assam.

22.3 Results and Discussion

Of the 440 community-level respondents who participated in the study, 36% respondents were found to be business-oriented people, 25% were students, and the remaining 38% were daily wage earners, housewives, retired persons and service holders. Out of these respondents, 52% live in nuclear families, 23% live in joint families, 16% are unmarried single, and the rest (9%) are couples with either one working adult or a dual career couple. On further investigation, it was found that, on average, the respondents had 3.5 dependents. The data indicated that the jobs undertaken by the respondents are not highly paid, thus indicating the economic struggle of these communities. All of this information was added to the SEMP, and based on this, the outline of surveys and workshops was designed.

During the survey, while discussing the problems that they are facing during the course of their stay in the community, a few major issues emerged that were faced by the respondents. These issues were Floods, Lack of green spaces and the rise of deforestation, Urban Sprawl: built-up areas, groundwater decline, reduced infiltration and lack of safe drinking water (Table 22.3). However, traffic and congestion, Air pollution, Safe shelter, poor healthcare and encroachment were also on the list of problems. While expressing the issues of floods, the respondents mentioned that the flash floods have made the lives of the people difficult. The study also revealed that 14% of the respondents mentioned that they faced issues related to very frequent floods and that it became difficult to walk, ride, or drive during these times. The majority of the respondents (63%) felt that the Government was primarily responsible for addressing these problems.

During the course of the questionnaire survey and workshops, it was found that the respondents were aware of the issues concerning their lives, however, the perception that the implementation of such practices is a costly affair made it a constraint to implement any changes at the local level. The community participants also understood the importance of implementing BGI to solve present-day issues (Table 22.4).

The participants also identified what they could do to solve those issues and hence identified the enablers of BGI to solve their flood-related issues. The results highlight the major enablers (Table 22.5) as mentioned: Constructed Wetlands, Green and Blue Roofs for rooftop agriculture, Rain Gardens, Downspout Disconnection for rainwater harvesting, Green Alleys, and Green School Yards.

Table 22.3 Table showing the responses on the major issues/problems

	Relevance				
Problems	Most				Least
Lack of safe drinking water	39	32	16	8	5
Groundwater decline	31	28	23	10	8
Poor status of wetlands and river	29	31	30	27	1
Lack of green spaces and rise of deforestation	25	41	24	8	2
Urban Sprawl: built-up area	21	30	37	9	4
Flood	13	22	20	20	25
Reduced infiltration	11	25	38	15	12
Air pollution	10	33	25	17	15
Traffic and congestion	9	27	25	23	17
Encroachment	7	24	37	14	19
Safe shelter	7	14	34	23	22
Poor healthcare	3	10	34	30	23

Table 22.4 To what extent do you think BGI implementation should be encouraged in the city for dealing with problems

	Relevance				
Issues	Most				Least
To bring sustainable development	45	36	13	4	2
To improve the green spaces	31	47	15	5	2
To deal with ecological problems	22	46	24	7	1
Improve water and air quality	21	47	24	7	1
To deal with urban flooding	15	50	23	5	7
To deal with problems of erosion	13	50	25	10	2

Table 22.5 According to the respondents, the possible enablers in the city of Guwahati to implement BGI

	Relevance				
Elements	Most				Least
Green school yards	27	30	24	11	8
Low impact development	22	16	36	17	9
Urban forest	11	34	40	14	1
Green alleys	11	23	31	25	10
Green and blue roofs for rooftop agriculture	11	28	30	15	17
Constructed wetlands	7	27	32	23	11
Rain gardens rooftop	6	24	31	21	18
Downspout disconnection for rainwater harvesting	5	25	40	14	17

Through the survey and the workshops, the participants started the discussion on the probable approaches that could be adopted for implementing BGI in their community (Table 22.6) and how the community members can actively participate in it. There was also an increase in interest from the community members in learning more about BGI (Table 22.7) which led to a successful interactive workshop on BGI.

Table 22.6 Table showing probable approaches to implement BGI in the study area

	Appropriateness				
Approaches	Most				Least
Wetlands and floodplain restoration and creation	26	43	22	7	2
Identification of vulnerable flood-prone areas	20	43	29	7	1
Floodplain restoration	18	32	26	10	14
Land preservation and restoration	18	49	21	10	3
Stream buffers	8	21	34	17	20

Table 22.7 Respondents' view on the statement: BGI

Statements	Strongly agree	Agree	Neutral	Disagree	Strongly disagree
Every person should give some of their time for the good of their local community	60	33	5	2	1
I would like to help my community maintain a system that prevents floods in our area	25	36	28	9	2
I would like to attend a seminar/workshop to learn what I can do to prevent flooding in our area	23	32	29	14	2
I do not feel that flooding is an important issue that will affect me	2	5	13	14	66
I do not feel that climate change is an important issue that will affect me	1	3	11	28	57

22.3.1 *Reflections on the Importance of Stakeholder Engagement and Two-Way Education*

The study found that 36% of the respondents are people who have stayed in the same community for more than 15 to 50 years, and 30% of the respondents are people who have stayed in the vicinity for more than 50 years and they have been an important part of the community as they are the direct receivers of any changes occurring locally. The SEMP model shows that stakeholders have different interests, powers, and influences in making contributions to the betterment of the community. The administrative stakeholders definitely have high power, but only with high interest in community development can climate change issues like urban floods be solved.

The study found that there are people with high interest but low power, and by targeting this segment of the stakeholders through community education and engagement in community activities, the stakeholders themselves can attain more power in the hope that they will be able to implement BGI in their locality, through positive community action. In this regard, respondents' views have been taken on whether they wish to learn more about BGI, and 32% have shown interest in helping the community through which they can maintain a system to prevent flooding and other issues that are being faced in daily lives which are affecting the climate and the livelihood of the people. Respondents also agreed that every person should give adequate time for the betterment of the local community. This contrasts with the views expressed during the earlier engagement, where respondents felt that BGI was primarily the government's responsibility. At the end of the workshop, the viewpoints

Fig. 22.8 Community people near Deepor Beel willing to make a change

and suggestions made by the community members were recorded in the form of visual representations (Figs. 22.8 and 22.9) so that the community members' outcomes could be fed back to them and they could also use it for future reference to empower themselves in the implementation of BGI in their community.

As stated by the community members, over the past two decades, their areas (Lokhra and Deepor Beel) have been degrading continuously due to chaotic rapid and unplanned development. The control of the damage being caused is beyond the scope of individuals and hence it was recommended by the community that the concerned government authorities pay immediate attention towards the major anthropogenic threats to mitigate the adverse impacts that are being faced by the community. The participants also showed interest and acceptance towards the concepts of blue-green infrastructure. They were very quick to recognize the problems and highly interested in solving the overall issues. Some of the major solutions they proposed were to incorporate bioretention areas in the road near the site to stop water flowing from the adjacent hilly roads, create more green infrastructures to increase the groundwater recharge table and conserve the existing blue-green infrastructures while avoiding further inappropriate land development (i.e., unregulated housing).

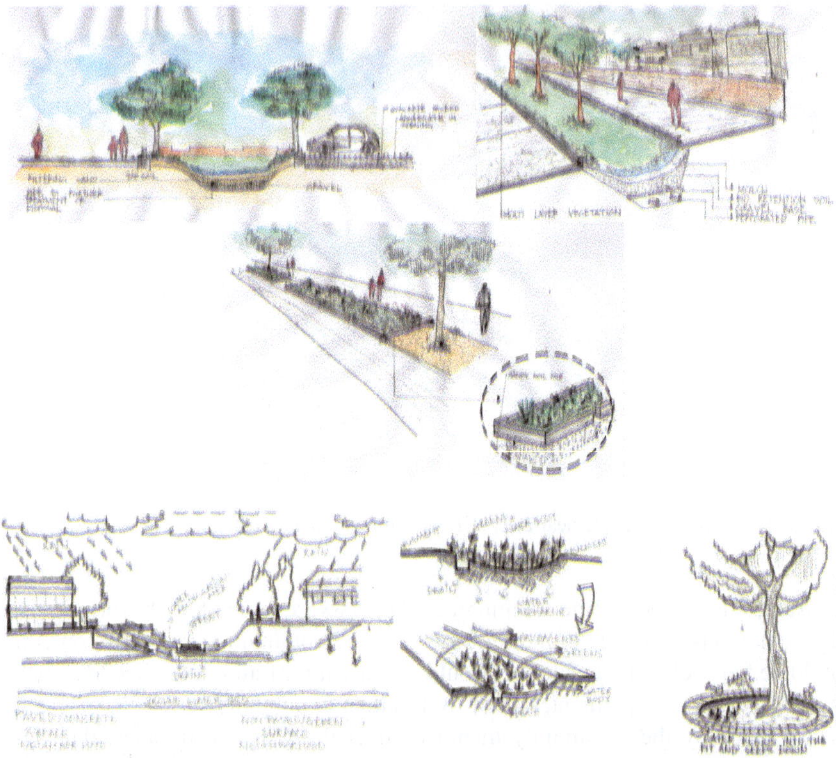

Fig. 22.9 Various possible solutions (as proposed by the residents of Lokhra)

22.4 Conclusions

Community education should not be a uni-directional exchange of information. This study not only provides important evidence of the benefits of supporting community education and awareness of flood risk and environmental degradation but also of the benefits of understanding diverse stakeholder needs. Providing the community with information as well as a platform to influence local decisions on BGI-related ideas and solutions led to a co-creation process that contributed towards removing potential local barriers and facilitated the design of scientifically informed, locally appropriate solutions to urban flood risk.

Due to the limitations of project scope and timeline, the data presented here represent only a preliminary 'snapshot' of the much longer process of developing and implementing BGI solutions. To fully understand the role of community education and engagement in enabling BGI, further longitudinal studies, covering the duration of the process from conception to installation and long-term maintenance, should be undertaken. Furthermore, although the initial study highlighted a lot of areas in Guwahati where such a kind of community engagement is required, the study was conducted only on two sites based on the project timeline and the number of local connections available with the ground team among all the sites.

The study's findings indicate that creating awareness amongst local stakeholders can help to nurture BGI approaches/solutions that are more likely to be appropriate and sustainable for the local context. The sense of community can be improved through well-planned and designed engagements and programs, where approaches to BGI can be viewed as a shared responsibility. A particularly interesting observation was that, through community engagement, projects can move the 'lower power, low interest' stakeholders to 'low power, high interest' (i.e., concerning the SEMP matrix) which can be key to moving towards community-based BGI. However, it is acknowledged that there is a gap that needs to be explored further: the gap between the roles of low- and high-power stakeholders. As an outcome of this study, it was found that when it comes to the implementation of BGI, the lack of funds will remain a major obstacle. Thus, it is noted that the stakeholders that have 'high power' and should have a 'high interest' (for instance, in this specific case, the Guwahati Metropolitan Development Authority, the Guwahati Development Department and the Guwahati Municipal Corporation) will be critical for enabling the (financial support) required to support locally effective and relevant BGI. Accordingly, BGI cannot simply be a 'top-down' undertaking of governments, national, regional or local agencies; to be sustainable and relevant, BGI will require the involvement of local stakeholders through new stewardship practices that are supported by Governmental funding and policies.

Acknowledgements This study was funded by UK NERC (Ref: 2021COPA&R35Marjoribanks).

References

Ashley R, Gersonius B, Digman C, Horton B, Smith B, Shaffer P (2018) Including uncertainty in valuing blue and green infrastructure for stormwater management. Ecosyst Serv 33:237–246

Attree P, French B, Milton B, Povall S, Whitehead M, Popay J (2011) The experience of community engagement for individuals: a rapid review of evidence. Health Soc Care Comm 19(3):250–260

Bosher L, Chmutina K (2017) Disaster risk reduction for the built environment. John Wiley & Sons, London

Bosher L, Dainty A, Carrillo P, Glass And J, Price A (2007) Integrating disaster risk management into construction: a UK perspective. Build Res Inform 35(2):163–177

Bozovic R, Maksimovic C, Mijic A, Smith KM, Suter I, Van Reeuwijk M (2017) Blue green solutions, a systems approach to sustainable, resilient and cost-efficient urban development. Imperial College, London

CEPT (2011) Carrying capacity-based urban development regulations. Guwahati Metropolitan Development Authority, Guwahati, India

UKGBC (2015) Demystifying green infrastructure. United Kingdom's Green Building Council, Report, London

Drosou N, Soetanto R, Hermawan F, Chmutina K, Bosher L, Hatmoko JUD (2019) Key factors influencing wider adoption of blue–green infrastructure in developing cities. Water 11(6):1234

Hamel P, Tan L (2022) Blue–green infrastructure for flood and water quality management in Southeast Asia: evidence and knowledge gaps. Environ Manage 69(4):699–718

Hemani S, Das AK (2016) City profile: Guwahati. Cities 50:137–157

Konadu DD, Fenner RA (2017) Catchment level water resource constraints on UK policies for low-carbon energy system transitions by 2030. Global Chall 1(5):1700006

Kumar A, Satheesh A, Mukherjee M, Naorem V, Ravan S, Schreyers L, Sudmeier-Rieux K, Walz Y, Mallawantantri A (2023) Harnessing geospatial data to design blue-green infrastructure for building regional and local scale disaster resilience. Fifth World Cong Disaster Manag V:10–30. https://doi.org/10.4324/9781003342090-3

Lamond J, Everett G (2019) Sustainable blue-green infrastructure: a social practice approach to understanding community preferences and stewardship. Landsc Urban Plan 191:103639

Liu HY, Jay M, Chen X (2021) The role of nature-based solutions for improving environmental quality, health and well-being. Sustainability 13(19):10950

Mills S, Waite C (2018) From big society to shared society? Geographies of social cohesion and encounter in the UK's national citizen service. Geografiska Annaler: Series b, Human Geog 100(2):131–148

Mishra HS, Bell S, Wilczyńska A, Balicka J (2021) Urban wetlands and stormwater management. In *Urban Blue Spaces* (pp. 336–371). Routledge

Mukherjee M, Satheesh A, Kumar A (2022) StEMAIRF-BGI as a tool for UHI mitigation using land use planning and designing. Global Urban Heat Island Mitigation 177–97. https://doi.org/10.1016/B978-0-323-85539-6.00002-0

Pawe CK, Saikia A (2020) Decumbent development: urban sprawl in the Guwahati Metropolitan Area, India. Singap J Trop Geogr 41(2):226–247

Sadeghi Z, Bin MH (2023) Enhancing recreational activities through planting design in neighbourhood parks. J Outdoor Recreat Tour 42:100591

Sánchez FG, Govindarajulu D (2023) Integrating blue-green infrastructure in urban planning for climate adaptation: lessons from Chennai and Kochi, India. Land Use Policy 124:106455

Soetanto R, Hermawan F, Drosou N, Bosher L, Hatmoko JUD (2022) Perceptions of social responsibility for community resilience to flooding: a comparison between communities in Indonesia and the UK. Water 14(3):433

Staddon C, De Vito L, Zuniga-Teran A, Schoeman Y, Hart A, Booth G (2018) Contributions of green infrastructure to enhancing urban resilience'. Environ Syst Dec 38:330–338. https://doi.org/10.1007/s10669-018-9702-9

TERI (2013) Risk assessment and review of prevailing laws, standards, policies and programmes to climate proof cities- Synthesis Report for Guwahati, The Energy Resources Institute (TERI), New Delhi

Chapter 23
Disruptive and Conventional Technologies in Disaster and Climate Education: Perspectives from China

Ztionhiying Zhao, Rong Huang, and Rajib Shaw

Abstract Risks and disasters alike from natural occurrences need decision-making processes that are well informed—coupled with a few other factors such as adoption of local knowledge and innovative approaches, as well as application. In China, local communities are seen to be instrumental in management of future and current disasters, as seen from their commitment to enhance and support technological developments and their accessibility through innovative techniques of preparedness, risk reduction, and resilience during recovery. Informed decision making and disaster risk management are processes that require research and development, technology adoption, application of new technologies, and promotion of approaches that are innovative and local knowledge to help in solving the arising complex issues posed by hazards. As such, there have been commitments to support and improve access to technologies and improve innovative approaches to reducing risk, preparedness, and resilient recovery within the local community. Risk reduction and preparedness in times of natural disasters therefore need to be supported with improved accessibility to technological developments since they are essential for disaster management and increasing resilience among communities in China. Dealing with disasters calls for a prompt and well-coordinated response that mitigates the adverse effects of catastrophes such as losses and injuries and lowers tension while offering a chance for faster recovery to normalcy after catastrophes. This chapter focuses on some of the key lessons and learning.

Keywords Disruptive technologies · Climate · Disaster · Conventional technology · Risk management · Policy · Development

Z. Zhao · R. Huang · R. Shaw (✉)
Graduate School of Media and Governance, Keio University, Fujisawa, Japan
e-mail: shaw@sfc.keio.ac.jp

A. Yildiz and R. Shaw (eds.), *Disaster and Climate Risk Education*, Disaster Risk Reduction, https://doi.org/10.1007/978-981-97-5987-3_23

23.1 Introduction

The concept and process of lowering the risks associated with disasters are best described through management and analysis of disaster triggers in a systematic way to lessen vulnerability and reduce exposure of property and people to hazards. Different researchers such as Torani et al. (2019) have indicated evidence of success in reduction of disaster risk through incorporating DRR into the formal education set up. This therefore promotes wise land management practices and improves readiness for catastrophic events among schools and communities at large. Considering recommendations such as 2015–2030 Sendai Framework for Disaster Risk Reduction (SFDRR), disaster risk reduction is instrumental in the quest for sustainable development. The SFDRR's priority 1 can be achieved successfully through accommodation of disaster risk knowledge entailing information on mitigation, response, disaster prevention, recovery, preparedness, and rehabilitation shall be quintessential. Emphasis on the importance of education has also been done in the Hyogo Framework for Action 2005–2015, which acknowledges the critical role education plays in Disaster Risk Reduction Education (Fu and Zhang 2024).

The complicated climatic and topographic conditions of China have made it on the receiving end of severe and frequent natural catastrophes. The natural disasters in China are distributed over a wide scope, come in diverse types, high frequencies, and are associated with great losses. The types of natural disasters that have previously hit China include landslides, typhoons, earthquakes, drought, sandstorms, forest fires, and intense sandstorms (Wang et al. 2023). The strategic location of China—meeting point of the Indian Ocean, Pacific and Eurasian waters—make it prone to earthquakes as the tectonic plates are in constant motion and still very active. Economic losses have crippled China's developments as seen from 324 billion Yuan spent on natural disasters as described by the Ministry of Emergency Management (MEM) in China. The face of adaptability to change in China is seen from the evolution process of the community when faced with natural disasters, as displayed by the far-fetched impacts on Chinese civilization. Considering the experience accumulated in DRRE by China, the country needs constant improvement and upgrading of systems and content within DRRE. Therefore, innovative strategies in the sector of disaster risk reduction education will prove handy in equipping China with the necessary resources for action during disaster.

In order to understand the dynamics and needs that led to the development of innovative technological strategies in the process of climate and disaster education, it is necessary to consider both evidence-based techniques and scientific knowledge. Understanding and managing disaster risks and events depends on scientific information such as structures and materials response to shock waves, and earth movements within the tectonic plates that cause buckling and separation (Hosseini and Izadkhah 2020). Equally, increased resilience and implementation procedures of risk reduction measures requires adoption of social sciences tools derived from conventional knowledge of cultural and social factors of the local communities that have been affected with similar disasters previously. Having a vast body of knowledge on disaster risk and management will be vital in ensuring the credibility and usability of the disaster risk education content delivered to individuals in the community. Progress has therefore been steered through technological inventions, and science, which are instrumental in provision of solutions for old and existing problems, providing new insights, and ensuring high standards are achieved in the development and implementation and policies.

Conventional technologies mainly entail the indigenous knowledge used by communities in risk reduction as well as dealing with natural disasters. The practices and techniques used by indigenous people is vital since the individuals have an advanced understanding of their local environs and surrounding conditions that has developed from habitation and adaptation over several generations, and subsequently imbedded in the community's way of life as a survival tactic (Cauton and Su 2020). Developing interest has been seen in the sector of the relationship between natural disasters and indigenous knowledge, highlighting the importance of the latter in improving policies of disaster risk reduction through incorporating early warning systems and disaster education in equal design. By assessing the new and conventional technologies in China, this chapter seeks does so provide valuable knowledge for tackling disasters that are easily transferable and adopted by other communities facing similar scenarios of disaster.

Subsequently, the integration of indigenous practices in technological and policy development shall encourage participation of the communities by empowering them in leading their fellows in disaster reduction activities in their local context, and having their traditional practices embedded in them, leading to higher adoption, implementation and success levels (Lambert and Scott 2019). Another importance of having the indigenous knowledge is that it provides a foundation for non-formal channels for dissemination of information pertaining disaster risk education when incorporated with advanced technological frameworks for disaster education. Subsequently, these efforts of disaster and climate education shall provide for minimal waste, improved outcomes, and reduced confusion in the events of disaster.

23.2 Literature Review

The magnitude and effects of natural and climatic disasters have been increasing substantially due to the increasing human activities and climate change condition; prompting communities globally to prepare, act and respond to disasters. Different methods of disaster risk reduction and education have been carried out and acknowledged for effectiveness in mitigating risk as defined through evidential measure (Klein et al. 2019). Especially the incorporation of indigenous conventional technology, which is ignored most of the time, when integrated with modern techniques of disaster mitigation and education allows for development of a holistic approach. Despite the technological advancements being realized in DRR education practices, calamities still strike and result in great losses. In order to achieve sustainable development, it is necessary therefore to guide the technocratic paradigm within the cultural and social dimension of care for the human environment. This section will define the different forms of conventional indigenous technology, and new disruptive technology in the disaster risk reduction sector followed by examples of different forms of conventional and disruptive technologies used in the provision of disaster and climate education.

23.3 Conventional Indigenous Technology

Conventional technology is commonly regarded as Indigenous knowledge, based on the characteristic principles sustainability and resilience to calamities, which defines why new paradigms of disaster risk reduction and climate education strategies are incorporating the knowledge (Munsaka and Dube 2018). Indigenous knowledge is constantly being incorporated widely in the scope of sectors such as ecology, land and natural resource management, biodiversity, health, environmental conservation, and education. Traditional societies, especially in China, live in harmony with nature since this is where they source their livelihood. They operate and live on local, traditional, or indigenous knowledge that is characteristic to each culture due to adaptation to surrounding conditions. The pattern of life is considered to be closely linked to the current ideas surrounding sustainable practices. Elements of balanced resources found in indigenous knowledge have recently gained recognition with their role in sustainable development. Indigenous knowledge is used in sectors of hazard identification through observation of natural phenomena such as observation of waves, plants, winds, celestial bodies, animals, and clouds.

Conventional or Indigenous knowledge is allured to contain a vast array of information on the natural set-up due to the centuries' old observation and experience with changing seasons and phenomena. The indigenous knowledge is sometimes challenged with the notion of being outdated and obsolete, but it is based on existing knowledge subjected to testing through trial and error before transmission to future generations through shared experiences practically or orally (Oktari et al. 2020). An example of conventional technology is seen in the Early Minoan civilization in the Ancient Mediterranean region, whose disaster risk reduction technique was anti-seismic architecture. The Minoan Civilization's Architectural concepts applied the use of different orientation and placement of wall elements so as to resist and withstand shock waves from all directions. Another distinctive feature is the near absence or absence of windows in the building's ground floor area. Lack of openings was done to ensure structural strength through elimination of voids that are points of weakness in the event of an earthquake. Also, lateral resistance was increased in the structures by reducing the overall size that served to reduce the structural weight of the building. Lastly the houses had minimal floors to keep the centre of gravity low in the event of a disaster.

23.4 New and Disruptive Technology

Recent developments indicate the great potential of disruptive technologies in offering solutions in parameters of mitigation and risk detection during an emergency. Disruptive technologies describe totally new high-performance technologies that have better deliverables as compared to the present market ready technologies (Cheikhrouhou et al. 2020). Disruptive technologies are characteristic in their advancements which give rise to a new market portion or new areas of research. Within the years, several technologies have been developed which went further and provided new opportunities in different research areas and industries. The technologies have been used throughout the process of minimizing disaster risk and the dissemination process of all the information in diverse setting to ensure the community is well versed with disruptive technologies adopted within parameters of risk reduction and mitigation.

The condition of climate change globally has called for the need of constant development and implementation of innovations and technology advancements that foster adoption of tactics that are more evidence-based within the realms of disaster risk reduction and education (Minges 2019). New and disruptive advancements are devised to counter and mitigate the evolving risks and the dynamic causes such as urbanization, climate change, environmental degradation, poverty, and population density. Technological innovations need to consider real life challenges in the face of natural disasters and calamities, and strategically apply the necessary tools to ensure efficiency. Existing studies provide solutions through individual developments and technological advancements, showing the need for an integration of these items in a fine mix of technologies through understanding the purpose of different items and their resilience potentials. This also allows for situational awareness and response in a manner tuned to the specific emergency scenario for example integrating social media data and other sources to sensors for sending out warnings for specific localities in times of emergencies. The process results in a holistic synthesis of technological resources through a comprehensive synthesis of locals and technological solutions.

Disruptive technologies are considered innovations with transformations that are dramatic in nature as seen from the alteration of operations and systems (Samarakkody et al. 2023). A wide array of technologies are currently being researched and used in the field of disaster reduction, resilience and education through adoption and improvement of indigenous sustainability measures and climate policies. In this context, the dramatic disruptions of the disruptive technologies are experienced in the climate change and mitigation processes that shift societal norms and challenge the status quoi. The conventional methods form part of the solutions, with disruptive technologies playing a significant role in addressing the climatic situation through radical innovations in sectors such as: tools and technologies for geodata collection and analysis including location-based services, mapping, cloud computing, data warehouses, geo-visualization, and sensor networks; public participation tools and technologies including web-based response participatory tools, living labs, crowdsourcing platforms, and social media; sectorial applications within cities, that alter operations in areas of transport, energy and environment; The different technologies at play include but are not limited to mobile phones, blockchain technology, big data, artificial intelligence, social media, and Internet of Things (Cheikhrouhou et al. 2020).

23.5 Examples of Technology Use in Climate and Disaster Education

23.6 Conventional Technology in Climate and Disaster Education

Oral traditions in climate and disaster education

Education on preparedness and risk mitigation mechanisms from the traditional perspective was done through oral traditions, whereby information was passed down in form of stories, songs, and folk tales (Cuaton and Su 2020). The training and education composed of water and food security protocols for emergencies; rescue, rebuilding and recovery strategies; environmental conservation techniques; and mutual support programs during times of recovery and relief. Knowledge on risk reduction and mitigation practices in the traditional set up includes information on customary practices, rules, beliefs and rituals related to environmental conservation and natural resource management. Reinforced respect to the environment in the traditional set up is seen from the diverse and dynamic use of an array of natural resources due to the resilience in their delivery of improved food security, calamity-resistant construction, and provision of coping mechanisms in the face of emergencies.

Community sensitization through administrators in provision of climate and disaster education

Another tactic that can be used for dissemination of indigenous knowledge is through; Provincial and District administrators at the local level since they are responsible for implementing, coordinating, developing, and teaching DRR programs, public awareness, emergency and early warning solutions, evacuation, rehabilitation, and reconstruction (Wang et al. 2023). Information, education, and communication materials can be distributed through the administrators, local community organizations, or even the council of elders. The integral role played by Sendai and Hyogo Frameworks in consideration of indigenous and local knowledge into DRR policy make it easier for local and national institutions to use the information in the design of curricula for DRR education programs.

Policy Development in climate and disaster education

Developing and adopting policies at the national level that mandate schools to add indigenous knowledge to the curricula in DRR courses can also prove effective in the delivery of disaster education from different cultures and integration of the same in school curricula (Klein et al. 2019). Learning from the indigenous people and strengthening their practices with scientific and technological prowess shall encourage a synergetic relationship where each part understands the role of the other such as traditional warning signs of blowing horns with modern preparedness and evacuation plans; developing and disseminating critical information to be incorporated in the respective education sector materials in regions where the indigenous knowledge is appropriate culturally, such as housing and infrastructure construction, shall aid both the locals and school communities in elevating their practices to be more accommodative for hazards and disasters; conducting dedicated community forums to promote awareness of locals on the potential and adoption of indigenous knowledge within the community environs.

23.7 New Technology in Climate and Disaster Education

Online Learning Platforms in climate disaster education

Online learning is one of the tools used in offering education within the parameters of technological interventions of disaster education (Amri et al. 2022). In order to ensure its effectiveness, online learning is undertaken in an active learning fashion, whereby the class activities are more student oriented through hands-on learning, paired discussions, group activities, and minimal use of traditional lectures. Online learning can also be undertaken through interactive websites such as 'Floodplain' that uses simulations and interactive maps in providing students with a deeper understanding for flood risks and aid in development of flood resilience plans. The online platform enables students to explore real-world disaster scenarios and familiarise with the potential impact of flooding in their locality and how to mitigate the same. (Rodrigues da Silva et al. 2023). In order for students to have a seamless online learning experience, the network infrastructure needs to be reliable so as to allow for practices such as downloading or accessing online learning material, and connecting with peers through social software. The network bandwidth within schools needs to be reliable to support a huge number of students studying simultaneously and offer flexible teaching and learning experiences for teachers and students alike with minimal interruption. Another example of an online learning platform for disseminating disaster education to kids is NASA's engaging website called 'NASA Climate Kids' (Diani 2021). The platform provides suitable digital learning resources in form of activites and games for all ages. The students are able to explore topics like extreme weather conditions, basics of climate change, and science behind changing weather patterns.

Different trends of education within the umbrella of ICT call for the need for adoption of due to the wide array of tools available like; e-books, small private online courses, animations, simulations, online micro-courses, and games, all of which have been formatted digitally to include graphics, video, audio, programs, and simulations of learning models (Abid et al. 2021). In order for students to have a seamless online learning experience, the network infrastructure needs to be reliable so as to allow for practices such as downloading or accessing online learning material, and connecting with peers through social software. The network bandwidth within schools needs to be reliable to support a huge number of students studying simultaneously and offer flexible teaching and learning experiences for teachers and students alike with minimal interruption.

Digital games and immersive technologies in climate disaster education

The gaming process is part of data visualization whereby it has a great potential for supporting decision-making protocols and increasing situational awareness. Data can be visualized through different methods such as through games, photos, info graphics, videos, or Virtual Reality (Anggaryani et al. 2023). The experiential environments presented by game-based approaches provide students with meaningful environments for exploring different problem-solving solutions in climate disaster scenarios (Aleksandra et al. 2018). The digital games are able to provide a holistic perception to achieve deep learning from serious games that are devised to be more complicated and challenging than simple educative games. An example is 'Disaster in My Backyard' game, whereby a disaster zone is simulated to present a real societal situation allowing students to work and apply theory. A simulated crisis environment enables students to develop skills and learn how to effectively apply theory in practice.

Chatbots in climate disaster education

Chatbots is one function of AI that could be used for disaster risk education purposes since it allows communication among several people in real time, at a fraction of the cost, and easily target the specific individuals in need of assistance as well. Artificial Intelligence (AI) is a theory that entails computer systems development to capacities of performing tasks that would otherwise need human intelligence in areas including decision making, translation of language, speech recognition, and visual perception (Arora et al. 2023). 'ChatGPT' or 'Bard' is one Chatbot example enhancing climate change mitigation effort through advanced language models. ChatGPT leverages machine learning and natural language processing thereby facilitating improved decision making, collaboration, and communication among stakeholders leading to accelerated mitigation strategies implementation (Rane et al. 2024). The model does this by effectively analysing extensive datasets while distilling key insights from climate science data and communicate effectively to the recipients. AI machines are created to adjust accordingly to new inputs and learn from experience, making them adaptable to all information systems as it is used in conjunction with big data. An example is the 'BCG's CO2 AI', that uses smart monitoring and machine learning to analyse the CO2 sources and footprints (Yayari et al. 2023). Data regarding emissions and greenhouse gases can further be used to create educational resources highlighting the sources of climate change and the importance of emission reduction strategies. Another application for AI in disaster management is the use of Piccard's AI flash flood alert system. This describes a fully automated end-to-end warning system used to predict floods occurrence in predetermined flood zones through AI technology (Torky et al. 2023). The real-time data and information can be integrated into online platforms or apps, allowing communities to be better prepared during flooding events. Applications of AI in the education sector are not so many yet, but there is evidence of high potential for AI to offer automated assessment of tests and customized teaching solutions using Chatbots. Some challenges, however, do exist concerning AI's dependence on high-quality datasets which are needed in large amounts and in real time so as to fuel the operations of the AI machines. Therefore, it needs proper data management protocols to cater for any arising changes. The autonomous decision-making process of the AI also raises the question on accountability when it delivers undesired or unforeseen decisions.

Cloud computing in climate disaster education

Cloud computing describes a computing protocol with the potential of providing disaster education through offering an array of IT Services connected through IP networks offering low-cost computing networks available for users on demand from any part of the country without having to install any software or hardware on the local devices (Lahoti 2018). The cloud is virtually unlimited in terms of resources by means of Web-GIS applications, geodata, storage, computing power, and infrastructure for platforms of information dissemination to individuals that will enable them be more aware of their environment and any looming dangers. An example of an online computing tool is the pan-African network called 'AfricaAdapt', utilising webinars, online resources, and mobile phone apps in the dissemination of best practices and knowledge regarding climate change and the necessary adaptations (Africa 2018). Concurrent cloud access from cloud computing also makes it efficient in community engagement and fosters a learning environment for locals whereby they can interact with the platform and get answers regarding emergencies, while the machines also harness information on the needs for locals to culture solutions to their specific needs. The systems working within the cloud computing platform, particularly IoT, AI and big data, have been considered vital in carrying out an informed and in-depth analysis of existing data concerning previous disasters so as to bring out any trends and key features of public safety and encourage education on the same.

Virtual Reality and Augmented Reality in climate disaster education

Data can be visualized through different methods such as through games, photos, info graphics, videos, or Virtual Reality (Anggaryani et al. 2023). All these can be cultured towards offering education to individuals in an effective manner through provision of visual insights. Virtual Reality (VR) is considered a form of Extended Reality (XR), which defines a computer generated environment that is three dimensional, interactive, and explorable by a person. AR, however, overlays virtual objects to the real world to enhance user perception. It has previously been used practically in a nursing school to provide disaster medicine training whereby students are put in a world with virtually simulated disaster scenarios that exposes them to a broad range of competencies and skills (Magi et al. 2023). Currently VR is already being implemented and used both as an awareness and training tool in disaster management protocols, although the technology required to film such scenarios is still more expensive and sophisticated than what is needed for regular photography.

23.8 Creating an Enabling Environment for Conventional and Disruptive Technology Adaptation

The damage and destruction from disasters have a lot of setbacks such as hampering economic activities and services, delaying long-term development opportunities, and disrupting efforts of poverty reduction in the region in question—China. It is essential, therefore, for China, which is prone to natural risks and disasters, to address the interrelated and complex affairs of disasters, climate change, sustainable development, and risk reduction education protocols (Wang et al. 2023). An enabling environment can be devised through tools and analyses that have provided significant help in offering solutions and making more data easily accessible. An enabling environment has been described in various ways by different scholars, and this section undertakes a deductive approach in determining the most appropriate concept as well as the specific applications in the case of disaster risk education, reduction and mitigation. In DRR, an enabling environment is mentioned to describe the participation and actions by different stakeholders and actors during the development and implementation processes that work to ensure success of the same (Sawaneh and Fan 2021). The enabling environment will be described briefly including the existing literature, specific roles played by different players, and the key actors involved in building resilience to disasters and ensure that development strategies work to reduce hazard vulnerability as well in-kind. Different steps towards creating an enabling environment include: policy development framework, ensured management and leadership commitment, reinforced institutional capacities, and integration of DRR protocols into project cycle, as well as advocacy both internally and externally.

Ensured support and commitment from management and leadership

The importance of leadership and management is seen in their vital role in mainstreaming operations, policies and functions (Septikasari et al. 2024). The ensured support and commitment will work to ensure characteristic levels of monitoring of progress and coordination of strategic processes across China so as to ensure the adoption of both conventional and disruptive technologies into the learning content of various curricula and education levels. In this case, the leadership and management of educational institutions are at the forefront of ensuring the adoption of various disaster reduction and mitigation information to relevant courses of education. Lack of commitment from leadership and policy makers in schools will channel down to the immediate level of teaching staff and further down to students who will not be receptive to such education. Therefore, due commitment both in teaching and practically showing examples in forms such as emergency drills shall be critical in reinforcing the information dissemination protocols.

Framework for strategy and policy development

One definition of the term 'policy' is 'commitment' or 'statement of intent' by a National Society. Its role is to provide guidance on what rule or principle is to guide the decision-making process so as to attain characteristic rational outcomes. It is necessary for the students and teachers to be involved in the process of policy development that will guide the dissemination of information (Fu and Zhang 2024).

The leadership and policy formulators and regulators need to mainstream the delivery and credibility of the information through policy formulation, and they will be held accountable of the same accordingly. The policies need to be formulated in accordance with clearly defined objectives of integrating both conventional and disruptive technologies in the education process of disaster risk and mitigation practices. These policies will be vital in providing managers and leaders with a mandate to follow when defining the curricula for disaster risk education.

Another sector that will strengthen the foundations of education within the realms of disaster mitigation is the legal ramification of technological innovations and research (Passadelli et al. 2024). For example having regulation and registration of drones ensures that there is a system of accountability in play which provides opportunities for information dissemination since standardization has been done and policies are in action. An argument is provided, however, on the regulations of using Big Data during the event of a disaster when it is urgently needed, calling for the need of sharing protocols to be present beforehand. In order to properly distribute information on technologies for disaster reduction, it shall be critical to have adequate capacities of ensuring that the management and planning operations realize success. Therefore the disaster agencies handling these technological advancements need to have well trained professionals who can clearly educate and inform the general public on how the technologies work and what is expected of them in the event of an emergency.

Reinforced institutional capacities

Upon implementing the mainstreaming process through management protocols, project planning, and formulating the proper curriculum –it is necessary to ensure the institutional capacities have the desired capacity to handle the dissemination of information to the respective students and through the appropriate channels, and with the intended learning materials (Amri et al. 2022). It is important to ensure the potential challenges to ownership rights that might exist are tackled through institutionalizing the learning content and ensure it is available to all. It is necessary also for the leadership management to understand the level of change to the learning environment and culture that adopting these disaster reduction education mechanisms will cause to the curriculum in its entirety.

Innovative financing models and approaches within the humanitarian sector can also be effective in the advancement of disaster risk education strategies (Munawar et al. 2021). Harnessing economic power from aid agencies and donors can provide help in defining financing models that is innovative such as impact investment, social enterprise, and transaction levies and taxes on services and goods. An example of this model is Forecast-based Financing (FbF), a system using weather and climate forecast tools to anticipate likely disasters in risk prone zones and mobilizing resources for preparation, risk mitigation, and education in early action strategies.

The reinforcement of institutions and policies also involves designation of a specific overseeing body to ensure definition of protocols, development of strategies and monitoring of progress of the strategies for information dissemination (Passadelli et al. 2024). The coordination of management protocols is necessary since the education has multiple sectors calling for multi-tiered engagement through involving the teachers so as to guide with their expertise on how best to offer sector-specific education for disaster risk mitigation and reduction. The teaching staff should, in turn, get training on delivery of education involving best practices and policies on disaster risk reduction to ensure proper understanding, implementation and efficient dissemination of the information to students who are the key stakeholders. Teaching staff briefings and regular communication on disaster risk reduction education importance can provide opportunities for delivering skills and knowledge to increase understanding. In order to ensure the institutions are effective in information delivery, regular assessments and drills shall be done to students within institutions to measure their understanding and application capacity in order to provide ways of strengthening and supporting institutional capacities through constant monitoring and evaluation.

Integration of DRR protocols into project cycle

Design and management of projects and programs are done through phases that are interrelated in a sequence that governs the project cycle management (PCM) (Gong et al. 2021). In order to mainstream the education process for adoption of conventional and disruptive technology into the disaster risk education curricula, it is essential for the key program developers and project managers to make it compulsory to address and consider disaster and climate change hazards and their mitigation processes. By considering the project cycle, the management personnel shall effectively design projects that are climate-smart and well informed on the needs of the hazard-prone zones.

For the disaster community in different backgrounds to clearly understand the best way for responsibly and aptly adapting emerging and disruptive technologies in crisis scenarios it is indispensable for them to undergo training on the same (Shoji et al. 2020). Different technologies would need to have manuals for users to familiarize with the basic operational requirements existing and how to deal with challenges such as coping with fake information on social media. It shall be necessary to expose the learning individuals to both the theoretical and practical aspects of deploying the technological innovations during training. For example, disaster management personnel should carry out drills on using new tools such as drones to locate needy individuals during floods before the actual need for evacuation after scanning inaccessible areas with drones arises.

Advocacy both internally and externally

Advocacy involves persuading individuals to adopt changes whether in practice, structures, policy or systems. It involves working with others, speaking for others, and supporting others to speak for themselves (Cabatay and Gonzales 2024). An example is seen within the laws of the Red Cross Red Crescent, whereby advocacy goes hand in hand with education and awareness raising as the three dimensions for strengthening interventions. Providing education and raising awareness can work to empower communities towards change so as to enjoy healthier, safer lives, whereas advocacy ensures the conditions with which they are in allow for the same in actuality. School and various learning institutions are involved in advocacy and awareness raising through social media also so as to promote diplomacy.

In order to alter advocacy internally, it shall be necessary change the thinking of individuals at the management level such as teachers who will be delivering the information. Externally, change will need to be experienced in the way businesses, organizations, and the government thinks, strategizes, and behaves in the parameters of disaster risk education (Wallengren-Lynch 2024). Having advocacy strategies both internally and externally will prove effective in eliminating discrimination of conventional knowledge, change public attitudes, and inspire community action regarding indigenous and new technologies, as well as offer better opportunities for understanding climate-related risks and how best they are addressed in the information dissemination process. The school administration shall ensure advocacy internally within the school, while government and non-government institutions shall reinforce the disaster risk education protocols externally.

23.9 Conclusion

In conclusion, this chapter analyses and provides the important role played by indigenous or conventional knowledge in the development of new technological inventions in the field of disaster risk climate education practices. Different forms of traditional disaster risk reduction and education techniques are first described, and their uses explained before their potential of incorporation with modern technologies is analysed. Additionally, this collection serves to show the importance of the indigenous knowledge as the backbone for development of disruptive technologies, curriculum development, and informed policy making. Also, the traditional knowledge has to be used daily and be recognized by the local community so as to be considered a tool for disaster risk reduction and education. Modern techniques are used document and record the effective use of traditional knowledge which entails an array of observational data derived from the natural phenomenon that can prove effective when merged with modern science practices of understanding and analysing natural hazards more precisely. This shows the need for modern technological data's combination with local context's understanding so as to improve people's lives and understand their needs. The nature of traditional knowledge interacting and interconnecting with animals, plants, humans and the environment at large sets up a balance for sustainable development and information dissemination in a holistic approach.

Application of technology during times of disaster in China has been done in creative ways using an array of monitoring and imagery tools. These technological advancements arose as a need to harness information and data that could previously not be accessed through conventional methods. However, some challenges still exist in the architecture for disaster response and education protocols in China such as lack of proper coordination between a complex series of departments, agencies, and ministries at the provincial, national and community levels. Therefore different perspectives have been put forward that indicate the potential of education and knowledge dissemination through both conventional and disruptive technologies, and the concerns of each. Different explanations have also been provided indicating the development process of achieving an enabling environment within the parameters of disaster risk reduction education and mitigation.

References

Abdel-Basset M, Mohamed R, Elhoseny M, Chang V (2020) Evaluation framework for smart disaster response systems in uncertainty environment. Mech Syst Signal Process 145:106941

Abid SK, Sulaiman N, Chan SW, Nazir U, Abid M, Han H, Ariza-Montes A, Vega-Muñoz A (2021) Toward an integrated disaster management approach: how artificial intelligence can boost disaster management. Sustainability 13:12560

Aleksandra Solinska-Nowak, Piotr Magnuszewski, Margot Curl, Adam French, Adriana Keating, Junko Mochizuki, Wei Liu, Reinhard Mechler, Michalina Kulakowska, Lukasz Jarzabek (2018) An overview of serious games for disaster risk management—prospects and limitations for informing actions to arrest increasing risk. Intern J Disast Risk Reduct 31:1013-1029. ISSN 2212-4209

Amri A, Lassa J, Sagala S, Hanifa R, Tebe Y, Kumar J (2022) Pathways to disaster risk reduction education integration in schools: insights from SPAB evaluation in Indonesia. Intern J Dis Risk Reduct 73:102860. https://doi.org/10.1016/j.ijdrr.2022.102860

Anggaryani Mita, Prastowo Tjipto, Suprapto Nadi, Lassa Jonatan, Madlazim Madlazim, Alifteria Faradila, Agusty Alifia, Lestari Nurita (2023) Virtual reality as experiential learning to promote STEM-DRR in tertiary education. ASM Sci J 18:1–12. https://doi.org/10.32802/asmscj.2023.1370

Arora Suchita, Kumar Sunil, Kumar Sandeep (2023) Artificial intelligence in disaster management: a survey. https://doi.org/10.1007/978-981-19-6634-7_56

Cabatay Mulina, Gonzales Heidi (2024) Disaster prevention and risk reduction education implementation in special education schools in indonesia: teachers' challenges, strategies, and recommendations. J Pendidikan Sains Indonesia 12:216–231. https://doi.org/10.24815/jpsi.v12i1.35191

Cuaton GP, Su Y (2020) Local-indigenous knowledge on disaster risk reduction: insights from the Mamanwa indigenous peoples in Basey, Samar after Typhoon Haiyan in the Philippines. Intern J Dis Risk Red 48:101596

Cheikhrouhou O, Koubaa A, Zarrad A (2020) A cloud based disaster management system. J Sens Actuator Netw 9:6

Comes T (2024) AI for crisis decisions. Ethics Inf Technol 26. https://doi.org/10.1007/s10676-024-09750-0

Desilia Nurul, Lassa Jonatan, Oktari Rina (2023) Integrating disaster education into school curriculum in Indonesia: a scoping review 5:263–274. https://doi.org/10.24815/ijdm.v6i2.34867

Diani G (2021) Popularizing scientific knowledge for children: a multimodal perspective. In *Multimodal Literacies Across Digital Learning Contexts* (pp. 118–134). Routledge

Africa I (2018) Use and communication of climate information to support uptake of adaptation actions in semi-arid regions in Africa and Asia

Fu Q, Zhang X (2024) Promoting community resilience through disaster education: review of community-based interventions with a focus on teacher resilience and well-being. PLoS ONE 19:e0296393. https://doi.org/10.1371/journal.pone.0296393

Gong Q, Yushan D, Guo F (2021) Disaster risk reduction education in school geography curriculum: review and outlook from a perspective of China. Sustainability 13:3963. https://doi.org/10.3390/su13073963

Madanian S, Johnson K, St Martin M, Sinha R, Cámara J, Parry D (2022) Adaptable socio-cyber physical systems for supporting disaster. Australas J Disaster Trauma Stud 26:221–234

Munawar Hafiz, Mojtahedi Moe, Hammad Ahmed, Kouzani Abbas, Mahmud MA (2021) Disruptive technologies as a solution for disaster risk management: a review. Sci Total Environ 806:151351. https://doi.org/10.1016/j.scitotenv.2021.151351

Minges M (2019) Disruptive technologies and their use in disaster risk reduction and management. Intern Telecom Union (ITU): Geneva, Switzerland

Hoffmann R, Blecha D (2020) Education and disaster vulnerability in southeast asia: evidence and policy implications. Sustainability 12:1401. https://doi.org/10.3390/su12041401

Hossin Md, Chen Lie, Asante Isaac, Asante Boadi, Evans, Stephen Yeboah (2023) Climate change and COP26: role of information technologies in disaster management and resilience. Environ Develop Sustain 1-27. https://doi.org/10.1007/s10668-023-04134-8

Hosseini KA, Izadkhah YO (2020) From "Earthquake and safety" school drills to "safe school-resilient communities": a continuous attempt for promoting community-based disaster risk management in Iran. Intern J Disast Red 45:101512

Klein JA, Tucker CM, Steger CE, Nolin A, Reid R, Hopping KA, Yager K (2019) An integrated community and ecosystem-based approach to disaster risk reduction in mountain systems. Environ Sci Policy 94:143–152

Lahoti S (2018) AI to the rescue: 5 ways machine learning can assist during emergency situations. Packt. 16 January. Available from https://hub.packtpub.com/ai-rescue-5-ways-machine-learning-can-assist-emergency-situations (accessed Feb 28, 2024)

Lambert S, Scott J (2019) International disaster risk reduction strategies and indigenous peoples. Intern Indig Policy J 10(2):1–21

Magi CE, Bambi S, Iovino P, El Aoufy K, Amato C, Balestri C, Rasero L, Longobucco Y (2023) Virtual reality and augmented reality training in disaster medicine courses for students in nursing: a scoping review of adoptable tools. Behav Sci (basel, Switzerland) 13(7):616. https://doi.org/10.3390/bs13070616

Munsaka E, Dube E (2018) The contribution of indigenous knowledge to disaster risk reduction activities in Zimbabwe: a big call to practitioners. Jàmbá: J Disaster Risk Stud 10(1):1–8

Oktari RS, Munadi K, Idroes R, Sofyan H (2020) Knowledge management practices in disaster management: Systematic review. Intern J Disaster Risk Red 51:101881

Passadelli Anthoula, Klonari Ekaterini, Ekaterini Nikolarea (2024) Disaster risk reduction education in geography at schools in Greece. Res J Educ 10:1–12. https://doi.org/10.32861/rje.101.1.12

Peker I, Ar I, Erol İ, Searcy C (2022) Leveraging blockchain in response to a pandemic through disaster risk management: an IF-MCDM framework. Oper Manag Res 16. https://doi.org/10.1007/s12063-022-00340-1

Phengsuwan J, Shah T, Thekkummal NB, Wen Z, Sun R, Pullarkatt D, Thirugnanam H, Ramesh MV, Morgan G, James P et al (2021) Use of social media data in disaster management: a survey. Fut Intern 13:46

Rane Nitin, Choudhary Saurabh, Rane Jayesh (2024) Contribution of ChatGPT and similar generative artificial intelligence for enhanced climate change mitigation strategies (January 2, 2024). Available at SRN: https://ssrn.com/abstract=4681720 or https://doi.org/10.2139/ssrn.4681720

Rodrigues da Silva A, Estima J, Marques J, Gamito I, Serra A, Moura L, Ferreira RM (2023) A web GIS platform to modeling, simulate and analyze flood events: the rivercure portal. ISPRS Int J Geo Inf 12(7):268

Samarakkody Aravindi, Dilanthi Amaratunga, Richard Haigh (2023) Technological innovations for enhancing disaster resilience in smart cities: a comprehensive urban scholar's analysis. Sustainability 15(15):12036. https://doi.org/10.3390/su151512036

Sarker MNI, Peng Y, Yiran C, Shouse RC (2020) Disaster resilience through big data: way to environmental sustainability. Int J Disaster Risk Reduct

Sawaneh IA, Fan L (2021) The mediating role of disaster policy implementation in disaster risk reduction and sustainable development in sierra leone. Sustainability 13(4):2112

Septikasari Zela, Retnawati Heri, Wilujeng Insih, Shaw Rajib, Handaka Irvan (2024) Needs assessment: strategy of integration disaster education at elementary schools in disaster prone areas. J Educ Learn (EduLearn) 18:244–252. https://doi.org/10.11591/edulearn.v18i1.20921

Shoji M, Takafuji Y, Harada T (2020) Behavioral impact of disaster education: evidence from a dance-based program in Indonesia. Intern J Disast Risk Red 45:101489

Torani S, Majd PM, Maroufi SS, Dowlati M, Sheikhi RA (2019) The importance of education on disasters and emergencies: a review article. J Educ Health Prom 24(8):85–97

Torky M, Gad I, Darwish A, Hassanien AE (2023) Artificial intelligence for predicting floods: a climatic change phenomenon. In *The Power of Data: Driving Climate Change with Data Science and Artificial Intelligence Innovations* (pp. 3–26). Cham: Springer Nature Switzerland. In areas

where drainage issues cannot be resolved, providing real time alert to residents and closing roads will provide sufficient information to reduce risks
Wallengren-Lynch M (2024) From compassion to action: school social workers at the forefront of emergency response. Nordic Soc Work Res. https://doi.org/10.1080/2156857X.2024.2309187
Wang Z, Han Z, Li Y (2023) The interplay between school preparedness and student's individual protective actions: the mediating role of disaster education. Sustainability 15:14888. https://doi.org/10.3390/su152014888
Winarni EW, Purwandari EP (2018) Disaster risk reduction for earthquake using mobile learning application to improve the students understanding in elementary school. Mediterr J Soc Sci 9(2):205–214
Yavari A, Mirza IB, Bagha H, Korala H, Dia H, Scifleet P, Sargent J, Tjung C, Shafiei M (2023) ArtEMon: artificial intelligence and internet of things powered greenhouse gas sensing for real-time emissions monitoring. Sensors 23(18):7971. https://doi.org/10.3390/s23187971

Correction to: Preparing for Disasters Through Citizenship Pedagogies

Hans Svennevig, Kaori Kitagawa, and Masayoshi Inoue

Correction to: Chapter 2 in: A. Yildiz and R. Shaw (eds.), *Disaster and Climate Risk Education*, Disaster Risk Reduction, https://doi.org/10.1007/978-981-97-5987-3_2

The original version of this chapter 2 authors names corrected and the references order has been updated. The book and the correction chapter have been updated with the changes.

The updated version of this chapter can be found at
https://doi.org/10.1007/978-981-97-5987-3_2

A. Yildiz and R. Shaw (eds.), *Disaster and Climate Risk Education*, Disaster Risk Reduction, https://doi.org/10.1007/978-981-97-5987-3_24

MIX
Papier aus verantwortungsvollen Quellen
Paper from responsible sources
FSC® C105338

If you have any concerns about our products,
you can contact us on
ProductSafety@springernature.com

In case Publisher is established outside the EU,
the EU authorized representative is:
Springer Nature Customer Service Center GmbH
Europaplatz 3, 69115 Heidelberg, Germany

Printed by Libri Plureos GmbH
in Hamburg, Germany